Studies in Analysis

ADVANCES IN MATHEMATICS
SUPPLEMENTARY STUDIES, VOLUME 4

ADVANCES IN
Mathematics
SUPPLEMENTARY STUDIES

Studies in Analysis

ADVANCES IN MATHEMATICS
SUPPLEMENTARY STUDIES, VOLUME 4

EDITED BY

Gian-Carlo Rota

Department of Mathematics
Massachusetts Institute of Technology
Cambridge, Massachusetts

With the Editorial Board
of *Advances in Mathematics*

ACADEMIC PRESS New York San Francisco London 1979
A Subsidiary of Harcourt Brace Jovanovich, Publishers

ACADEMIC PRESS, INC.
111 Fifth Avenue, New York, New York 10003

United Kingdom Edition published by
ACADEMIC PRESS, INC. (LONDON) LTD.
24/28 Oval Road, London NW1 7DX

Library of Congress Cataloging in Publication Data

Main entry under title:

Studies in analysis.

(Advances in mathematics : Supplementary studies ; v. 4)
Includes bibliographies.
1. Mathematical analysis--Addresses, essays, lectures. 2. Functional analysis--Addresses, essays, lectures. I. Rota, Gian-Carlo, 1932–
II. Series.
QA300.S9 515 78-21276
ISBN 0-12-599150-9

PRINTED IN THE UNITED STATES OF AMERICA

79 80 81 82 83 84 9 8 7 6 5 4 3 2 1

Contents

Manifolds of C^r Mappings and Applications to Differentiable Dynamical Systems

John Franks

Positive Definite Functions on Abelian Semigroups without Zero

Paul Ressel

Invariant Weights on Decomposable von Neumann Algebras

Nghiem Dang-Ngoc

Robert R. Kallman

List of Contributors

Numbers in parentheses indicate the pages on which the authors' contributions begin.

CHRISTOPHER I. BYRNE† (83), Department of Mathematics, University of Utah, Salt Lake City, Utah 84112

NGHIEM DANG-NGOC (311), Laboratoire de Probabilités, Université de Paris, Paris, France

HUGO BEIRÃO DA VEIGA (241), Instituto de Fisica e Matemática, Lisbon, Portugal

JOHN FRANKS (271), Department of Mathematics, Northwestern University, Evanston, Illinois 60201

JOHN A. GOSSELIN (223), Department of Mathematics, University of Georgia, Athens, Georgia 30601

DAVID HANDELMAN (171), Mathematics Department, University of Ottowa, Ottowa, Ontario, Canada

M. P. HEBLE (135), Department of Mathematics, University of Toronto, Toronto, Canada

NORMAN E. HURT‡ (83), Department of Mathematics, Cleveland State University, Cleveland, Ohio 44115

ROBERT R. KALLMAN (321), Department of Mathematics, University of Florida, Gainesville, Florida 32601

JOSÉ G. LLAVONA (197), Facultad de Ciencias Matemáticas, Universidad Complutense, Madrid, Spain

JORGE MUJICA§ (1), Department of Mathematics, University of Rochester, Rochester, New York 14650

PAUL RESSEL (291), Institut für Mathematische Statistik, Westfälische Wilhems-Universität, Münster, Germany

MARC A. RIEFFEL (43), Department of Mathematics, University of California at Berkeley, Berkeley, California 94720

D. SUNDARARAMAN (233), The Ramanujan Institute, University of Madras, Madras, India

KAZUHIRO YAMAMOTO (123), Department of Mathematics, Hokkaido University, Sapporo, Japan

† Present address: Harvard University, Cambridge, Massachusetts.

‡ Present address: Mechanics Research, Inc., McLean, Virginia 22101.

§ Present address: Instituto de Matemática, Universidad Estadual de Campinas, Campinas, S.P. Brazil.

Preface

The supplementary volumes of the journal *Advances in Mathematics* are issued from time to time to facilitate publication of papers already accepted for publication in the journal. The volumes will deal in general—but not always—with papers on related subjects, such as algebra, topology, foundations, etc., and are available individually and independently of the journal.

STUDIES IN ANALYSIS
ADVANCES IN MATHEMATICS SUPPLEMENTARY STUDIES, VOL. 4

Spaces of Germs of Holomorphic Functions

BY JORGE MUJICA†

Department of Mathematics
University of Rochester
Rochester, New York

INTRODUCTION

In the theory of holomorphic mappings one is often interested in functions that are locally holomorphic around a given set X. It is then convenient to develop an explicit local theory, which leads to the notion of a holomorphic germ on X. The set $\mathscr{H}(X)$ of all holomorphic germs on X can be viewed as a ring or as a topological vector space. When X is a point in C^n, or more generally in a complex locally convex space E, then the algebraic properties of the ring $\mathscr{H}(X)$ have been extensively studied, leading to many applications in analytic geometry; see, for instance, [19, 20, 36, 37]. On the other hand, when X is any compact set in E, then the vector space $\mathscr{H}(X)$ can be endowed with a natural locally convex topology, whose properties have been studied by many authors, leading to various applications on the global behavior of holomorphic functions; see, for instance, [5, 6, 15, 18, 25, 26]. This chapter is concerned with the second aspect of the local theory.

Let $\mathscr{H}(U)$ denote the vector space of all holomorphic functions on an open subset U of a complex locally convex space E. The study of topologies on $\mathscr{H}(U)$ was begun by Nachbin [30, 31] and Alexander [1]. It was motivated by the fact that the compact open topology is not the only natural topology on $\mathscr{H}(U)$ when E is infinite-dimensional; we refer to Nachbin [32, 33] for an account of natural topologies on $\mathscr{H}(U)$. Motivated by the notion of analytic functionals ported by compact sets, which were studied by Martineau [25], Nachbin [30, 31] introduced and studied the compact ported topology τ_ω on $\mathscr{H}(U)$, where τ_ω is useful in the study of holomorphic continuation; see, Dineen [10, 12] and Matos [27].

The study of the locally convex space $(\mathscr{H}(U), \tau_\omega)$ is by no means straightforward when E is infinite-dimensional. In this chapter we endow the space

† Present address: Instituto de Matemática, Universidade Estadual de Campinas, Campinas, S. P. Brazil.

ISBN 0-12-599150-9

of germs $\mathscr{H}(K)$, with $K \subset E$ compact, with a natural locally convex topology, and study its properties. In the infinite-dimensional case the study of $\mathscr{H}(K)$ is much easier than that of $(\mathscr{H}(U), \tau_\omega)$, and we use the topological properties of the spaces $\mathscr{H}(K)$, with $K \subset U$, to obtain information about $(\mathscr{H}(U), \tau_\omega)$. For instance, we prove that $(\mathscr{H}(U), \tau_\omega)$ is complete whenever U is an open subset of a Banach space, thus answering a question raised by Nachbin [32]. We now describe in more detail the contents of this chapter.

In Section 1 we collect some results on topological vector spaces which are often used in subsequent sections.

In Section 2 we introduce the locally convex space $\mathscr{H}(K)$ as the inductive limit of the spaces $(\mathscr{H}(U), \tau_\omega)$, where U varies over all open neighborhoods of K. Alternatively, $\mathscr{H}(K)$ can be viewed as the inductive limit of the Banach spaces $\mathscr{H}^\infty(U)$ of all bounded holomorphic functions on U. Then $\mathscr{H}(K)$ is a Hausdorff, barreled, bornological space, and when E is semimetrizable, then $\mathscr{H}(K)$ is a (DF)-space. All of these results are immediate extensions of the case where E is a Banach space, which was studied by Chae [5, 6].

In Section 3 we show that whenever E is semimetrizable then $\mathscr{H}(K) = \varinjlim \mathscr{H}^\infty(U)$ is a regular inductive limit, i.e., every bounded subset of $\mathscr{H}(K)$ is contained and bounded in some $\mathscr{H}^\infty(U)$. In the case where E is a Banach space, this result was obtained independently by Chae [6] and Hirschowitz [22], but their proofs are rather complicated, and actually were originally incomplete. Our proof is much simpler and relies on a theorem of Grothendieck on (DF)-spaces. Next we impose one further restriction on E, in terms of the spaces $\mathscr{P}(^mE)$ of continuous m-homogeneous polynomials on E. We say that E satisfies condition (B) if its topology is defined by a directed family of seminorms $\{\alpha\}$ such that for each $m \in N$, each Banach space $\mathscr{P}(^mE_\alpha)$ has the induced topology of $\mathscr{P}(^mE)$, where $\mathscr{P}(^mE)$ is endowed with the inductive limit topology of the Banach spaces $\mathscr{P}(^mE_\alpha)$. This condition was considered first by Barroso [4]. We show that whenever E is a semimetrizable space satisfying condition (B), then the inductive limit $\mathscr{H}(K) = \varinjlim \mathscr{H}^\infty(U)$ is Cauchy regular, i.e., given $\mathscr{X} \subset \mathscr{H}(K)$ bounded, there exists $U \supset K$ such that $\mathscr{X}$ is contained and bounded in $\mathscr{H}^\infty(U)$, and furthermore, each net $\{f_\alpha\} \subset \mathscr{X}$ is $\mathscr{H}(K)$-Cauchy if and only if it is $\mathscr{H}^\infty(U)$-Cauchy. The Cauchy regularity of $\mathscr{H}(K) = \varinjlim \mathscr{H}^\infty(U)$ was established by Chae [6] in the case where E is a Banach space.

In Section 4 we give examples and counterexamples of semimetrizable spaces satisfying condition (B). This class includes all seminormed spaces, all countable products of seminormed spaces, and several function spaces. We show that a Fréchet space satisfies condition (B) whenever it satisfies condition (C) studied by Nachbin [34]. Conversely, we prove that every distinguished Frechet space satisfying condition (B) also satisfies condition (C).

In Section 5 we show that $(\mathscr{H}(U), \tau_\omega)$ is complete whenever U is an open subset of a semimetrizable space satisfying condition (B). This answers a question raised by Nachbin [32]. To prove that $(\mathscr{H}(U), \tau_\omega)$ is complete we express $(\mathscr{H}(U), \tau_\omega)$ as the projective limit of a family of complete (DF)-spaces, each of which is a vector subspace of some $\mathscr{H}(K)$, with $K \subset U$.

In Section 6 we define $K \subset U$ to be U-Runge if the canonical image of $\mathscr{H}(U)$ in $\mathscr{H}(K)$ is sequentially dense. We say that U has the Runge property if every compact subset of U is contained in another one which is U-Runge. We prove that $(\mathscr{H}(U), \tau_\omega)$ is the projective limit of the spaces $\mathscr{H}(K)$, with $K \subset U$, whenever E is a semimetrizable space satisfying condition (B) and $U \subset E$ has the Runge property. In the case where E is a Banach space, Chae [5, 6] obtained the same conclusion under a somewhat different and apparently stronger Runge property. We show, however, that both Runge properties coincide. We also give examples of open sets with the Runge property.

In Section 7 we show that $(\mathscr{H}(U), \tau_\omega)$ is a multiplicatively locally convex algebra, i.e., τ_ω is defined by the continuous seminorms p such that $p(fg) \leqslant p(f)p(g)$ for all $f, g \in \mathscr{H}(U)$. This answers a question raised by Matos [27]. This result has applications to the construction of the envelope of holomorphy of U. The methods used in this section are similar to those of Section 5.

1. Preliminaries on Topological Vector Spaces

For the convenience of the reader, we devote this preliminary section to collecting some definitions and results from the theory of topological vector spaces which are used throughout this chapter.

Definition 1.1. For each $\alpha \in A$, A any set, let $\pi_\alpha : E \to E_\alpha$ be a linear mapping of a vector space E into a locally convex space E_α. The *projective topology* on E is the smallest topology on E for which all mappings π_α are continuous.

The projective topology on E is locally convex, and a mapping $\phi : X \to E$ of a topological space X into E is continuous if and only if $\pi_\alpha \circ \phi$ is continuous for every α.

Definition 1.2. For each $\alpha \in A$, A a directed set, let E_α be a locally convex space, and assume that whenever $\alpha \leqslant \beta$ there exists a continuous linear mapping $\pi_{\alpha\beta} : E_\beta \to E_\alpha$ such that

(a) $\pi_{\alpha\alpha}$ is the identity for each α;
(b) $\pi_{\alpha\beta} \circ \pi_{\beta\gamma} = \pi_{\alpha\gamma}$ whenever $\alpha \leqslant \beta \leqslant \gamma$.

Then the family of spaces and linear mappings $\{E_\alpha, \pi_{\alpha\beta}\}$ is called a *projective system.* The subspace

$$E = \{\{x_\alpha\} \in \Pi E_\alpha : \pi_{\alpha\beta}(x_\beta) = x_\alpha \text{ for } \alpha \leqslant \beta\}$$

of ΠE_α, with the induced topology, is called the *projective limit* of $\{E_\alpha, \pi_{\alpha\beta}\}$, and we write $E = \varprojlim E_\alpha$.

If π_α denotes the restriction to E of the canonical projection $\Pi E_\alpha \to E_\alpha$ then E has the *projective topology* with respect to the mappings $\pi_\alpha : E \to E_\alpha$.

PROPOSITION 1.1. *The projective limit of a family of complete locally convex spaces is complete.*

For a proof see [38, p. 52, Proposition 5.3].

cs(E) denotes the set of all continuous seminorms on a locally convex space E. For $\alpha \in \mathrm{cs}(E)$, E_α denotes the vector space E seminormed by α, and E/α denotes the associated normed space $E_\alpha/\alpha^{-1}(0)$.

PROPOSITION 1.2. *If E is a complete Hausdorff locally convex space, then $E = \varprojlim E/\alpha$, where α varies in* $\mathrm{cs}(E)$.

For a proof see [38, p. 53, Proposition 5.4].

DEFINITION 1.3. For each $\alpha \in A$, A any set, let $\pi_\alpha : E_\alpha \to E$ be a linear mapping of a locally convex space E_α into a vector space E. The *inductive topology* on E is the largest locally convex topology on E for which all mappings π_α are continuous.

A linear mapping $\phi : E \to F$ of E into a locally convex space F is *continuous* if and only if $\phi \circ \pi_\alpha$ is continuous for every α.

If every E_α is *barreled*, so is E. If every E_α is *bornological*, so is E.

DEFINITION 1.4. Let a vector space E be the union of a family of locally convex spaces $\{E_\alpha\}$, directed under inclusion, and such that each inclusion mapping $E_\alpha \hookrightarrow E_\beta$ is continuous. Then E, endowed with the inductive topology with respect to the inclusion mappings $E_\alpha \to E$, is called the *inductive limit* of the subspaces $\{E_\alpha\}$, and we write $E = \varinjlim E_\alpha$.

DEFINITION 1.5. Let $E = \varinjlim E_\alpha$ be an inductive limit of subspaces.

(a) $E = \varinjlim E_\alpha$ is a *strict* inductive limit if E_α has the induced topology of E_β whenever $E_\alpha \subset E_\beta$.

(b) $E = \varinjlim E_\alpha$ is a *regular* inductive limit if every bounded subset of E is contained and bounded in some E_α.

(c) $E = \varinjlim E_\alpha$ is a *Cauchy regular* inductive limit if given $B \subset E$ bounded, there exists α such that B is contained and bounded in E_α, and furthermore, every net $\{x_\alpha\} \subset B$ is *E-Cauchy* if and only if it is E_α-Cauchy.

PROPOSITION 1.3. *Let $E = \lim E_n$ be the strict inductive limit of an increasing sequence of subspaces. Then*:

(*a*) *Each E_n has the induced topology of E.*

(*b*) *If E_n is closed in E_{n+1} for each n, then $E = \varinjlim E_n$ is a Cauchy regular inductive limit.*

(*c*) *If each E_n is Hausdorff and complete, then E is Hausdorff and complete.*

For a proof see [38, pp. 58–59, Propositions 6.4–6.6].

PROPOSITION 1.4. *Let F be a Hausdorff locally convex space which is the union of an increasing sequence of Fréchet spaces $\{F_n\}$ and assume that each inclusion mapping $F_n \hookrightarrow F$ is continuous. Let $\phi: E \to F$ be a continuous linear mapping of a Fréchet space E into F. Then there exists n and a continuous linear mapping $\phi_n: E \to F_n$ such that the following diagram is commutative*:

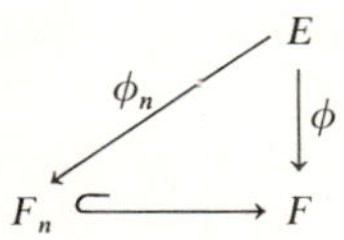

For a proof see [17, p. 16, Theorem A or 18, p. 268, Theorem 1].

DEFINITION 1.6. (a) An (*LF*)-*space* is the inductive limit of an increasing sequence of Fréchet spaces.

(b) A locally convex space is *ultrabornological* if it has the inductive topology with respect to a family of Banach spaces.

PROPOSITION 1.5. *Let $\phi: F \to E$ be a continuous injective linear mapping of an* (LF)-*space F onto a Hausdorff ultrabornological space E. Then ϕ is a homeomorphism.*

For a proof see [17, p. 17, Theorem B or 18, p. 271, Theorem 2].

DEFINITION 1.7. A locally convex space E is a (*DF*)-*space* if:

(a) E has a fundamental sequence of bounded sets.

(b) Every strongly bounded countable union of equicontinuous subsets of E' is equicontinuous.

Every infrabarreled space satisfies condition (b) of definition 1.7. Hence every normed space is a (DF)-space. It is also clear that the strong dual of a (DF)-space is a Fréchet space.

PROPOSITION 1.6. *The strong dual of a metrizable locally convex space is a* (DF)*-space.*

For a proof see [16, p. 61, Theorem 1 or 18, p. 302, Theorem 1].

PROPOSITION 1.7. *A quasi-complete* (DF)*-space is complete.*

For a proof see [16, p. 77, Corollary 2 or 18, p. 311, Corollary 2].

PROPOSITION 1.8. *Let E be the inductive limit of an increasing sequence of* (DF)*-spaces E_n. Then E is a* (DF)*-space and every bounded subset of E is contained in the E-closure of a bounded subset of some E_n.*

For a proof see [16, p. 78, Theorem 9 or 18, p. 312, Proposition 5].

2. THE SPACE OF GERMS $\mathscr{H}(K)$

Let K be a compact subset of a complex locally convex space E. In this section we introduce the space $\mathscr{H}(K)$ of all holomorphic germs on K and study its elementary properties. We begin by recalling some basic facts on holomorphic functions. Our basic reference is Nachbin [31] in the case where E is a Banach space and Barroso [4] in the general case.

DEFINITION 2.1. Let $U \subset E$ be open. A function $f: U \to C$ is *holomorphic* if for each $\xi \in U$ there exists a sequence $\{P_m\}$ with $P_m \in \mathscr{P}({}^mE)$, such that

$$f(x) = \sum_{m=0}^{\infty} P_m(x - \xi)$$

uniformly on a neighborhood of ξ.

The sequence $\{P_m\}$ is unique and we write $P_m = (1/m!)\hat{d}^m f(\xi)$.

DEFINITION 2.2. The function $f: U \to C$ is *G-holomorphic* if its restriction to each complex line is holomorphic.

The following result will be often used without further reference. For a proof see [35, p. 25, Theorem 1.2.8].

PROPOSITION 2.1. *Given $f: U \to C$, the following conditions are equivalent: (a) f is holomorphic, (b) f is G-holomorphic and continuous, and (c) f is G-holomorphic and locally bounded.*

$\mathscr{H}(U)$ denotes the vector space of all holomorphic functions on U. Nachbin [30, 31] introduced the compact ported topology τ_ω on $\mathscr{H}(U)$ as follows.

DEFINITION 2.3. A seminorm p on $\mathscr{H}(U)$ is *ported* by a compact subset K of U if for each V open with $K \subset V \subset U$ there exists $c(V) > 0$ such that

$$p(f) \leqslant c(V) \sup_{x \in V} |f(x)|$$

for all $f \in \mathscr{H}(U)$. τ_ω denotes the locally convex topology defined by all such seminorms.

PROPOSITION 2.2. *$\mathscr{H}^\infty(U)$ is a Banach space.*

Proof. Let $\{f_n\}$ be a Cauchy sequence in $\mathscr{H}^\infty(U)$. Define f on U as the pointwise limit of f_n. Then f is G-holomorphic and bounded on U. Thus $f \in \mathscr{H}^\infty(U)$. Clearly $f_n \to f$ uniformly on U. Q.E.D.

DEFINITION 2.4. Given $K \subset E$ compact, we consider the set

$$\bigcup_{U \supset K} \mathscr{H}(U)$$

Two functions in this set are said to be *equivalent* if they coincide on some open neighborhood of K. Each equivalence class is said to be the *germ* of a holomorphic function on K or, for short, a *holomorphic germ* on K. $\mathscr{H}(K)$ denotes the set of all holomorphic germs on K.

The canonical mappings $\mathscr{H}(U) \to \mathscr{H}(K)$, with $U \supset K$, induce a well-defined vector space structure on $\mathscr{H}(K)$.

The vector space $\mathscr{H}(K)$ is endowed with the inductive topology with respect to the canonical linear mappings

$$(\mathscr{H}(U), \tau_\omega) \to \mathscr{H}(K)$$

with $U \supset K$.

The topology of $\mathscr{H}(K)$ does not change if we restrict $U \supset K$ in such a way that each connected component of U meets K. By uniqueness of holomorphic continuation, the canonical mapping $\mathscr{H}(U) \to \mathscr{H}(K)$ is then injective, and we identify $\mathscr{H}(U)$ with a vector subspace of $\mathscr{H}(K)$. Under this identification we have $\mathscr{H}(K) = \bigcup \mathscr{H}(U)$ and $\mathscr{H}(K)$ is the inductive limit of the subspaces $(\mathscr{H}(U), \tau_\omega)$.

PROPOSITION 2.3. $\mathscr{H}(K) = \varinjlim \mathscr{H}^\infty(U)$.

Proof. K being compact, $\mathscr{H}(K) = \bigcup \mathscr{H}^\infty(U)$. The continuity of the identity mapping.

$$\varinjlim \mathscr{H}^\infty(U) \to \varinjlim(\mathscr{H}(U), \tau_\omega)$$

follows from the continuity of the inclusion mappings

$$\mathscr{H}^\infty(U) \hookrightarrow (\mathscr{H}(U), \tau_\omega).$$

Conversely let $p \in \text{cs}(\varinjlim \mathscr{H}^\infty(U))$. To show that $p \in \text{cs}(\varinjlim(\mathscr{H}(U), \tau_\omega))$, it suffices to prove that $p \in \text{cs}(\mathscr{H}(U), \tau_\omega)$ for $U \supset K$ fixed. Let V be open with $K \subset V \subset U$. Since $p \in \text{cs}(\mathscr{H}^\infty(V))$, there exists $c(V) \geqslant 0$ such that

$$p(f) \leqslant c(V) \sup_{x \in V} |f(x)|$$

for all $f \in \mathscr{H}^\infty(V)$, and hence for all $f \in \mathscr{H}(U)$. Thus p is a seminorm on $\mathscr{H}(U)$, ported by K. Q.E.D.

COROLLARY. *$\mathscr{H}(K)$ is barreled and bornological.*

We set some notation. For $\alpha \in \text{cs}(E)$, $s > 0$, and $\xi \in E$ we set

$$B_{\alpha,s}(\xi) = \{x \in E: \alpha(x - \xi) < s\}$$

$$\bar{B}_{\alpha,s}(\xi) = \{x \in E: \alpha(x - \xi) \leqslant s\}$$

For $X \subset E$ we set

$$X_{\alpha,s} = B_{\alpha,s}(X) = \bigcup_{\xi \in X} B_{\alpha,s}(\xi)$$

$$\bar{X}_{\alpha,s} = \bar{B}_{\alpha,s}(X) = \bigcup_{\xi \in X} \bar{B}_{\alpha,s}(\xi)$$

Remark. K being compact, for each $U \supset K$ open there exists $\alpha \in \text{cs}(E)$ and V α-open with $K \subset V \subset U$. Hence there exists $s > 0$ such that $K \subset K_{\alpha,s} \subset V \subset U$. In particular we obtain

$$\mathscr{H}(K) = \varinjlim_{\alpha,s}(\mathscr{H}(K_{\alpha,s}), \tau_\omega) = \varinjlim_{\alpha,s} \mathscr{H}^\infty(K_{\alpha,s})$$

PROPOSITION 2.4. *If E is a semimetrizable locally convex space, then $\mathscr{H}(K)$ is the inductive limit of an increasing sequence of Banach spaces. In particular, $\mathscr{H}(K)$ is a* (DF)*-space.*

Proof. Let $\{\alpha_n\}$ be an increasing sequence of seminorms defining the topology of E and let $\{s_n\}$ be a sequence of positive numbers, decreasing to zero. By the previous remark we then obtain

$$\mathscr{H}(K) = \varinjlim \mathscr{H}^{\infty}(K_{\alpha_n,s_n})$$

The last assertion follows from Proposition 1.8. Q.E.D.

COROLLARY. *If E is a semimetrizable locally convex space, then the strong dual $\mathscr{H}(K)'$ of $\mathscr{H}(K)$ is a Fréchet space.*

LEMMA 2.1. *For each $m \in N$ and $B \subset E$ bounded, the mapping*

$$p: f \to \sup_{\substack{\xi \in K \\ h \in B}} |\hat{d}^m f(\xi)(h)|$$

defines a continuous seminorm on $\mathscr{H}(K)$.

Proof. We show that the restriction of p to each $\mathscr{H}^{\infty}(K_{\alpha,s})$ is continuous. Let $f \in \mathscr{H}^{\infty}(K_{\alpha,s})$. By Cauchy inequalities [4, p. 530, Proposition 1.5] we obtain

$$\begin{aligned} \sup_{\substack{\xi \in K \\ h \in B}} |\hat{d}^m f(\xi)(h)| &\leqslant \sup_{\substack{\xi \in K \\ h \in B}} \|\hat{d}^m f(\xi)\|_{\alpha} [\alpha(h)]^m \\ &\leqslant C^m (m!/s^m) \sup_{x \in K_{\alpha,s}} |f(x)| \end{aligned}$$

where $C = \sup_{h \in B} \alpha(h)$. Q.E.D.

PROPOSITION 2.5. *$\mathscr{H}(K)$ is Hausdorff.*

Proof. Let $f \in \mathscr{H}(K)$ and assume that $p(f) = 0$ for every $p \in \mathrm{cs}(\mathscr{H}(K))$. From Lemma 2.1 we obtain in particular that $\hat{d}^m f(\xi) = 0$ for all $\xi \in K$ and $m \in N$. It follows that $f = 0$ on a neighborhood of K. Q.E.D.

All of the preceding results about $\mathscr{H}(K)$ are immediate extensions of the case where E is a Banach space, which was studied by Chae [5, 6]. We consider now a situation which is vacuous when E is a seminormed space. Barroso [4] noticed that every holomorphic germ on K is a holomorphic germ on K_{α}, for some $\alpha \in \mathrm{cs}(E)$, where K_{α} denotes the set K, regarded as a compact subset of the seminormed space K_{α}. We show that $\mathscr{H}(K)$ is actually the inductive limit of the subspaces $\mathscr{H}(K_{\alpha})$, where α runs over all continuous seminorms of E.

For $U \subset E$, α-open, U_{α} denotes the set U as an open subset of E_{α} and U/α denotes the canonical image $\pi_{\alpha}(U)$ of U in E/α. We have a similar notation for $K \subset E$ compact.

LEMMA 2.2. *For each $\alpha \in \text{cs}(E)$ and $U \subset E$, α-open, the Banach spaces $\mathscr{H}^\infty(U)$, $\mathscr{H}^\infty(U_\alpha)$, and $\mathscr{H}^\infty(U/\alpha)$ are isometrically isomorphic.*

Proof. Let $f \in \mathscr{H}^\infty(U)$. Then f is G-holomorphic and bounded on U_α. Thus f is α-continuous on U, i.e., $f \in \mathscr{H}^\infty(U_\alpha)$. This shows $\mathscr{H}^\infty(U) = \mathscr{H}^\infty(U_\alpha)$.

We next define f^* on U/α by $f^*(\pi_\alpha(x)) = f(x)$. Then f^* is well defined, for f is α-continuous. From

$$f^*(\pi_\alpha(\xi) + \lambda\pi_\alpha(h)) = f(\xi + \lambda h)$$

it follows that f^* is G-holomorphic and, being bounded, is holomorphic. Thus $\mathscr{H}^\infty(U_\alpha) = \mathscr{H}^\infty(U/\alpha)$. Q.E.D.

From Lemma 2.2 and the remark preceding Proposition 2.4 we obtain

PROPOSITION 2.6. $\mathscr{H}(K) = \varinjlim \mathscr{H}(K_\alpha) = \varinjlim \mathscr{H}(K/\alpha)$.

3. BOUNDED SUBSET OF $\mathscr{H}(K)$

In this section we show that $\mathscr{H}(K) = \varinjlim \mathscr{H}^\infty(U)$ is a regular inductive limit whenever E is semimetrizable. Furthermore, if we impose one further condition on E, which we call condition (B), then we can prove that $\mathscr{H}(K) = \varinjlim \mathscr{H}^\infty(U)$ is actually Cauchy regular.

When E is semimetrizable, the regularity of $\mathscr{H}(K) = \varinjlim \mathscr{H}^\infty(K_{\alpha,s})$, and in fact its Cauchy regularity, would follow at once from Proposition 1.3 if $\mathscr{H}(K) = \varinjlim \mathscr{H}^\infty(K_{\alpha,s})$ were a strict inductive limit. However, this is not the case.

PROPOSITION 3.1. *Neither of the inductive limits $\mathscr{H}(K) = \varinjlim \mathscr{H}^\infty(K_{\alpha,s})$ or $\mathscr{H}(K) = \varinjlim(\mathscr{H}(K_{\alpha,s}), \tau_\omega)$ is strict.*

Proof. Fix α, β, s, t with $\alpha \leqslant \beta$ and $s > t$. Then we certainly have that $K_{\alpha,s} \not\subset \hat{K}_{\beta,t}$, where $\hat{K}_{\beta,t}$ denotes the closed, convex, balanced hull of $K_{\beta,t}$. Pick $x_0 \in K_{\alpha,s}$ such that $x_0 \notin \hat{K}_{\beta,t}$. By the Hahn–Banach theorem (see [14, p. 31, Proposition 2.4]) there exists $\phi \in E'$ with

$$\phi(x_0) = 1, \qquad |\phi(x)| \leqslant \sigma < 1 \quad \text{for} \quad \text{all } x \in \hat{K}_{\beta,t}$$

If $f_n = \phi^n$, then $f_n \to 0$ uniformly on $K_{\beta,t}$ but $f_n(x_0) = 1$ for all n. Thus neither $\mathscr{H}^\infty(K_{\alpha,s})$ has the induced topology of $\mathscr{H}^\infty(K_{\beta,t})$ nor $(\mathscr{H}(K_{\alpha,s}), \tau_\omega)$ has the induced of $(\mathscr{H}(K_{\beta,t}), \tau_\omega)$ Q.E.D.

Remark. We do not know whether the inductive limit $\mathscr{H}(K) = \varinjlim \mathscr{H}(K_\alpha)$ is strict.

Chae [6] and Hirschowitz [22] proved independently that $\mathscr{H}(K) = \varinjlim \mathscr{H}^\infty(U)$ is a regular inductive limit whenever E is a Banach space. Actually Chae proved Cauchy regularity. However, their proofs are rather complicated, and in fact were originally incomplete. In [28] we gave a much simpler proof, based on a theorem of Grothendieck on (DF)-spaces: Proposition 1.8. Our proof also showed Cauchy regularity. Here we apply the same technique to show that the inductive limit $\mathscr{H}(K) = \varinjlim \mathscr{H}^\infty(U)$ is regular whenever E is semimetrizable. However, we are unable to prove Cauchy regularity in this case. The following lemma is the heart of the proof.

LEMMA 3.1. *The closed unit ball of $\mathscr{H}^\infty(K_{\alpha,s})$ is closed in $\mathscr{H}(K)$.*

Proof. Let $\mathscr{B}$ denote the closed unit ball of $\mathscr{H}^\infty(K_{\alpha,s})$. We first prove the following:

(*) If a net $\{f_\lambda\} \subset \mathscr{B}$ is $\mathscr{H}(K)$-Cauchy, then $f_\lambda - f_\mu \to 0$ uniformly on each compact subset of $K_{\alpha,s}$.

Let $L \subset K_{\alpha,s}$ be compact. Then $L \subset K_{\alpha,t}$ for some $t < s$. Since $\{f_\lambda\}$ is $\mathscr{H}(K)$-Cauchy, Lemma 2.1 implies that

(1) $\sup_{\xi \in K,\, x \in L} |(1/m!)\hat{d}^m(f_\lambda - f_\mu)(\xi)(x - \xi)| \to 0$

and since $\{f_\lambda\} \subset \mathscr{B}$, the Cauchy inequalities imply that

(2) $\sup_{\xi \in K} \|(1/m!)\hat{d}^m f_\lambda(\xi)\|_\alpha \leqslant 1/s^m$.

Using (2) we obtain, for every $k \in N$,

$$\begin{aligned}
\sup_{x \in L} |(f_\lambda - f_\mu)(x)| &\leqslant \sum_{m=0}^{\infty} \sup_{\substack{\xi \in K\\ x \in L}} |(1/m!)\hat{d}^m(f_\lambda - f_\mu)(\xi)(x - \xi)| \\
&\leqslant \sum_{0}^{k} \sup_{\substack{\xi \in K\\ x \in L}} |(1/m!)\hat{d}^m(f_\lambda - f_\mu)(\xi)(x - \xi)| \\
&\quad + \sum_{k+1}^{\infty} \sup_{\substack{\xi \in K\\ x \in L}} \|(1/m!)\hat{d}^m(f_\lambda - f_\mu)(\xi)\|_\alpha [\alpha(x - \xi)]^m \\
&\leqslant \sum_{0}^{k} \sup_{\substack{\xi \in K\\ x \in L}} |(1/m!)\hat{d}^m(f_\lambda - f_\mu)(\xi)(x - \xi)| + \sum_{k+1}^{\infty} (2/s^m)t^m
\end{aligned}$$

Given $\epsilon > 0$ we first choose k so that

$$\sum_{k+1}^{\infty} (2/s^m)t^m = 2 \sum_{k+1}^{\infty} (t/s)^m \leqslant \epsilon/2$$

and then by (1) we can find λ_0 such that

$$\sum_0^k \sup_{\substack{\xi \in K \\ x \in L}} |(1/m!)\hat{d}^m(f_\lambda - f_\mu)(\xi)(x - \xi)| \leqslant \epsilon/2 \qquad \text{for} \quad \lambda, \mu \geqslant \lambda_0$$

Thus $\sup_{x \in L} |(f_\lambda - f_\mu)(x)| \leqslant \epsilon$ for $\lambda, \mu \geqslant \lambda_0$ proving $(*)$.

Now let $f \in \mathscr{H}(K)$-closure of $\mathscr{B}$. Then there exists a net $\{f_\lambda\} \subset \mathscr{B}$ with $f_\lambda \to f$ in $\mathscr{H}(K)$. Thus $\{f_\lambda\}$ is $\mathscr{H}(K)$-Cauchy and, therefore, by $(*)$, $f_\lambda - f_\mu \to 0$ uniformly on each compact subset of $K_{\alpha,s}$. We define g on $K_{\alpha,s}$ as the pointwise limit of $\{f_\lambda\}$. The function g is well defined and $f_\lambda \to g$ uniformly on each compact subset of $K_{\alpha,s}$. Hence $g \in \mathscr{B}$. To conclude the proof we show that $f = g$ on a neighborhood of K. It suffices to show that:

(3) $\hat{d}^m f(\xi) = \hat{d}^m g(\xi)$ for all $\xi \in K$, $m \in N$.

Fix $h \in E$ and choose $\rho > 0$ such that $\{\xi + \zeta h : \zeta \in C, |\zeta| \leqslant \rho\} \subset K_{\alpha,s}$. Then, using the Cauchy integral formula [4, p. 530, Proposition 1.4] we obtain

$$|\hat{d}^m(f_\lambda - g)(\xi)(h)| \leqslant (m!/s^m) \sup_{|\zeta| = \rho} |(f_\lambda - g)(\xi + \zeta h)|$$

and the right-hand side goes to zero because the set $\{\xi + \zeta h : |\zeta| = \rho\}$ is compact. Thus

(4) $\hat{d}^m f_\lambda(\xi)(h) \to \hat{d}^m g(\xi)(h)$.

Since $f_\lambda \to f$ in $\mathscr{H}(K)$, Lemma 2.1 implies that

(5) $\hat{d}^m f_\lambda(\xi)(h) \to \hat{d}^m f(\xi)(h)$.

Now (3) follows from (4) and (5). Q.E.D.

THEOREM 3.1. *Let E be a semimetrizable locally convex space and let $K \subset E$ be compact. Then $\mathscr{H}(K) = \varinjlim \mathscr{H}^\infty(K_{\alpha,s})$ is a regular inductive limit, i.e., every bounded subset of $\mathscr{H}(K)$ is contained and bounded in some $\mathscr{H}^\infty(K_{\alpha,s})$.*

Proof. $\mathscr{H}(K)$ is the inductive limit of an increasing sequence of Banach spaces:

$$\mathscr{H}(K) = \varinjlim \mathscr{H}^\infty(K_{\alpha_n,s_n})$$

Therefore, by Proposition 1.8, given $\mathscr{X} \subset \mathscr{H}(K)$ bounded there exists n and k such that $\mathscr{X}$ is contained in the $\mathscr{H}(K)$-closure of $k\mathscr{B}_n$, where $\mathscr{B}_n$ denotes the closed unit ball of $\mathscr{H}(K_{\alpha_n,s_n})$. By Lemma 3.1., $k\mathscr{B}_n$ is closed in $\mathscr{H}(K)$. Hence $\mathscr{X} \subset k\mathscr{B}_n$. Q.E.D.

COROLLARY. *Let $\mathscr{X} \subset \mathscr{H}(K)$, where K is a compact subset of a semimetrizable locally convex space E. Then the following are equivalent:*

(a) *$\mathscr{X}$ is bounded in $\mathscr{H}(K)$.*
(b) *$\mathscr{X}$ is contained and bounded in $\mathscr{H}^{\infty}(U)$ for some $U \supset K$.*
(c) *$\mathscr{X}$ is contained and bounded in $(\mathscr{H}(U), \tau_{\omega})$ for some $U \supset K$.*
(d) *$\mathscr{X}$ is contained and bounded in $\mathscr{H}(K_{\alpha})$ for some $\alpha \in \text{cs}(E)$.*

As an application of Theorem 3.1 we show that $\mathscr{H}(K)$ is not metrizable whenever K is a compact subset of a semimetrizable locally convex space. We need a lemma.

LEMMA 3.2. *Whenever $\alpha \leqslant \beta$ and $s > t$ we have that*

$$\mathscr{H}^{\infty}(K_{\alpha,s}) \subsetneq \mathscr{H}^{\infty}(K_{\beta,t})$$

Proof. As in the proof of Proposition 3.1 we can find $\phi \in E'$ with

$$\phi(x_0) = 1, \qquad |\phi(x)| \leqslant \sigma < 1 \qquad \text{for} \quad \text{all } x \in \hat{K}_{\beta,t}$$

where $x_0 \in K_{\alpha,s}$, $x_0 \notin \hat{K}_{\beta,t}$. We define $f \in \mathscr{H}^{\infty}(K_{\beta,t})$ by

$$f(x) = 1/[1 - \phi(x)] = \sum_{m=0}^{\infty} [\phi(x)]^m$$

Certainly f has no holomorphic extension to x_0. Thus $f \notin \mathscr{H}^{\infty}(K_{\alpha,s})$. Q.E.D.

PROPOSITION 3.2. *Let E be a semimetrizable locally convex space and let $K \subset E$ be compact. Then $\mathscr{H}(K)$ is not metrizable.*

Proof. We write

$$\mathscr{H}(K) = \varinjlim \mathscr{H}^{\infty}(K_{\alpha_n,s_n})$$

where $\alpha_n \leqslant \alpha_{n+1}$ and $s_n > s_{n+1}$ for all n. By Lemma 3.2 we can find a sequence $\{f_n\} \subset \mathscr{H}(K)$ such that

$$f_n \notin \mathscr{H}^{\infty}(K_{\alpha_n,s_n}), \qquad f_n \in \mathscr{H}^{\infty}(K_{\alpha_{n+1},s_{n+1}})$$

Assume $\mathscr{H}(K)$ was metrizable. Then by the Mackey countability condition (see [23, p. 116, Proposition 3]), we could find $\{\lambda_n\}$, $\lambda_n > 0$, such that $\{\lambda_n f_n\}$ is bounded in $\mathscr{H}(K)$. Then by Theorem 3.1 we could find j so that $\{\lambda_n f_n\} \subset \mathscr{H}^{\infty}(K_{\alpha_j,s_j})$. Hence $\{f_n\} \subset \mathscr{H}^{\infty}(K_{\alpha_j,s_j})$, a contradiction. Q.E.D.

We do not know if any of the inductive limits $\mathscr{H}(K) = \varinjlim \mathscr{H}^{\infty}(U)$, $\mathscr{H}(K) = \varinjlim (\mathscr{H}(U), \tau_{\omega})$, or $\mathscr{H}(K) = \varinjlim \mathscr{H}(K_{\alpha})$ are regular when E is not semimetrizable. We have shown that all of the previous inductive limits are regular whenever E is semimetrizable. However, we are unable to prove Cauchy regularity in this case. In order to obtain Cauchy regularity we impose a further restriction on E.

For any locally convex space E we have

$$\mathscr{P}(^m E) = \bigcup_\alpha \mathscr{P}(^m E_\alpha)$$

Each $\mathscr{P}(^m E_\alpha)$ is a Banach space and $\mathscr{P}(^m E)$ is endowed with the inductive limit topology

$$\mathscr{P}(^m E) = \varinjlim \mathscr{P}(^m E_\alpha)$$

LEMMA 3.3. *If* $\gamma \in \mathrm{cs}[\mathscr{P}(^m E)]$, *then*

$$p(f) = \sup_{\xi \in K} \gamma[\hat{d}^m f(\xi)]$$

defines a continuous seminorm on $\mathscr{H}(K)$.

Proof. We show that the restriction of p to each $\mathscr{H}^\infty(K_{\alpha,s})$ is continuous. Since the restriction of γ to $\mathscr{P}(^m E_\alpha)$ is continuous, there exists $C_\alpha > 0$ such that

$$\gamma(P) \leqslant C_\alpha \|P\|_\alpha \qquad \text{for} \quad \text{all } P \in \mathscr{P}(^m E_\alpha)$$

Therefore for any $f \in \mathscr{H}^\infty(K_{\alpha,s})$ we have

$$p(f) = \sup_{\xi \in K} \gamma[\hat{d}^m f(\xi)] \leqslant C_\alpha \sup_{\xi \in K} \|\hat{d}^m f(\xi)\|_\alpha \leqslant (C_\alpha m!/s^m) \sup_{x \in K_{\alpha,s}} |f(x)| \quad \text{Q.E.D.}$$

DEFINITION 3.1. A locally convex space E is said *to satisfy condition* (*B*) if its topology can be defined by a directed family of seminorms $\{\alpha\}$ such that each $\mathscr{P}(^m E_\alpha)$ has the induced topology of $\mathscr{P}(^m E)$, for every $m \in N$.

Condition (B) was introduced by Barroso (see [4, p. 543, theorem 3.1]). Certainly every seminormed space satisfies condition (B). We give more examples in Section 4.

LEMMA 3.4. *Assume E satisfies condition* (*B*). *Then for every net* $\{f_\lambda\} \subset \mathscr{H}^\infty(K_{\alpha,s})$ *which is* $\mathscr{H}(K)$-*Cauchy we have*

$$\sup_{\xi \in K} \|\hat{d}^m (f_\lambda - f_\mu)(\xi)\|_\alpha \to 0$$

for each $m \in N$.

Proof. Since E satisfies condition (B), there exists $\gamma \in \mathrm{cs}[\mathscr{P}(^m E)]$ such that

$$\|P\|_\alpha \leqslant \gamma(P) \qquad \text{for all } P \in \mathscr{P}(^m E_\alpha)$$

Hence

$$\sup_{\xi \in K} \|\hat{d}^m(f_\lambda - f_\mu)(\xi)\|_\alpha \leqslant \sup_{\xi \in K} \gamma[\hat{d}^m(f_\lambda - f_\mu)(\xi)]$$

and the right-hand side goes to zero, by Lemma 3.3. Q.E.D.

LEMMA 3.5. *Let E be any locally convex space satisfying condition (B), let $K \subset E$ be compact, and let $\mathscr{X} \subset \mathscr{H}^\infty(K_{\alpha,s})$ be bounded. Then every net $\{f_\lambda\} \subset \mathscr{X}$ which is $\mathscr{H}(K)$-Cauchy is also $\mathscr{H}^\infty(K_{\alpha,t})$-Cauchy for every $t < s$.*

Proof. $\mathscr{X}$ being a bounded subset of $\mathscr{H}^\infty(K_{\alpha,s})$ there exists $C > 0$ such that

$$\sup_{x \in K_{\alpha,s}} |f_\lambda(x)| \leqslant C \qquad \text{for every } \lambda$$

and so, by Cauchy inequalities, we have

(1) $\sup_{\xi \in K} \|(1/m!)\hat{d}^m f_\lambda(\xi)\|_\alpha \leqslant C/s^m$ for every λ.

On the other hand, by Lemma 3.4,

(2) $\sup_{\xi \in K} \|(1/m!)\hat{d}^m(f_\lambda - f_\mu)(\xi)\|_\alpha \to 0$.

Using (1) we obtain for each $k \in N$:

$$\begin{aligned}\sup_{x \in K_{\alpha,t}} |(f_\lambda - f_\mu)(x)| &\leqslant \sum_{m=0}^{\infty} \sup_{\xi \in K} \|(1/m!)\hat{d}^m(f_\lambda - f_\mu)(\xi)\|_\alpha t^m \\ &\leqslant \sum_{0}^{k} \sup_{\xi \in K} \|(1/m!)\hat{d}^m(f_\lambda - f_\mu)(\xi)\|_\alpha t^m + \sum_{k+1}^{\infty} (2C/s^m)t^m\end{aligned}$$

Given $\epsilon > 0$ we first choose $k \in N$ so that

$$\sum_{k+1}^{\infty} (2C/s^m)t^m = 2C \sum_{k+1}^{\infty} (t/s)^m \leqslant \epsilon/2$$

and then, by (2), we can find λ_0 such that

$$\sum_{0}^{k} \sup_{\xi \in K} \|(1/m!)\hat{d}^m(f_\lambda - f_\mu)(\xi)\|_\alpha t^m \leqslant (\epsilon/2) \qquad \text{for } \lambda, \mu \geqslant \lambda_0$$

Thus

$$\sup_{x \in K_{\alpha,t}} |(f_\lambda - f_\mu)(x)| \leqslant \epsilon \qquad \text{for} \quad \lambda, \mu \geqslant \lambda_0 \qquad \text{Q.E.D.}$$

From Theorem 3.1 and Lemma 3.5 we obtain at once the next theorem.

THEOREM 3.2. *Let E be a semimetrizable locally convex space satisfying condition (B), and let $K \subset E$ be compact. Then $\mathscr{H}(K) = \varinjlim \mathscr{H}^\infty(K_{\alpha,s})$ is a Cauchy regular inductive limit, i.e., given $\mathscr{X} \subset \mathscr{H}(K)$ bounded, there exists $\alpha \in \mathrm{cs}(E)$ and $s > 0$ such that*:

(a) $\mathscr{X}$ is contained and bounded in $\mathscr{H}^\infty(K_{\alpha,s})$,
(b) Every net $\{f_\lambda\} \subset \mathscr{X}$ is $\mathscr{H}(K)$-Cauchy if and only if it is $\mathscr{H}^\infty(K_{\alpha,s})$-Cauchy.

COROLLARY 1. *Let E be a semimetrizable locally convex space satisfying condition (B) and let $K \subset E$ be compact. Then the inductive limits $\mathscr{H}(K) = \varinjlim(\mathscr{H}(K_{\alpha,s}), \tau_\omega)$ and $\mathscr{H}(K) = \varinjlim \mathscr{H}(K_\alpha)$ are both Cauchy regular.*

COROLLARY 2. *Let E be a semimetrizable locally convex space satisfying condition (B) and let $K \subset E$ be compact. Then every bounded subset of $\mathscr{H}(K)$ is metrizable.*

COROLLARY 3. *Let E be a semimetrizable locally convex space satisfying condition (B), let $K \subset E$ be compact, and let $\mathscr{X} \subset \mathscr{H}(K)$. Then the following are equivalent*:

(a) *$\mathscr{X}$ is compact in $\mathscr{H}(K)$.*
(b) *$\mathscr{X}$ is contained and compact in $\mathscr{H}^\infty(U)$ for some $U \supset K$.*
(c) *$\mathscr{X}$ is contained and compact in $(\mathscr{H}(U), \tau_\omega)$ for some $U \supset K$.*
(d) *$\mathscr{X}$ is contained and compact in $\mathscr{H}(K_\alpha)$ for some $\alpha \in \mathrm{cs}(E)$.*

COROLLARY 4. *Let E be a semimetrizable locally convex space satisfying condition (B) and let $K \subset E$ be compact. Then $\mathscr{H}(K)$ satisfies the Mackey convergence condition, i.e., given a null sequence $\{f_n\} \subset \mathscr{H}(K)$ there exists a sequence $\{\lambda_n\}$, with $\lambda_n > 0$ and $\lambda_n \to \infty$, such that $\{\lambda_n f_n\}$ is still a null sequence.*

Proof. Let $\{f_n\}$ be a null sequence in $\mathscr{H}(K)$. By Theorem 3.2 $\{f_n\}$ is a null sequence in $\mathscr{H}^\infty(U)$ for some $U \supset K$. Therefore there exists $\{n_j\} \subset N$, $\{n_j\}$ strictly increasing, such that

$$\sup_{x \in U} |f_n(x)| \leqslant 1/j^2 \qquad \text{for} \quad n \geqslant n_j$$

If we define $\lambda_n = j$ for $n_j \leqslant n < n_{j+1}$, then

$$\sup_{x \in U} |\lambda_n f_n(x)| \leqslant 1/j \qquad \text{for} \quad n_j \leqslant n < n_{j+1}$$

and $\{\lambda_n f_n\}$ is a null sequence in $\mathscr{H}^\infty(U)$, hence in $\mathscr{H}(K)$. Q.E.D.

The most important consequence of Theorem 3.2 is the following.

THEOREM 3.3. *Let E be a semimetrizable locally convex space satisfying condition (B) and let $K \subset E$ be compact. Then $\mathscr{H}(K)$ is complete.*

Proof. It is clear from Theorem 3.2 that every closed bounded subset of $\mathscr{H}(K)$ is complete. Thus $\mathscr{H}(K)$ is quasi-complete, and hence complete, by Proposition 1.7. Q.E.D.

Theorem 3.2 and its consequences were established by Chae [6] in the case where E is a Banach space. We have shown that Theorem 3.2 is still valid when E is a semimetrizable locally convex space satisfying condition (B). This class of spaces is studied in detail in the next section.

4. THE CONDITION (B)

The condition (B) introduced in Section 3 does not seem easy to check in specific cases. In this section we establish the relationship of condition (B) to several other conditions which have been studied in greater detail. We show, in particular, that a distinguished Fréchet space satisfies condition (B) if and only if it satisfies condition (C) studied by Nachbin in [34]. The class of semimetrizable locally convex spaces satisfying condition (B) includes all countable products of seminormed spaces and several function spaces, like the space of continuous functions $\mathscr{C}(X; C)$, the space of locally integrable pth power functions $\mathscr{L}^p_{\text{loc}}(X; \mu; C)$, and the space of continuously m-differentiable functions $\mathscr{C}^m(U; C)$. On the other hand, the space $\mathscr{H}(U)$ of holomorphic functions on a domain U in C^n is a distinguished Fréchet space not satisfying condition (B).

DEFINITION 4.1. A locally convex space E is said to *satisfy condition (A)* if its topology can be defined by a directed family of seminorms $\{\alpha\}$ such that for every α there exists a bounded set $B_\alpha \subset E$ whose image in E/α is an O-neighborhood.

This condition was introduced by Aron [3].

PROPOSITION 4.1. *Let E be a semimetrizable locally convex space satisfying condition (A). Then there exists an increasing sequence of seminorms* $\{\alpha_n\}$ *defining the topology of E, and a bounded set* $B \subset E$ *whose image in each* E/α_n *is an* O*-neighborhood.*

Proof. There exists $\{\alpha_n\}$ increasing and defining the topology of E, and $\{B_n\}$, $B_n \subset E$ bounded, such that $\pi_n\ (B_n)$ is an O-neighborhood in E/α_n. By the Mackey countability condition (see [23, p. 116, proposition 3), there exists $\{\lambda_n\}$, $\lambda_n > 0$, such that $B = \bigcup \lambda_n B_n$ is a bounded subset of E. Thus $\pi_n(B) \supset \lambda_n \pi_n(B_n)$, and $\pi_n(B)$ is an O-neighborhood in E/α_n for every n.

Q.E.D.

PROPOSITION 4.2. *Let E be any locally convex space. The following are equivalent*:

(C_1) *The topology of E can be defined by a directed family of seminorms* $\{\alpha\}$ *such that for every* α, *the mapping* $\pi_\alpha: E \to E/\alpha$ *is open.*

(C_2) *The topology of E can be defined by a directed family of seminorms* $\{\alpha\}$ *such that, whenever* $\alpha \leqslant \beta$, *the mapping* $\pi_{\alpha\beta}: E/\beta \to E/\alpha$ *is open.*

(C_3) *The topology of E can be defined by a directed family of seminorms* $\{\alpha\}$ *such that, whenever* $\alpha \leqslant \beta$, *there exists a bounded subset of* E/β *whose image in* E/α *is an* O*-neighborhood.*

Proof. $(C_1) \Rightarrow (C_2)$. Whenever $\alpha \leqslant \beta$ the following diagram is commutative.

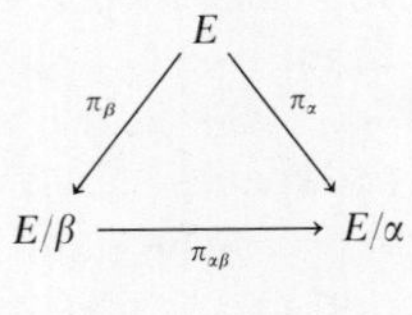

Let $U \subset E/\beta$ be open. Then π_β being continuous, $\pi_\beta^{-1}(U) \subset E$ is open. Also π_α being open, $\pi_\alpha[\pi_\beta^{-1}(U)] \subset E/\alpha$ is open. But

$$\pi_\alpha[\pi_\beta^{-1}(U)] = \pi_{\alpha\beta}(U)$$

Thus $\pi_{\alpha\beta}$ is open.

$(C_2) \Rightarrow (C_1)$. Fix α. Let U be a basic O-neighborhood in $E: U = \pi_\beta^{-1}(V)$, where V is an O-neighborhood in E/β. Choose γ with $\gamma \geqslant \alpha$, $\gamma \geqslant \beta$. Then the mappings

$$\pi_{\alpha\gamma}: E/\gamma \to E/\alpha \qquad \text{and} \qquad \pi_{\beta\gamma}: E/\gamma \to E/\beta$$

are both continuous and open. Then $\pi_{\beta\gamma}$ being continuous, $\pi_{\beta\gamma}^{-1}(V)$ is an

O-neighborhood in E/γ. Also $\pi_{\alpha\gamma}$ being open, $\pi_{\alpha\gamma}[\pi_{\beta\gamma}^{-1}(V)]$ is an O-neighborhood in E/α. But

$$\pi_{\alpha\gamma}[\pi_{\beta\gamma}^{-1}(V)] = \pi_{\alpha}[\pi_{\beta}^{-1}(V)] = \pi_{\alpha}(V).$$

Hence $\pi_{\alpha}(U)$ is an O-neighborhood in E/α, and π_{α} is open.

$(C_2) \Rightarrow (C_3)$. Given $\alpha \leqslant \beta$, we let B denote the closed unit ball of E/β. Then $\pi_{\alpha\beta}$ being open, $\pi_{\alpha\beta}(B)$ is an O-neighborhood in E/α.

$(C_3) \Rightarrow (C_2)$. Given $\alpha \leqslant \beta$, there exists a bounded subset B of E/β whose image in E/α is an O-neighborhood. Let U be any O-neighborhood in E/β. Since U absorbs B, $\pi_{\alpha\beta}(U)$ is an O-neighborhood in E/α, and $\pi_{\alpha\beta}$ is open, Q.E.D.

DEFINITION 4.2. A locally convex space E is said to *satisfy condition* (C) if it satisfies any of the equivalent conditions of Proposition 4.2.

This condition was introduced by Nachbin [34]. The following result is also due to Nachbin [34].

PROPOSITION 4.3. *Let E be a locally convex space which has a continuous norm. Then E satisfies condition* (C) *if and only if E is normable.*

Proof. Suppose E has a continuous norm γ and satisfies condition (C). Then there exists $\alpha \in \text{cs}(E)$ such that $\gamma \leqslant C\alpha$ for some $C > 0$ and the mapping $E \to E/\alpha$ is open. γ being a norm, so is α. Hence the identity mapping $E \to E_{\alpha}$ is continuous and open, i.e., a homeomorphism. Q.E.D.

DEFINITION 4.3. A locally convex space E is said to *satisfy condition* (N), or to be *N-complete*, if its topology can be defined by a directed family of seminorms $\{\alpha\}$ such that each E_{α} is complete.

This condition has been studied by Nachbin [34] and Barroso [4]. The following theorem gives the connections between conditions (A), (C), (N), and (B).

THEOREM 4.1. *Let E be a locally convex space.*

(a) *For E arbitrary, condition* (A) *implies condition* (B), *condition* (A) *implies condition* (C), *and condition* (N) *implies condition* (C).

(b) *For E Fréchet, conditions* (A), (N), *and* (C) *are equivalent, and each of them implies condition* (B).

(c) *For E Fréchet and distinguished, conditions* (A), (N), (C), *and* (B) *are equivalent.*

We give the proof of the theorem in a number of lemmas.

LEMMA 4.1. *For an arbitrary locally convex space E, condition (A) implies condition (B).*

Proof. Assume E satisfies condition (A) and fix α, m. There exists a bounded set $B_\alpha \subset E$ whose image $\pi_\alpha(B_\alpha)$ is an O-neighborhood in E/α. We may as well assume that

$$\pi_\alpha(B_\alpha) \supset \{x \in E/\alpha; \alpha(x) \leqslant 1\} \qquad (*)$$

For $P \in \mathscr{P}(^m E)$ we define

$$\gamma(P) = \sup_{x \in B_\alpha} |P(x)|$$

$B_\alpha \subset E$ being bounded, $C_\beta = \sup_{x \in B_\alpha} \beta(x) < \infty$ for every β. Thus for $P \in \mathscr{P}(^m E_\beta)$ we have

$$\gamma(P) \leqslant \sup_{x \in B_\alpha} \|P\|_\beta [\beta(x)]^m = C_\beta{}^m \|P\|_\beta$$

and γ is a continuous seminorm on $\mathscr{P}(^m E) = \varinjlim \mathscr{P}(^m E_\beta)$. Let $x \in E$ with $\alpha(x) \leqslant 1$. Then by $(*)$ we have that $\pi_\alpha(x) \in \pi_\alpha(B_\alpha)$, i.e., $\pi_\alpha(x) = \pi_\alpha(y)$ for some $y \in B_\alpha$. Hence $\alpha(x - y) = 0$ and therefore $P(x) = P(y)$ for all $P \in \mathscr{P}(^m E_\alpha)$. Therefore for all $P \in \mathscr{P}(^m E_\alpha)$, we have

$$\|P\|_\alpha = \sup_{\alpha(x) \leqslant 1} |P(x)| \leqslant \sup_{y \in B_\alpha} |P(y)| = \gamma(P)$$

and $\mathscr{P}(^m E_\alpha)$ has the induced topology of $\mathscr{P}(^m E)$. Q.E.D.

LEMMA 4.2. *For an arbitrary locally convex space E, condition (A) implies condition (C).*

The proof is similar to the proof of $(C_3) \Rightarrow (C_2)$ in Proposition 4.2.

LEMMA 4.3. *For an arbitrary locally convex space E, condition (N) implies condition (C).*

Proof. For $\alpha \leqslant \beta$ the mapping $\pi_{\alpha\beta}: E/\beta \to E/\alpha$ is continuous and surjective, and E/β and E/α are Banach spaces. By the open mapping theorem, $\pi_{\alpha\beta}$ is open. Q.E.D.

LEMMA 4.4. *For E Fréchet, condition (C) implies condition (A).*

Proof. Let E be a Fréchet space satisfying condition (C). Then there exists an increasing sequence of seminorms $\{\alpha_j\}$ defining the topology of E, and such that for every $j \geqslant 2$ there exists a bounded set $A_j \subset E/\alpha_j$ whose image in E/α_{j-1} is an O-neighborhood. After replacing each A_j by a suitable multiple, we may assume that:

(1) $\pi_{j-1,j}(A_j) \supset A_{j-1}$ for every $j \geqslant 3$.

Fix n and define

(2) $B_n = \bigcap_{j>n} \pi_j^{-1}(A_j)$.

Since $\pi_j(B_n) \subset A_j$ for all $j > n$, B_n is a bounded subset of E. We claim that

(3) $\pi_{n+1}(B_n) = A_{n+1}$.

We need only show that $A_{n+1} \subset \pi_{n+1}(B_n)$. Let $x_{n+1} \in A_{n+1}$. Applying (1) and induction we can find $x_j \in A_j$ ($j \geqslant n + 2$) such that

(4′) $x_j = \pi_{j,\,j+1}(x_{j+1})$ for $j > n$.

We also define $x_j \in E/\alpha_j$ for $1 \leqslant j \leqslant n$ by

(4″) $x_j = \pi_{j,\,j+1}(x_{j+1})$ for $1 \leqslant j \leqslant n$.

Thus we have

(4) $x_j = \pi_{jk}(x_k)$ for $j < k$.

Since E is complete, $E = \varprojlim E/\alpha_j$ by Proposition 1.2 and there exists $x \in E$ with $\pi_j(x) = x_j$ for all j. Since $\pi_j(x) = x_j \in A_j$ for all $j > n$, $x \in B_n$. Thus

$$x_{n+1} = \pi_{n+1}(x) \in \pi_{n+1}(B_{n+1})$$

and $A_{n+1} \subset \pi_{n+1}(B_{n+1})$, proving (3).

Using (3) we have

$$\pi_n(B_n) = \pi_{n,\,n+1} \circ \pi_{n+1}(B_n) = \pi_{n,\,n+1}(A_{n+1})$$

But $\pi_{n,\,n+1}(A_{n+1})$ is by hypothesis an O-neighborhood in E/α_n. Thus $\pi_n(B)$ is an O-neighborhood in E/α_n and E satisfies condition (A). Q.E.D.

LEMMA 4.5. *For E Fréchet, condition (C) implies condition (N).*

Proof. Let E be a Fréchet space satisfying condition (C). Since E is Fréchet and $E \to E/\alpha_n$ is open, E/α_n is complete. One can prove this fact as follows. Since the mapping $E \to E/\alpha_n$ is open, the mapping $E \to \widehat{E/\alpha_n}$ of E into the completion of E/α_n is nearly open. By the open mapping theorem (see [14, p. 41, proposition 1.3] or [18, p. 69, lemma]), the mapping $E \to \widehat{E/\alpha_n}$ is surjective. Hence $E/\alpha_n = \widehat{E/\alpha_n}$. Q.E.D.

LEMMA 4.6. *For E Fréchet and distinguished, condition (B) implies condition (C).*

Proof. Let E be a distinguished Fréchet space satisfying condition (B). Since E is metrizable and distinguished, the strong dual E_β' is complete and bornological (see [23, p. 223, proposition 6 and p. 289, theorem 1). Therefore E_β' is the inductive limit of a family of Banach spaces (see [38, p. 63, proposition 8.4]). Thus E_β' is ultrabornological.

Since E satisfies condition (B), there exists an increasing sequence of seminorms $\{\alpha_n\}$ defining the topology of E such that, for every m, each $\mathscr{P}(^mE_n)$ has the induced topology of $\mathscr{P}(^mE)$. We apply condition (B) for $m = 1$ only. Thus we consider the dual E' endowed with the inductive limit topology

$$E_i' = \varinjlim (E_{\alpha_n})_\beta'$$

Since each inclusion mapping $(E_{\alpha_n})_\beta' \hookrightarrow E_\beta'$ is continuous, so is the identity mapping $E_i' \to E_\beta'$. Thus E_i' is an (LF)-space and E_β' is ultrabornological. By Proposition 1.5, the identity mapping $E_i' \to E_\beta'$ is a homeomorphism. Thus each $(E/\alpha_n)_\beta' = (E_{\alpha_n})_\beta'$ has the induced topology of E_β'.

Clearly the topology induced by E_β' on $(E/\alpha_n)'$ coincides with the topology of uniform convergence on subsets of the form $\pi_n(B)$, with $B \subset E$ bounded. Let V_n denote the closed unit ball of E/α_n. Then there exists $B_n \subset E$ bounded, convex, and balanced such that

$$\sup_{x \in V_n} |\phi(x)| \leqslant \sup_{x \in \pi_n(B_n)} |\phi(x)| \qquad \text{for all } \phi \in (E/\alpha_n)'$$

Thus

$$V_n^0 \supset [\pi_n(B_n)]^0$$

and

$$V_n = V_n^{00} \subset [\pi_n(B_n)]^{00} = \overline{\pi_n(B_n)}$$

by the bipolar theorem. Let U be an O-neighborhood in E. We may assume $U \supset B_n$ and then

$$\overline{\pi_n(U)} \supset \overline{\pi_n(B_n)} \supset V_n$$

Thus $\overline{\pi_n(U)}$ is an O-neighborhood in E/α_n and the mapping $\pi_n : E \to E/\alpha_n$ is nearly open. By the open mapping theorem (see [14, p. 41, Proposition 1.3] or [18, p. 69, lemma]), π_n is open, and E satisfies condition (C). Q.E.D.

The proof of Theorem 4.1 is now complete.

Remark. From the proof of Lemma 4.6 we see that for a distinguished Fréchet space, condition (B) for $m = 1$ implies condition (B) for all m.

Next we give examples and counterexamples of semimetrizable convex spaces satisfying condition (B). All of these examples were already considered by Nachbin [34] when studying condition (C).

EXAMPLE 4.1. The product of an arbitrary family of seminormed spaces satisfies condition (A), and hence conditions (B) and (C). (If the family is countable, then the product is semimetrizable.)

Proof. Let $E = \prod_{\gamma \in \Gamma} E_\gamma$, where E_γ is seminormed by γ. Then

$$\alpha_I(x) = \sup_{\gamma \in I} \gamma(x_\gamma) \qquad (I \subset \Gamma \text{ finite})$$

gives a directed set of seminorms defining the topology of E. Let

$$B = \{x \in E : \gamma(x_\gamma) \leqslant 1 \text{ for all } \gamma \in \Gamma\}$$

Then B is a bounded subset of E and $\pi_{\alpha_I}(B)$ is the closed unit ball of E/α_I.
Q.E.D.

EXAMPLE 4.2. Let X be a completely regular topological space, and let $E = \mathscr{C}(X; C)$ be the space of all continuous C-valued functions on X, endowed with the compact open topology. Then E satisfies condition (A), and hence conditions (B) and (C). (If X is denumerable at infinity, then E is metrizable.)

Proof. The topology of E is defined by the directed family of seminorms

$$\alpha_k : f \to \sup_{x \in K} |f(x)|$$

with $K \subset X$ compact. Let $B = \{f \in E : |f(x)| \leqslant 1 \text{ for all } x \in X$. Thus B is a bounded subset of E. Fix $K \subset X$ compact and let $\pi_k : E \to E/\alpha_k$ denote the canonical mapping. We claim that

$$\pi_K(B) \supset \{f \in E/\alpha_K : \alpha_K(f) \leqslant \tfrac{1}{2}\}$$

Let $f \in E$ with $|f| \leqslant \frac{1}{2}$ on K. Choose $U \supset K$ open with $|f| \leqslant 1$ on U. With X being completely regular, there exists $\phi : X \to [0, 1]$ continuous with $\phi = 1$ on K, $\phi = 0$ off U. Let $g = \phi \cdot f$. Then $g = f$ on K, $|g| \leqslant |f| \leqslant 1$ on U, and $g = 0$ off U. Therefore $\alpha_K(f - g) = 0$ and $g \in B$. Thus $\pi_K(f) = \pi_K(g) \in \pi_K(B)$.
Q.E.D.

EXAMPLE 4.3. Let X be a locally compact space and let μ be a Borel measure on X. For $1 \leqslant p \leqslant \infty$, let $E = \mathscr{L}^p_{\text{loc}}(X; \mu; C)$ be the vector space of all measurable functions $f : X \to C$ such that $|f|^p$ is locally integrable on X. Then E is endowed with the topology defined by the directed family of seminorms.

$$\alpha_K : f \to \left(\int_K |f|^p \, d\mu \right)^{1/p}$$

with $K \subset X$ compact. Then E satisfies condition (A) and hence conditions (B) and (C). (If X is denumerable at infinity, then E is metrizable.)

Proof. Let $B = \{f \in E : \alpha_L(f) \leqslant 1 \text{ for all } L \subset X \text{ compact}\}$. Then B is a bounded subset of E. We claim that

$$\pi_K(B) \supset \{f \in E/\alpha_K : \alpha_K(f) \leqslant 1\}$$

Given $f \in E$ with $\alpha_K(f) \leqslant 1$, we define g on X by $g = f$ on K, $g = 0$ off K. Then $g \in E$, $\alpha_K(f - g) = 0$, and $\alpha_L(g) \leqslant \alpha_K(f) \leqslant 1$ for every $L \subset X$ compact. Thus $\pi_K(f) = \pi_K(g) \in \pi_K(B)$. Q.E.D.

For our next example we need a lemma.

LEMMA 4.7. *Let E, F be Banach spaces, let $U \subset E$ be open, and let $f: U \to F$ be continuously m-differentiable. For each k, $0 \leqslant k \leqslant m$, and $[\xi, \xi + t] = \{\xi + \lambda t, 0 \leqslant \lambda \leqslant 1\} \subset U$ we have that*

$$D^k f(\xi + t) = \sum_{j=0}^{m-k} \frac{1}{j!} D^{k+j} f(\xi) t^j + \left(\int_0^1 \frac{m(1-\lambda)^{m-1}}{m!} [D^m f(\xi + \lambda t) - D^m f(\xi)] \, d\lambda \right) t^{m-k}$$

where $D^k f(\xi)$ denotes the kth derivative of f at ξ.

Proof. By Taylor's formula (see [9, p. 190]), we have

$$f(\xi + t) = \sum_{j=0}^{m-1} \frac{1}{j!} D^j f(\xi) t^j + \left(\int_0^1 \frac{(1-\lambda)^{m-1}}{(m-1)!} D^m f(\xi + \lambda t) \, d\lambda \right) t^m$$

But since $\int_0^1 m(1-\lambda)^{m-1} \, d\lambda = 1$, we obtain at once

$$f(\xi + t) = \sum_{j=0}^{m} \frac{1}{j!} D^j f(\xi) t^j + \left(\int_0^1 \frac{m(1-\lambda)^{m-1}}{m!} [D^m f(\xi + \lambda t) - D^m f(\xi)] \, d\lambda \right) t^m \qquad (*)$$

This proves the lemma for $k = 0$. To prove it for $k = 1, 2, \ldots, m$ we simply apply $(*)$ to $D^k f$, for $D^k f \in \mathscr{C}^{m-k}(U; F_k)$, where $F_0 = F$, $F_1 = \mathscr{L}(E; F_0), \ldots, F_k = \mathscr{L}(E; F_{k-1})$. Q.E.D.

EXAMPLE 4.4. Let $U \subset R^n$ be open and let $E = \mathscr{C}^m(U; C)$ denote the space of continuously m-differentiable functions $f: U \to C$. For $p = (p_1, \ldots, p_n) \in$

N^n with $|p| = p_1 + \cdots + p_n \leqslant m$ we write

$$\partial^p f = \partial^{|p|} f / \partial x_1^{p_1} \cdots \partial x_n^{p_n}$$

E is endowed with the topology defined by the seminorms.

$$f \to \sup_{\xi \in K} |\partial^p f(\xi)|$$

with $|p| \leqslant m$ and $K \subset U$ compact. Then E is a Fréchet space satisfying condition (N), and hence satisfying conditions (A), (B), and (C).

Proof. We show that E satisfies condition (N). For $k = 0, 1, \ldots, m$ we have

$$\frac{1}{k!} D^k f(\xi) t^k = \sum_{|p|=k} \frac{1}{p_1! \cdots p_n!} \partial^p f(\xi) t^k \tag{$*$}$$

The seminorms

$$\alpha_k : f \to \sum_{|p| \leqslant m} \frac{1}{p_1! \cdots p_n!} \sup_{\xi \in K} |\partial^p f(\xi)|,$$

with $K \subset U$ form a directed family defining the topology of E.

Fix $K \subset U$ compact and let $\{f_\nu\}$ be a Cauchy sequence in E_{α_K}. Then $\alpha_K(f_\nu - f_\mu) \to 0$ and $(*)$ implies that

$$\sup_{\xi \in K} \|D^k(f_\nu - f_\mu)(\xi)\| \to 0$$

Therefore for each $k = 0, 1, \ldots, m$, there exists $g_k \in \mathscr{C}(K; F_k)$ such that $D^k f_\nu \to g_k$ uniformly on K. We will find $f \in \mathscr{C}^m(U; C)$ such that $D^k f = g_k$ on K for $0 \leqslant k \leqslant m$.

By Lemma 4.7 we have that

$$D^k f_\nu(\xi + t) = \sum_{j=0}^{m-k} \frac{1}{j!} D^{k+j} f_\nu(\xi) t^j + \left(\int_0^1 \frac{m(1-\lambda)^{m-1}}{m!} [D^m f_\nu(\xi + \lambda t) - D^m f_\nu(\xi)\, d\lambda] \right) t^{m-k}$$

for each λ, whenever $[\xi, \xi + t] \subset U$. Letting $\nu \to \infty$ we obtain, whenever $[\xi, \xi + t] \subset K$,

$$g_k(\xi + t) = \sum_{j=0}^{m-k} \frac{1}{j!} g_{k+j}(\xi) t^j + \left(\int_0^1 \frac{m(1-\lambda)^{m-1}}{m!} [g_m(\xi + \lambda t) - g_m(\xi)]\, d\lambda \right) t^{m-k}$$

Since g_m is uniformly continuous on K, for any $\epsilon > 0$ there exists $\delta > 0$ such that whenever $[\xi, \xi + t] \subset K$ and $\|t\| < \delta$ we have

$$\|g_k(\xi + t) - \sum_{j=0}^{m-k} \frac{1}{j!} g_{k+j}(\xi)t^j\| \leqslant \epsilon \|t\|^{m-k}$$

Thus we may apply Whitney's extension theorem (see [43, p. 65 theorem I]) and find $f \in \mathscr{C}^m(U; C)$ with $D^k f = g_k$ on K for $0 \leqslant k \leqslant m$. Therefore $f_\nu \to f$ in E_{α_k} and E_{α_k} is complete. Q.E.D.

Finally we give a counterexample.

EXAMPLE 4.5. Let U be a connected open subset of C^n and let E be $\mathscr{H}(U)$, endowed with the compact open topology. Then E is a distinguished Fréchet space with a continuous norm. Since E is not normable, E does not satisfy any of the conditions (A), (N), (C), or (B).

Proof. E is a Montel space, hence the strong dual E_β' is also Montel, in particular, barreled. Thus E is distinguished.

If K is a compact subset of U with nonempty interior, then

$$f \to \sup_{x \in K} |f(x)|$$

defines a continuous norm on E. Q.E.D.

5. COMPLETENESS OF $(\mathscr{H}(U), \tau_\omega)$

The question of completeness of $(\mathscr{H}(U), \tau_\omega)$ was raised first by Nachbin [32]. Partial affirmative answers were given by Dineen, Aron, and Chae. Dineen first proved the completeness of $(\mathscr{H}(E), \tau_\omega)$ when E is a Banach space [11], and later he proved the completeness of $(\mathscr{H}(E), \tau_\omega)$ when E is a countable product of Banach spaces [13]. Aron [2] proved that $(\mathscr{H}(U), \tau_\omega)$ is complete whenever U is a ξ-balanced open subset of a Banach space. Dineen and Aron used a polynomial expansion approach, which seems to work only when U is balanced with respect to one of its points. Chae [5, 6] proved that $(\mathscr{H}(U), \tau_\omega)$ is complete when E is a Banach space and $U \subset E$ is an open set having a certain Runge approximation property. Chae used a germ approach: he showed that $(\mathscr{H}(U), \tau_\omega)$ is the projective limit of the spaces $\mathscr{H}(K)$ with $K \subset U$. In this section we extend this results by proving that $(\mathscr{H}(U), \tau_\omega)$ is complete whenever U is an arbitrary open subset of a semimetrizable locally convex space satisfying condition (B). Our approach is similar to that of Chae: we show that $(\mathscr{H}(U), \tau_\omega)$ is the projective limit of a family of complete (DF)-spaces, each of which is a vector subspace of some $\mathscr{H}(K)$, with $K \subset U$.

We first introduce some notation. Let E be any locally convex space and let $K \subset E$ be compact. For any $V \supset K$ open all of whose connected components meet K, we let $H_V{}^K$ denote the image of the injective mapping $\mathscr{H}(V) \to \mathscr{H}(K)$. If we endow each $H_V{}^K$ with the unique norm making the bijection $\mathscr{H}^\infty(V) \to H_V{}^K$ an isometry, we have

$$\mathscr{H}(K) = \varinjlim_V H_V{}^K$$

Let M be any vector subspace of $\mathscr{H}(K)$, and define $M_V = M \cap H_V{}^K$ and $\tilde{M} = \bigcup_V \tilde{M}_V$, where $\tilde{M}_V$ is the closure of M_V in $H_V{}^K$. In a diagram we have

$$\begin{array}{ccccc} M & \hookrightarrow & \tilde{M} & \hookrightarrow & \mathscr{H}(K) \\ \uparrow & & \uparrow & & \uparrow \\ M_V & \hookrightarrow & \tilde{M}_V & \hookrightarrow & H_V{}^K \end{array}$$

Then $\tilde{M}_V$ is the completion of the vector subspace M_V of the Banach space $H_V{}^K$. We endow M and $\tilde{M}$ with the inductive limit topologies.

$$M = \varinjlim M_V, \qquad \tilde{M} = \varinjlim_V \tilde{M}_V$$

Lemma 5.1. *Let E be any locally convex space, let $K \subset E$ be compact, and let M be a vector subspace of $\mathscr{H}(K)$. Then with the preceding notation we have*:

(a) *M is sequentially dense in $\tilde{M}$.*
(b) *M has the induced topology of $\tilde{M}$.*

Proof. (a) is clear. We show (b). It suffices to show that each continuous seminorm on M extends to a continuous seminorm on $\tilde{M}$. Given $p \in \mathrm{cs}(M)$, let p_V denote the restriction of p to M_V. Then $p_V \in \mathrm{cs}(M_V)$. Since $\tilde{M}_V$ is the completion of M_V, p_V extends uniquely to a continuous seminorm $\tilde{p}_V$ on $\tilde{M}_V$. We define $\tilde{p}$ on $\tilde{M}$ by $\tilde{p} = \tilde{p}_V$ on each $\tilde{M}_V$. Then $\tilde{p}$ is well defined and continuous on $\tilde{M}$. Q.E.D.

Lemma 5.2. *Let E be a semimetrizable locally convex space satisfying condition* (B), *let $K \subset E$ be compact, and let M be a vector subspace of $\mathscr{H}(K)$. Then*:

(a) *The inductive limits $M = \varinjlim M_V$ and $\tilde{M} = \varinjlim \tilde{M}_V$ are both Cauchy regular.*
(b) *$\tilde{M}$ is complete. Thus $\tilde{M}$ is the completion of M.*

Proof. (a) Since $\mathscr{H}(K) = \varinjlim H_V{}^K$ is Cauchy regular, by Theorem 3.2, it follows at once from the diagram

$$\begin{array}{ccc} M & \hookrightarrow & \mathscr{H}(K) \\ \uparrow & & \uparrow \\ M_V & \hookrightarrow & H_V{}^K \end{array}$$

that $M = \varinjlim M_V$ is also Cauchy regular. Here we use the fact $M \cap H_V{}^K = M_V$. We do not know where $\tilde{M} \cap H_V{}^K$ is contained in $\tilde{M}_W$ for some $W \supset K$, and therefore we cannot apply the same idea to show that $\tilde{M} = \varinjlim \tilde{M}_V$ is Cauchy regular.

To show that $\tilde{M} = \varinjlim \tilde{M}_V$ is Cauchy regular, we proceed as follows. Let $\{\alpha_n\}$ be an increasing sequence of seminorms defining the topology of E, let $\{s_n\}$ be a sequence of positive numbers decreasing to zero, and let $V_n = K_{\alpha_n, s_n}$. We write $H_n{}^K$ for $H_{V_n}^K$ and $\tilde{M}_n$ for $\tilde{M}_{V_n}$. Then we have

$$\mathscr{H}(K) = \varinjlim H_n{}^K, \qquad \tilde{M} = \varinjlim \tilde{M}_n$$

We begin by proving the following:

(i) Let $B \subset \tilde{M}_n$ be bounded. Then every net $\{f_\lambda\} \subset B$ which is $\tilde{M}$-Cauchy is also $\tilde{M}_{n+1}$-Cauchy.

By Lemma 3.5 the following is true.

(ii) Let $B \subset H_n{}^K$ be bounded. Then every net $\{f_\lambda\} \subset B$ which is $\mathscr{H}(K)$-Cauchy is also H_{n+1}^K-Cauchy.

Using (ii), (i) follows at once from the diagram:

$$\begin{array}{ccccc} H_n{}^K & \hookrightarrow & H_{n+1}^K & \hookrightarrow & \mathscr{H}(K) \\ \uparrow & & \uparrow & & \uparrow \\ \tilde{M}_n & \hookrightarrow & \tilde{M}_{n+1} & \hookrightarrow & \tilde{M} \end{array}$$

Next we show the following.

(iii) Let $B \subset \tilde{M}_n$ be bounded. Then the $\tilde{M}$-closure of B is contained and bounded in $\tilde{M}_{n+1}$.

Let $f \in \tilde{M}$-closure of B. Then there exists a net $\{f_\lambda\} \subset B$ with $f_\lambda \to f$ in $\tilde{M}$. Then $\{f_\lambda\}$ is $\tilde{M}$-Cauchy and hence $\tilde{M}_{n+1}$-Cauchy by (i). Then $\tilde{M}_{n+1}$ being complete, there exists $g \in \tilde{M}_{n+1}$ such that $f_\lambda \to g$ in $\tilde{M}_{n+1}$, and hence in $\tilde{M}$. Also $\tilde{M}$ being Hausdorff, $f = g$. Thus $f \in \tilde{M}_{n+1}$-closure of B. We have proved that

$$\tilde{M}\text{-closure of } B \subset \tilde{M}_{n+1}\text{-closure of } B$$

and the set on the right-hand side is bounded in $\tilde{M}_{n+1}$. This shows (iii).

Now we can prove that $\tilde{M} = \varinjlim \tilde{M}_V$ is Cauchy regular. Let B be a bounded subset of $\tilde{M}$. By Proposition 1.8 there exists n such that B is contained in the $\tilde{M}$-closure of a bounded subset of $\tilde{M}_n$. By (iii), B is then contained and bounded in $\tilde{M}_{n+1}$. By (i) every net $\{f_\lambda\} \subset B$ which is $\tilde{M}$-Cauchy is also $\tilde{M}_{n+2}$-Cauchy. Thus $\tilde{M} = \varinjlim \tilde{M}_V$ is Cauchy regular.

(b) Since each $\tilde{M}_V$ is complete and $\tilde{M} = \varinjlim \tilde{M}_V$ is Cauchy regular, it follows at once that $\tilde{M}$ is quasi-complete. By Proposition 1.7, $\tilde{M}$ is complete.

Q.E.D.

We introduce some notation. Let E be any locally convex space and let $U \subset E$ be open. For each $K \subset U$ compact, we let $\mathscr{H}^K(U)$ denote the image of the canonical mapping $\mathscr{H}(U) \to \mathscr{H}(K)$, endowed with the inductive topology introduced at the beginning of this section:

$$\mathscr{H}^K(U) = \varinjlim_V \mathscr{H}_V{}^K(U)$$

We also consider the inductive limit

$$\tilde{\mathscr{H}}^K(U) = \varinjlim \tilde{\mathscr{H}}_V{}^K(U)$$

also defined at the beginning of this section. When E is a semimetrizable locally convex space satisfying condition (B), then $\tilde{\mathscr{H}}^K(U)$ is the completion of $\mathscr{H}^K(U)$, by Lemma 5.2.

LEMMA 5.3. *Let E be any locally convex space and let $U \subset E$ be open. Then the Nachbin topology τ_ω on $\mathscr{H}(U)$ coincides with the projective topology with respect to the canonical mappings*:

$$\pi_K : \mathscr{H}(U) \to \mathscr{H}^K(U)$$

with $K \subset U$ compact.

Proof. The projective topology on $\mathscr{H}(U)$ is defined by the seminorms $p_K \circ \pi_K$, where p_K varies over all continuous seminorms on $\mathscr{H}^K(U)$ and K varies over all compact subsets of U. We claim that each $p_K \circ \pi_K$ is a continuous seminorm on $(\mathscr{H}(U), \tau_\omega)$. Let V be open with $K \subset V \subset U$, and assume that each connected component of V meets K. Since the restriction of p_K to $H_V{}^K$ is continuous, there exists $c(V) > 0$ such that

$$p_K \circ \pi_K(f) \leqslant c(V) \sup_{x \in V} |\pi_K(f)(x)| = c(V) \sup_{x \in V} |f(x)|$$

for all $f \in \mathscr{H}(U)$ with $\sup_{x \in V} |f(x)| < \infty$. Then we obviously have that

$$p_K \circ \pi_K(f) \leqslant c(V) \sup_{x \in V} |f(x)|$$

for all $f \in \mathscr{H}(U)$. Thus $p_K \circ \pi_K$ is ported by K.

Conversely, let p be a seminorm on $\mathscr{H}(U)$, ported by $K \subset U$. If $f, g \in \mathscr{H}(U)$ are such that $\pi_K(f) = \pi_K(g)$, then f and g coincide on an open neighborhood V of U, and it follows that $p(f) = p(g)$ for

$$p(f - g) \leqslant c(V) \sup_{x \in V} |f(x) - g(x)| = 0$$

Thus we may write $p = p_K \circ \pi_K$, where p_K is a well-defined seminorm on $\mathscr{H}^K(U)$. Since p is ported by K, the restriction of p_K to each $\mathscr{H}_V{}^K(U)$ is continuous, and p_K is a continuous seminorm on $\mathscr{H}_V{}^K(U)$. Q.E.D.

LEMMA 5.4. *Let E be any locally convex space and let $U \subset E$ be open. Then the Nachbin topology τ_ω on $\mathscr{H}(U)$ coincides with the projective topology with respect to the canonical mappings*

$$\mathscr{H}(U) \to \tilde{\mathscr{H}}^K(U)$$

with $K \subset U$ compact.

This follows at once from Lemmas 5.3 and 5.1.

LEMMA 5.5. *Let E be any semimetrizable locally convex space and let $U \subset E$ be open. Then*

$$(\mathscr{H}(U), \tau_\omega) = \varprojlim_{K \subset U} \mathscr{H}^K(U)$$

Proof. Clearly the spaces $\mathscr{H}^K(U)$, with $K \subset U$, form a projective system. By Lemma 5.3 it suffices to show that

$$\mathscr{H}(U) = \varprojlim_{K \subset U} \mathscr{H}^K(U) \qquad \text{algebraically}$$

If $f \in \mathscr{H}(U) \to f_K \in \mathscr{H}^K(U)$ is the canonical mapping, then

$$\varprojlim_{K \subset U} \mathscr{H}^K(U) = \{\{f_K\} \in \prod_{K \subset U} \mathscr{H}^K(U) : f_L | K = f_K \text{ for } K \subset L\}$$

We show that the canonical mapping

$$f \in \mathscr{H}(U) \to \{f_K\} \in \varprojlim_{K \subset U} \mathscr{H}^K(U)$$

is surjective. Given $\{f_K\} \in \varprojlim \mathscr{H}^K(U)$, we define f on U by $f = f_K$ on each $K \subset U$. Then f is well defined, G-holomorphic on U, and continuous on compact subsets of U. Then E being semimetrizable, f is continuous on U. Thus $f \in \mathscr{H}(U)$. Q.E.D.

LEMMA 5.6. *Let E be any semimetrizable locally convex space and let $U \subset E$ be open. Then*

$$(\mathscr{H}(U), \tau_\omega) = \varprojlim_{K \subset U} \tilde{\mathscr{H}}^K(U)$$

Proof. The same proof of Lemma 5.5. shows that

$$\mathscr{H}(U) = \varprojlim_{K \subset U} \tilde{\mathscr{H}}^K(U) \quad \text{algebraically}$$

Combining this with Lemma 5.4 we obtain the desired result. Q.E.D.

From Lemmas 5.6 and 5.2 we obtain at once the next theorem.

THEOREM 5.1. *Let E be a semimetrizable locally convex space satisfying condition (B) and let U be any open subset of E. Then $(\mathscr{H}(U), \tau_\omega)$ is complete.*

Dineen [13] has also shown that $(\mathscr{H}(E), \tau_\omega)$ is complete whenever E is the direct sum of a sequence of finite-dimensional spaces. This result cannot be obtained by our methods, for such an E is not in general metrizable.

We have shown that τ_ω is the projective topology on $\mathscr{H}(U)$ with respect to the canonical mappings $\mathscr{H}(U) \to \tilde{\mathscr{H}}^K(U)$. We do not know whether each $\mathscr{H}^K(U)$ has the induced topology of $\mathscr{H}(K)$. If this were the case, then τ_ω would be the projective topology on $\mathscr{H}(U)$ with respect to the canonical mappings $\mathscr{H}(U) \to \mathscr{H}(K)$. We do not know whether this is true in general.

6. THE RUNGE PROPERTY

We define $K \subset U$ to be U-Runge if the canonical image of $\mathscr{H}(U)$ in $\mathscr{H}(K)$ is sequentially dense. We say that U has the Runge property if every compact subset of U is contained in another one which is U-Runge. Assuming that E is a semimetrizable locally convex space satisfying condition (B) and that $U \subset E$ has the Runge property, we show that $(\mathscr{H}(U), \tau_\omega)$ is the projective limit of the spaces $\mathscr{H}(K)$, with $K \subset U$. Chae [5, 6] obtained the same conclusion when E is a Banach space and $U \subset E$ satisfies a somewhat different and apparently stronger Runge property. We show, however, that both Runge properties coincide. We conclude this section by giving examples of open sets with the Runge property. Every ξ-balanced open set has the Runge property. The image under a biholomorphic mapping of an open set with the Runge property also has the Runge property. Finally, every Reinhardt domain containing the origin in a Banach space with an unconditional basis also has the Runge property. It has been conjectured by Chae [5, 6] that every open set has the Runge property.

DEFINITION 6.1. Let U be an open subset of a locally convex space E.

(a) A compact subset K of U is said to be *U-Runge* if the canonical image of $\mathscr{H}(U)$ in $\mathscr{H}(K)$ is sequentially dense.

(b) U is said to have the *Runge property* if every compact subset of U is contained in another one which is U-Runge.

LEMMA 6.1. *Let E be a semimetrizable locally convex space satisfying condition (B), let $K \subset E$ be compact, and let M be a vector subspace of $\mathscr{H}(K)$. Then, with the notation of Section 5, we have*:

(a) $\tilde{M}$ is the sequential closure of M in $\mathscr{H}(K)$.

(b) If M is sequentially dense in $\mathscr{H}(K)$, then, for each $V \supset K$ open, there exists W open with $K \subset W \subset V$, such that $H_V{}^K \subset \tilde{M}_W$. In particular, $\tilde{M} = \mathscr{H}(K)$ as topological vector spaces.

Proof. (a) Clearly we have $\subset$. We show $\supset$. Let $f \in \mathscr{H}(K)$ and let $\{f_j\} \in M$ with $f_j \to f$ in $\mathscr{H}(K)$. $\{f_j\} \cup \{f\}$ being bounded in $\mathscr{H}(K)$, there exists $V \supset K$ such that $\{f_j\} \cup \{f\} \subset H_V{}^K$ and $f_j \to f$ in $H_V{}^K$. Since $\{f_j\} \subset M \cap H_V{}^K = M_V$, we get that $f \in \tilde{M}_V \subset \tilde{M}$.

(b) Let $\{V_n\}$ be a fundamental sequence of open neighborhoods of K. The inclusion mappings $H_n{}^K \hookrightarrow \mathscr{H}(K)$ and $\tilde{M}_j \hookrightarrow \mathscr{H}(K)$ are continuous. If M is sequentially dense in $\mathscr{H}(K)$, then $\tilde{M} = \mathscr{H}(K)$ as sets, by (a), and hence

$$\mathscr{H}(K) = \bigcup \tilde{M}_j$$

Therefore, by Proposition 1.4, for each n there exists j such that $H_n{}^K \subset \tilde{M}_j$.

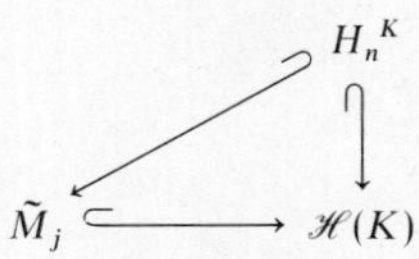

Thus for each $V \supset K$ there exists W with $K \subset W \subset V$ such that $H_V{}^K \subset \tilde{M}_W$.

Q.E.D

THEOREM 6.1. *Let E be a semimetrizable locally convex space satisfying condition (B). If $U \subset E$ is an open set with the Runge property, then*

$$(\mathscr{H}(U), \tau_\omega) = \varprojlim_{K \subset U} \mathscr{H}(K)$$

Proof. By Lemma 5.6,

$$(\mathscr{H}(U), \tau_\omega) = \varprojlim_{K \subset U} \tilde{\mathscr{H}}^K(U)$$

Moreover, since U has the Runge property, we may assume that in the preceding projective limit K varies over all compact subsets of U which are U-Runge. But for any such K, Lemma 6.1 implies that

$$\tilde{\mathscr{H}}^K(U) = \mathscr{H}(K)$$

as topological vector spaces. It follows that

$$(\mathscr{H}(U), \tau_\omega) = \varprojlim_{K \subset U} \mathscr{H}(K) \qquad \text{Q.E.D.}$$

THEOREM 6.2. *Let E be a semimetrizable locally convex space satisfying condition (B), let $U \subset E$ be open, and let $K \subset U$ be compact. Then the following are equivalent*:

(a) *K is U-Runge, i.e., $\mathscr{H}^K(U)$ is sequentially dense in $\mathscr{H}(K)$.*

(b) *Every bounded subset of $\mathscr{H}(K)$ is contained in the $\mathscr{H}(K)$-closure of a bounded subset of $\mathscr{H}^K(U)$.*

(c) *Given $\mathscr{X} \subset \mathscr{H}(K)$ bounded, there exists W open with $K \subset W \subset U$ such that $\mathscr{X} \subset \mathscr{H}(W)$ and for each $f \in \mathscr{X}$ there exists $\{f_j\} \subset \mathscr{H}(U)$ with $f_j \to f$ in $(\mathscr{H}(W), \tau_\omega)$.*

(d) *Given $\mathscr{X} \subset \mathscr{H}(K)$ bounded, there exists W open with $K \subset W \subset U$ such that $\mathscr{X} \subset \mathscr{H}^\infty(W)$ and for each $f \in \mathscr{X}$ there exists $\{f_j\} \subset \mathscr{H}(U) \cap \mathscr{H}^\infty(W)$* with $f_j \to f$ in $\mathscr{H}^\infty(W)$.

(e) Given V open with $K \subset V \subset U$, there exists W open with $K \subset W \subset V$, such that for each $f \in \mathscr{H}^\infty(V)$ there exists $\{f_j\} \subset \mathscr{H}(U) \cap \mathscr{H}^\infty(W)$ with $f_j \to f$ in $\mathscr{H}^\infty(W)$.

Proof. We write $M = \mathscr{H}^K(U)$ and use the notation of Section 5.

(a) $\Rightarrow$ (e). By (a), M is sequentially dense in $\mathscr{H}(K)$. Therefore, by Lemma 6.1., for each V open with $K \subset V \subset U$, there exists W open with $K \subset W \subset V$ such that $H_V{}^K \subset \tilde{M}_W$. This is precisely (e).

(e) $\Rightarrow$ (d). Given $\mathscr{X} \subset \mathscr{H}(K)$ bounded, by Theorem 3.1 there exists V open with $K \subset V \subset U$ such that $\mathscr{X} \subset \mathscr{H}^\infty(V)$. Now apply (e).

(d) $\Rightarrow$ (c) and (c) $\Rightarrow$ (a) are obvious. This shows that (a) $\Leftrightarrow$ (c) $\Leftrightarrow$ (d) $\Leftrightarrow$ (e).

(e) $\Rightarrow$ (b). Given $\mathscr{X} \subset \mathscr{H}(K)$ bounded, by Theorem 3.1 there exists V open with $K \subset V \subset U$ such that $\mathscr{X}$ is contained and bounded in $H_V{}^K$. By (e) there exists W open with $K \subset W \subset V$ such that $H_V{}^K \subset \tilde{M}_W$. Thus $\mathscr{X}$ is contained and bounded in $\tilde{M}_W$. Hence there exists $\mathscr{Y} \subset M_W$ bounded such that $\mathscr{X}$ is contained in the $H_W{}^K$-closure of $\mathscr{Y}$. Thus $\mathscr{Y}$ is a bounded subset of M and $\mathscr{X}$ is contained in the $\mathscr{H}(K)$-closure of $\mathscr{Y}$.

(b) ⇒ (a) Let $f \in \mathscr{H}(K)$. By (b) there exists $\mathscr{Y} \subset M$ bounded such that $f \in \mathscr{H}(K)$-closure of $\mathscr{Y}$. By Theorem 3.2 there exists V open with $K \subset V \subset U$, such that $\{f\} \cup \mathscr{Y} \subset H_V{}^K$ and $f \in H_V{}^K$-closure of $\mathscr{Y}$. Hence there exists a sequence $\{f_j\} \subset \mathscr{Y} \subset M$ with $f_j \to f$ in $H_V{}^K$, hence in $\mathscr{H}(K)$. Q.E.D.

In the case where E is a Banach space, Chae [5, 6] defined K to be U-Runge by condition (e) of Theorem 6.2. In the case where $E = C^n$, Martineau [25] defined K to be U-Runge by condition (b) and proved the equivalence (b) ⇔ (c).

Now we give some examples of open sets with the Runge property.

EXAMPLE 6.1. Let U be a ξ-balanced open subset of a locally convex space E. Then U has the Runge property.

Proof. It suffices to show that every ξ-balanced compact subset K of U is U-Runge. Given $f \in \mathscr{H}(K)$ there exists V open and ξ-balanced with $K \subset V \subset U$ such that $f \in \mathscr{H}(V)$. Then the Taylor series of f at ξ converges to f in $\mathscr{H}^\infty(V)$, hence in $\mathscr{H}(K)$. Q.E.D.

To give our second example we need some preliminary material. We recall the definition of a holomorphic mapping between locally convex spaces.

DEFINITION 6.2. Let E, F be locally convex spaces and let $U \subset E$ be open. A mapping $f: U \to F$ is *holomorphic* if for each $\xi \in U$ there exists a sequence $\{P_m\}$, with $P_m \in \mathscr{P}(^mE; F)$ such that for each $\beta \in \mathrm{cs}(F)$, there exists an open neighborhood V of ξ with $V \subset U$ such that

$$\sup_{x \in V} \beta\left[f(x) - \sum_{m=0}^{k} P_m(x - \xi) \right] \to 0$$

when $k \to \infty$.

The following result can be found in [21, p. 79, th. 3].

PROPOSITION 6.1. *Let E, F be locally convex spaces, let $U \subset E$ be open, and let $f \in \mathscr{H}(U; F)$. Let $V \subset F$ be open and let $g \in \mathscr{H}(V)$. If $f(U) \subset V$, then $g \circ f \in \mathscr{H}(U)$.*

In [21] it is required that F be quasi-complete. This additional hypothesis is necessary there for the polynomial coefficients P_m are allowed to take values in the completion $\hat{F}$ of F. But with our definition, $P_m \in \mathscr{P}(^mE; F)$, and F need not be quasi-complete.

LEMMA 6.2. *Let U, V be open sets in locally convex spaces, let ϕ* be a biholomorphic mapping of U onto V, let $K \subset U$ be compact, and let $L = \phi(K)$, Then $\mathscr{H}(K)$ and $\mathscr{H}(L)$ are isomorphic as topological vector spaces.

Proof. For each U' open with $K \subset U' \subset U$, set $V' = \phi(U')$. Then the linear mapping

$$g \in \mathscr{H}^{\infty}(V') \to g \circ \phi \in \mathscr{H}^{\infty}(U')$$

is an isometry. Thus we get an isomorphism $\mathscr{H}(L) \to \mathscr{H}(K)$ such that the diagram

$$\begin{array}{ccc} \mathscr{H}(L) & \longrightarrow & \mathscr{H}(K) \\ \uparrow & & \uparrow \\ \mathscr{H}^{\infty}(V') & \longrightarrow & \mathscr{H}^{\infty}(U') \end{array}$$

is commutative for each U' open with $K \subset U' \subset U$. Q.E.D.

EXAMPLE 6.2. Let U, V be open sets in locally convex spaces and let ϕ be a biholomorphic mapping of U onto V. If U has the Runge property, then so does V.

Proof. By Lemma 6.2, if $K \subset U$ is U-Runge, then $L = \phi(K)$ is V-Runge. Q.E.D.

For a third example we need some results of Katz [24]. Let E denote a complex Banach space with an unconditional basis $\{e_n\}$ and let $\{\phi_n\}$ denote the associated sequence of coefficient functionals. Thus for every $x \in E$ we have

$$x = \sum_{n=1}^{\infty} \varphi_n(x) e_n$$

and the series converges unconditionally.

DEFINITION 6.3. $X \subset E$ is a *Reinhardt set* with respect to $\{e_n\}$ if

$$\sum_{n=1}^{\infty} \lambda_n \phi_n(x) e_n \in X$$

whenever $x \in X$ and $\{\lambda_n\} \subset C$ with $|\lambda_n| = 1$.

DEFINITION 6.4. For $X \subset E$, $X_{\mathscr{R}}$ denotes its Reinhardt hull:

$$X_{\mathscr{R}} = \left\{ \sum_{n=1}^{\infty} \lambda_n \phi_n(x) e_n : x \in X, \{\lambda_n\} \subset C, |\lambda_n| = 1 \right\}$$

Clearly $X_{\mathscr{R}}$ is the smallest Reinhardt set containing X.

LEMMA 6.3. *Let $U \subset E$ be a domain containing the origin and let $J \subset U$ be compact. Then there exists K compact, connected, and containing the origin with $J \subset K \subset U$.*

LEMMA 6.4. *If $X \subset E$ is compact, then $X_{\mathscr{R}}$ is compact. If $X \subset E$ is connected, then $X_{\mathscr{R}}$ is connected.*

LEMMA 6.5. *If X is a Reinhardt set, then X_s is a Reinhardt open set for each $s > 0$.*

PROPOSITION 6.2. *Let V be a Reinhardt domain containing the origin. Then there exists W open, balanced, and containing V such that (V, W) is an extension pair.*

Proofs of these results can be found in [24]. Now we have:

EXAMPLE 6.3. Let E be a Banach space with an unconditional basis and let $U \subset E$ be a Reinhardt domain containing the origin. Then U has the Runge property.

Proof. Let $J \subset U$ be compact. By Lemma 6.3. we can find K compact, connected, and containing the origin with $J \subset K \subset U$. By Lemma 6.4, $L = K_{\mathscr{R}}$ is a compact, connected Reinhardt set in U that contains the origin. We show that L is U-Runge. Let $f \in \mathscr{H}(L)$. Then $f \in \mathscr{H}(L_s)$ for some $s > 0$. By Lemma 6.5, L_s is a Reinhardt domain containing the origin. By Proposition 6.2 there exists W open, balanced, and containing L_s such that f extends to an $\tilde{f} \in \mathscr{H}(W)$. W being balanced, the Taylor series of $\tilde{f}$ at 0 converges to $\tilde{f}$ uniformly on a neighborhood of L. Thus the Taylor series of f at 0 converges to f in $\mathscr{H}(L)$ and L is U-Runge. Q.E.D.

7. MULTIPLICATIVE LOCAL CONVEXITY OF $\mathscr{H}(K)$ AND $(\mathscr{H}(U), \tau_\omega)$

Matos [27] considered on $\mathscr{H}(U)$ the topology $\tau_{\omega a}$ defined by all τ_ω-continuous seminorms p such that

$$p(fg) \leqslant p(f)p(g)$$

for all f, $g \in \mathscr{H}(U)$. Matos used the spectrum of the topological algebra $(\mathscr{H}(U), \tau_{\omega a})$ to give a construction of the envelope of holomorphy of U. Similar constructions of the envelope of holomorphy of U, but using other topologies on $\mathscr{H}(U)$, have been given by Alexander [1], Cœuré [7], and Schottenloher [39–41]. Actually, none of the constructions in [1], [7], or [27] was completely satisfactory, and it was Schottenloher in [39] who gave first a satisfactory construction of the envelope of holomorphy via the spectrum.

In this section we show that the topology $\tau_{\omega a}$ defined by Matos coincides with the Nachbin topology τ_ω. Our methods are similar to those of section 5. We begin with:

THEOREM 7.1. *Let E be a semimetrizable locally convex space and let $K \subset E$ be compact. Then $\mathscr{H}(K)$ is a multiplicatively locally convex topological algebra; that is, its topology is defined by the continuous seminorms p such that*

$$p(fg) \leqslant p(f)p(g)$$

for all $f, g \in \mathscr{H}(K)$.

Proof. It sufficies to show that the convex O-neighborhoods $\mathscr{U}$ with $\mathscr{U} \cdot \mathscr{U} \subset \mathscr{U}$ form an O-neighborhood base in $\mathscr{H}(K)$.

Let $\{V_n\}$ be a decreasing fundamental sequence of neighborhoods of K. For $n \in N$ and $0 < \epsilon_n \leqslant 1$ we write

$$\mathscr{U}_{n,\epsilon_n} = \left\{ f \in \mathscr{H}^\infty(V_n) : \sup_{x \in V_n} |f(x)| \leqslant \epsilon_n \right\}$$

For each n, as ϵ_n varies, the sets $\mathscr{U}_{n,\epsilon_n}$ form an O-neighborhood base in $\mathscr{H}^\infty(V_n)$. Since

$$\mathscr{H}(K) = \varinjlim \mathscr{H}^\infty(V_n),$$

the sets

$$\mathscr{U}_\epsilon = \text{convex, balanced hull of } \bigcup_{n=1}^{\infty} \mathscr{U}_{n,\epsilon_n}$$

form an O-neighborhood base in $\mathscr{H}(K)$, when the sequence $\epsilon = \{\epsilon_n\}$ varies. We show that $\mathscr{U}_\epsilon \cdot \mathscr{U}_\epsilon \subset \mathscr{U}_\epsilon$ for each $\epsilon = \{\epsilon_n\}$.

Certainly

$$\mathscr{U}_\epsilon = \{\text{finite sums } \textstyle\sum \lambda_n f_n : \sum |\lambda_n| \leqslant 1,\ f_n \in \mathscr{U}_{n,\epsilon_n}\}$$

Let $f, g \in \mathscr{U}_\epsilon$. Then

$$f = \sum \lambda_n f_n \qquad \text{with} \quad \sum |\lambda_n| \leqslant 1, \qquad f_n \in \mathscr{U}_{n,\epsilon_n}$$

and

$$g = \sum \mu_j g_j \qquad \text{with} \quad \sum |\mu_j| \leqslant 1, \qquad g_j \in \mathscr{U}_{j,\epsilon_j}$$

Therefore,

$$fg = \sum_{n,j} \lambda_n \mu_j f_n g_j$$

with

$$\sum_{n,j} |\lambda_n \mu_j| = \sum |\lambda_n| \cdot \sum |\mu_j| \leqslant 1$$

and

$$f_n g_j \in \begin{cases} \mathscr{U}_{j,\epsilon_j} & \text{for} \quad n \leqslant j \\ \mathscr{U}_{n,\epsilon_n} & \text{for} \quad n \geqslant j \end{cases}$$

It follows that $fg \in \mathscr{U}_\epsilon$. Q.E.D.

To prove that $(\mathscr{H}(U), \tau_\omega)$ is a multiplicatively locally convex topological algebra, we need the following lemma.

LEMMA 7.1. *Let E be a semimetrizable locally convex space, let $K \subset E$ be compact, and let M be a subalgebra of $\mathscr{H}(K)$. Then M, endowed with the inductive topology introduced in Section 5, is a multiplicatively locally convex topological algebra.*

Proof. We use the notation of the proof of Theorem 7.1 and also the notation of Section 5.

For each n, as ϵ_n varies, the sets

$$\mathscr{V}_{n,\epsilon_n} = \left\{ f \in M_n : \sup_{x \in V_n} |f(x)| \leqslant \epsilon_n \right\} = M_n \cap \mathscr{U}_{n,\epsilon_n}$$

form an O-neighborhood base in M_n. Since

$$M = \varinjlim M_n$$

the sets

$$\mathscr{V}_\epsilon = \text{convex, balanced hull of } \bigcup_{n=1}^{\infty} \mathscr{V}_{n,\epsilon_n}$$

form an O-neighborhood base in M, when the sequence $\epsilon = \{\epsilon_n\}$ varies. We have that

$$\mathscr{V}_\epsilon = \{ \text{finite sums} \sum \lambda_n f_n : \sum |\lambda_n| \leqslant 1, \, f_n \in \mathscr{V}_{n,\epsilon_n} \}$$

and we show as in the proof of Theorem 7.1 that $\mathscr{V}_\epsilon \cdot \mathscr{V}_\epsilon \subset \mathscr{V}_\epsilon$. Q.E.D.

THEOREM 7.2. *Let E be any semimetrizable locally convex space and let $U \subset E$ be open. Then $(\mathscr{H}(U), \tau_\omega)$ is a multiplicatively locally convex topological algebra; that is, its topology is defined by the continuous seminorms p such that*

$$p(fg) \leqslant p(f)p(g)$$

for all $f, g \in \mathscr{H}(U)$.

Proof. For each $K \subset U$ compact, let $\mathscr{H}^K(U)$ denote the image of the canonical mapping $\mathscr{H}(U) \to \mathscr{H}(K)$, endowed with the inductive topology of Section 5. By Lemma 7.1 each $\mathscr{H}^K(U)$ is a multiplicatively locally convex topological algebra. Using Lemma 5.3 we obtain that $(\mathscr{H}(U), \tau_\omega)$ is also a multiplicatively locally convex topological algebra. Q.E.D.

ACKNOWLEDGMENTS

This chapter is a modified and expanded version of the author's doctoral dissertation at the University of Rochester. The author is deeply grateful to his thesis advisor, Professor Leopoldo Nachbin, as well as to Professors Richard Aron and Martin Schottenloher, for their advice and help. This thesis was written during the academic year 1973–1974 while the author was on leave of absence from the University of Rochester at the University of Kentucky. Some results from this chapter were announced without proofs in [29].

REFERENCES

1. H. ALEXANDER, Analytic functions on Banach spaces, Thesis, Univ. of California, Berkeley, 1968.
2. R. ARON, Holomorphy types for open subsets of a Banach space, *Studia Math.* **45** (1973), 273–289.
3. R. ARON, unpublished work.
4. J. A. BARROSO, Topologias nos espaços de aplicações holomorfas entre espaços localmente convexos, *Anais Acad. Brasil. Ciencias* **43** (1971), 527–546.
5. S. B. CHAE, Sur les espaces localement convexes de germes holomorphes, *C. R. Acad. Sci. Paris* **271** (1970), 990–991.
6. S. B. CHAE, Holomorphic germs on Banach spaces, *Ann. Inst. Fourier Grenoble* **21** (1971), 107–141.
7. G. COEURÉ, Fonctions plurisousharmoniques sur les spaces vectoriels topologiques et applications a l'étude des fonctions analytiques, *Ann. Inst. Fourier Grenoble* **20** (1969), 361–432.
8. G. COEURÉ, "Analytic Functions and Manifolds in Infinite Dimensional Spaces," North-Holland Publ., Amsterdam, 1974.
9. J. DIEUDONNÉ, "Foundations of Modern Analysis," Vol. I, Academic Press, New York, 1969.

10. S. Dineen, Topologie de Nachbin et prolongement analytique en dimension infinie, *C. R. Acad. Sci. Paris* **271** (1970), 643–644.
11. S. Dineen, Holomorphy types on a Banach space, *Studia Math.* **39** (1971), 241–288.
12. S. Dineen, Holomorphic functions on (c_0, X_b)-modules, *Math. Ann.* **196** (1972), 106–116.
13. S. Dineen, Holomorphic functions on locally convex topological vector spaces: I. Locally convex topologies on $\mathscr{H}(U)$, *Ann. Inst. Fourier Grenoble* **23** (1973), 19–54.
14. K. Floret and J. Wloka, "Einfuhrung in die Theorie der lokalkonvexen Räume," Springer-Verlag, Berlin and New York, 1968.
15. A. Grothendieck, Sur certains espaces de fonctions holomorphes, *J. Reine Angew. Math.* **192** (1953), 35–64, 77–95.
16. A. Grothendieck, Sur les espaces (F) et (DF), *Summa Brasil. Math.* **3** (1954), 57–122.
17. A. Grothendieck, "Produits tensoriels topologiques et spaces nucléaires," Amer. Math. Soc., Providence, Rhode Island, 1955.
18. A. Grothendieck, "Espaces vectoriels topologiques," 2nd ed., Soc. Mat. Sao Paulo, 1958.
19. R. Gunning and H. Rossi, "Analytic Functions of Several Complex Variables," Prentice–Hall, Englewood Cliffs, New Jersey, 1965.
20. M. Hervé, "Several Complex Variables, Local Theory," Oxford Univ. Press, London and New York and Tata Inst. of Fundamental Res., Bombay, 1963.
21. M. Hervé, "Analytic and Plurisubharmonic Functions," Springer-Verlag, Berlin and New York, 1971.
22. A. Hirschowitz, Bornologie des espaces de fonctions analytiques en dimension infinie, *Séminaire Lelong 1970*, pp. 21–33, Springer-Verlag, Berlin and New York, 1971.
23. J. Horváth, "Topological Vector Spaces and Distributions," Vol. I, Addison-Wesley, Reading, Massachusetts, 1966.
24. G. Katz, Analytic Continuation in Infinite Dimension and Reinhardt Sets, Thesis, Univ. of Rochester, Rochester, New York, 1974.
25. A. Martineau, Sur les fonctionnelles analytiques et la transformation de Fourier–Borel, *J. Anal. Math.* **11** (1963), 1–164.
26. A. Martineau, Sur la topologie des espaces de fonctions holomorphes, *Math. Ann.* **163** (1966), 62–88.
27. M. Matos, "Holomorphic Mappings and Domains of Holomorphy," Centro Brasileiro de Pesquisas Físicas, Rio de Janeiro, 1970.
28. J. Mujica, On a proposition of Chae and Hirschowitz, *Symp. Infinite Dimensional Function Theory*, Northern Illinois Univ., De Kalb, Illinois, 1973.
29. J. Mujica, On the Nachbin topology in spaces of holomorphic functions, *Bull. Amer. Math. Soc.*, **81** (1975), 904–906.
30. L. Nachbin, On the topology of the space of all holomorphic functions on a given open subset, *Indag. Math.* **29** (1967), 366–368.
31. L. Nachbin, "Topology on Spaces of Holomorphic Mappings," Springer-Verlag, Berlin and New York, 1969.
32. L. Nachbin, Concerning Spaces of Holomorphic Mappings, Rutgers Univ., New Brunswick, New Jersey, 1970.
33. L. Nachbin, Sur les espaces vectoriels topologiques d'applications continues, *C. R. Acad. Sci. Paris* **271** (1970), 596–598.
34. L. Nachbin, Uniformité d'holomorphie et type exponentiel, *Seminaire Lelong 1970*, pp. 216–224, Springer-Verlag, Berlin and New York, 1971.
35. Ph. Noverraz, "Pseudo-convexité, convexité polynomiale et domains d'holomorphie en dimension infinie," North–Holland Publ., Amsterdam, 1973.
36. J. P. Ramis, "Sous-ensembles Analytiques d'une Varieté Banachique Complexe," Springer-Verlag, Berlin and New York, 1970.

37. J. P. RAMIS, Some recent results and open problems in infinite dimensional analytic geometry, *Proc. Infinite Dimensional Holomorphy, Univ. of Kentucky, 1973*, pp. 186–195, Springer-Verlag, Berlin and New York, 1974.
38. H. H. SCHAEFER, "Topological Vector Spaces," Springer-Verlag, Berlin and New York, 1971.
39. M. SCHOTTENLOHER, Uber analytische Fortsetzung in Banachraumen, *Math. Ann.* **199** (1972), 313–336.
40. M. SCHOTTENLOHER, Analytic continuation and regular classes in locally convex Hausdorff spaces, *Portugal. Math.* **33** (1974), 219–250.
41. M. SCHOTTENLOHER, Riemann domains: basic results and open problems, *Proc. Infinite Dimensional Holomorphy, Univ. of Kentucky. 1973*, pp. 196–212. Springer-Verlag, Berlin and New York, 1974.
42. M. SCHOTTENLOHER, Das Leviproblem in unendlichdimensionalen Raumen mit Schauderzerlegung, Habilitation Thesis, Ludwig Maximilians Univ. München, 1974.
43. H. WHITNEY, Analytic extensions of differentiable functions defined in closed sets, *Trans. Amer. Math. Soc.* **36** (1934), 63–89.

References Added in Proof. Since this was written, the following intimately related articles have appeared:

44. P. AVILES AND J. MUJICA, Holomorphic germs and homogeneous polynomials on quasinormable metrizable spaces, *Rend. Mat.* **10** (1977), 117–127.
45. K.-D. BIERSTEDT AND R. MEISE, Nuclearity and the Schwartz property in the theory of holomorphic functions on metrizable locally convex spaces, *in* "Infinite Dimensional Holomorphy and Applications," pp. 93–129, North–Holland Publ., Amsterdam, 1977.

AMS (MOS) subject classifications: primary, 46E10; secondary, 46E25, 32E25, 46A15, 32E30.

STUDIES IN ANALYSIS
ADVANCES IN MATHEMATICS SUPPLEMENTARY STUDIES, VOL. 4

Unitary Representations of Group Extensions; an Algebraic Approach to the Theory of Mackey and Blattner[†]

BY MARC A. RIEFFEL

Department of Mathematics
University of California at Berkeley
Berkeley, California

In [32] Mackey used his theory of induced representations for separable locally compact groups [30] to attack the problem of describing the unitary representation theory of a separable locally compact group G in terms of the representation theories of a normal subgroup N and of the quotient group G/N. In the case in which N is nicely embedded in G, he obtained far-reaching generalizations of results obtained earlier for finite groups by Clifford [12]. Mackey's main theorems were generalized to nonseparable groups by Blattner [5, 6] with weakened hypotheses concerning how N is embedded in G. This necessitated introducing methods of proof which were considerably less measure-theoretic than those of Mackey. In [36] a theory of induced representations for C^*-algebras was introduced which has Mackey's definition for groups as a special case. This theory for C^*-algebras can involve no measure theory, and so required techniques that are different from those of either Mackey or Blattner. In this chapter we prove a version of the central theorem of Mackey's analysis for group extensions by using the approach introduced in [36]. In particular, our proof is fairly algebraic, and involves no measure theory beyond the elementary facts about Haar measure needed, for example, to convolve continuous functions of compact support and to construct the integrated forms of unitary representations. Our version of Mackey's theorem is not quite as strong as Blattner's, in large part because some of Blattner's hypotheses are explicitly and essentially measure-theoretic. But our version is applicable in many of the situations that ordinarily arise, such as the case in which the normal subgroup is "regularly embedded." The rather algebraic nature of the proof given here makes some of the mechanisms involved in the situation

[†] This research was partially supported by National Science Foundation Grant GP-30798X.

ISBN 0-12-599150-9

somewhat more transparent. Also, some of the techniques used and intermediate results obtained are of independent interest. For example, all of Section 3 develops general information relating ideals and quotients of C^*-algebras to the imprimitivity bimodules introduced in [36], while in other places results on weak containment (Proposition 4.1) and on the ideal centers of C^*-algebras (Proposition 5.3) are obtained. (See *Note Added in Proof*, p. 80.)

The chapter is organized in the following way. In Section 1 basic definitions are given and our version of Mackey's theorem is stated. Section 2 is devoted to constructing the C^*-algebras involved in the proof of the main theorem and to establishing a crucial relation between them. Section 3, which can be read independently, was just described. In Section 4 the material of Section 3 is applied to the situation involved in the main theorem, while in Section 5 systems of imprimitivity are tied into the situation by means of ideal centers of C^*-algebras. The proof of the main theorem is then concluded in Section 6. Various alternative hypotheses (related to work of Effros [17] and Glimm [24]) under which the main theorem remains true are gathered together in Section 7. Finally, in Section 8 we sketch for the reader's convenience for remaining part of Mackey's analysis, which does not involve the inducing process, but is concerned with projective representations. We make a few minor improvements of the proofs, but essentially all the material of this last section already exists in various places in the literature.

1. The Statement of the Main Theorem

In order to be able to state our main theorem we need to know what it means for a representation of a C^*-algebra to live on a given subset of the primitive ideal space of the algebra, at least when the subset is nice. This can be done in terms of the projection-valued measure introduced by Glimm [25] (see also the appendix of [23]), but since we will only need to consider subsets which are locally closed (equivalently, open in their closure, see p. 41 of [7]), it is easier to work directly with the ideals involved.

Let A be a C^*-algebra. We will let $\mathrm{Prim}(A)$ denote the primitive ideal space of A with the hull–kernel topology [14]. By a Hermitian A-module we will mean (as in [36, 37]) the Hilbert space of a nondegenerate $*$-representation of A. We will usually work with Hermitian modules rather than the corresponding representations. Let V be a Hermitian A-module, let C be a closed subset of $\mathrm{Prim}(A)$, and let $\ker(C)$ denote the intersection of the primitive ideals in C. In analogy with the case of a commutative C^*-algebra and representations thereof defined by measures, we say that V *lives on* C if $\ker(C)V = \{0\}$, so that V can be viewed as a Hermitian module over

$A/\ker(C)$. In fact, this is the starting point for the definition of Glimm's projection-valued measure. Suppose now that E is a subset of Prim(A) which is locally closed, so that $E = \bar{E} - E'$, where $\bar{E}$ is the closure of E, $E' = \bar{E} - E$, and E' is a closed set. We would like to say what it means for a Hermitian A-module V to live on E. For this it must certainly live on $\bar{E}$, so that it can be viewed as a module over $A/\ker(\bar{E})$. Now by Proposition 3.2.1 of [14], Prim($A/\ker(\bar{E})$) is naturally homeomorphic to $\bar{E}$. Thus we can view E as an open subset of Prim($A/\ker(\bar{E})$), with complement E'. Again, working in analogy with the commutative case, we see that to say that V lives on E should mean that as a module over $\ker(E')$ it is nondegenerate (and so Hermitian), that is, $\ker(E')V = V$. (Throughout this chapter whenever we juxtapose a subset of an algebra with a subset of a module this will denote the closed linear span of the products of corresponding elements.)

DEFINITION 1.1. Let A be a C^*-algebra, let V be a Hermitian A-module, and let E be a locally closed subset of Prim(A). Let $\bar{E}$ denote the closure of E in Prim(A) and let $E' = \bar{E} - E$. Then we will say that V *lives* on E, or that E *carries* V, if $\ker(\bar{E})V = \{0\}$ and V is nondegenerate as a $\ker(E')$-module, so that V is a Hermitian module over $\ker(E')/\ker(\bar{E})$.

Thus the category of all Hermitian A-modules that live on E is naturally isomorphic with the category of all Hermitian modules over $\ker(E')/\ker(\bar{E})$.

If H is any locally compact group, we let $C^*(H)$ denote its group C^*-algebra [14], and we will write Prim(H) instead of Prim($C^*(H)$). By a unitary H-module we will mean the Hilbert space of a strongly continuous unitary representation of G, as in [36]. Suppose now that G is a locally compact group and that N is a normal subgroup of G. (Subgroups will always be assumed to be closed.) Then the inner automorphisms of G carry N into itself, and so G acts as a group of automorphisms of N, and so of all structures naturally associated to N. In particular, G acts as a group of $*$-automorphisms of $C^*(N)$, and this action is easily seen to be strong operator-continuous. As a result, G acts as a topological transformation group on Prim(N) (see [20, 25]). In particular, if $J \in$ Prim(N), we can let GJ denote the orbit of J in Prim(N) under the action of G and we can let G_J denote the stability subgroup of G at J. Since the primitive ideal space of any C^*-algebra is a T_0 topological space [14], G_J will be a closed subgroup of G by Lemma 1 of [6]. There is, of course, a natural bijection of G/G_J onto GJ, which will be continuous, but is not, in general, a homeomorphism.

We now state our version of the main theorem of Mackey's analysis [32] for group extensions.

THEOREM 1.1. (The Main Theorem). *Let G be a locally compact group and let N be a normal subgroup of G. Let $J \in$ Prim(N), let GJ denote the orbit* *of J in* Prim(N) *under the action of G, and let G_J denote the stability subgroup of J. We make the following hypotheses*:

(1) $\{J\}$ *is locally closed in* Prim(N).

(2) *GJ is locally closed in* Prim(N), *and the canonical map of G/G_J onto GJ is a homeomorphism.*

Then the process of inducing representations from G_J to G establishes an equivalence of the category of unitary G_J-modules whose restrictions to N live on $\{J\}$ with the category of unitary G-modules whose restrictions to N live on GJ. This equivalence will preserve weak containment of representations.

We remark that Mackey stated and proved his original version of this theorem for projective representations, but the theorem for projective representations can be easily deduced from that for ordinary representations, as discussed, for example, in [3], and so we concern ourselves only with ordinary representations. In Section 7 we shall examine various other conditions which ensure that the hypotheses of Theorem 1.1 are satisfied.

We also remark that the statement about weak containment has as a special case one of the main results (Proposition 5) of [16]. This result was also known to Fell (as yet unpublished, but see 6.4 of [2]). See also [39].

Our method of proving Theorem 1.1 will be to find (in the next section) C^*-algebras whose categories of Hermitian modules are isomorphic with the categories of unitary modules mentioned in the theorem, and then to show (in Sections 3–6) that from the inducing process one obtains an imprimitivity bimodule [36] between these C^*-algebras, so that their categories of modules are equivalent by Theorem 6.23 of [36].

2. The C^*-Algebras

For any locally compact group H let $C_c(H)$ denote the space of continuous functions of compact support on H. Then on $C_c(H)$ we have the inductive limit topology, the supremum norm $\| \ \|_\infty$, and the L^1-norm, $\| \ \|_1$ (for left Haar measure on H). In addition, $C_c(H)$ with convolution and its usual involution [14] is a (dense) subalgebra of $C^*(H)$, and so can be equipped with the norm from that algebra.

Suppose now that H is a subgroup of the locally compact group G. Then, as discussed in Section 4 of [36], both $C_c(G)$ and $C_c(H)$ can be viewed as subalgebras of the algebra $M(G)$ of finite measures on G (by viewing their elements as densities against the left Haar measures of each group), and in $M(G)$ the convolution of an element of $C_c(G)$ on the left or right by an element of $C_c(H)$ will again be an element of $C_c(G)$. For future use we state the formulas for these convolutions in terms of left Haar measures, where Δ denotes the modular function for G. For $f \in C_c(G)$ and $\varphi \in C_c(H)$ we have

$$\varphi * f(y) = \int_H \varphi(t) f(t^{-1}y)\, dt, \qquad f * \varphi(y) = \int_H f(yt^{-1}) \Delta(t^{-1}) \varphi(t)\, dt$$

for all $y \in G$.

For any C^*-algebra A we will let $M(A)$ denote the double centralizer algebra of A [9]. Then, according to Proposition 4.1 of [36], the left and right actions of $C_c(H)$ on $C_c(G)$ previously defined extend to give a $*$-homomorphism of $C^*(H)$ into $M(C^*(G))$. It is not known whether this homomorphism is injective in general (see the comments after Proposition 4.1 of [36]). However, in the case of interest to us, in which H is normal, it is an easy consequence of Theorem 4.5 of [21] that this homomorphism is injective if G is separable. Moreover, J. M. G. Fell has pointed out (personal communication) that this is also true in the nonseparable case as well. Actually, a new proof of this fact will emerge later from our general theory (Proposition 4.1).

Let H still be a subgroup of G which need not be normal in G. We will not introduce any notation for the homomorphism of $C^*(H)$ into $M(C^*(G))$, but for $d \in C^*(H)$ and $a \in C^*(G)$ we will simply write da or ad for the action of d on the left or right of a under this homomorphism. We remark that since $C_c(H)$ contains an approximate identity for the inductive limit topology for its action on $C_c(G)$, $C^*(G)$ will be an essential (i.e., nondegenerate) $C^*(H)$-module for both the left and the right actions of $C^*(H)$ on $C^*(G)$.

Suppose now that E is a locally closed subset of $\operatorname{Prim}(H)$. Let $E' = \bar{E} - E$, so that E' is closed. Let W be a unitary G-module. We would like to reformulate the statement that the restriction W_H of W to H lives on E. To begin, we must have $\ker(\bar{E})W_H = \{0\}$. This will hold if and only if

$$C^*(G)\ker(\bar{E})C^*(G)W = \{0\}.$$

In addition, we must have $\ker(E')W_H = W_H$. This will hold if and only if

$$C^*(G)\ker(E')C^*(G)W = W.$$

Notice that $C^*(G)\ker(E')C^*(G)$ and $C^*(G)\ker(\bar{E})C^*(G)$ are ideals in $C^*(G)$. (When we say an "ideal" in a C^*-algebra we will always mean a two-sided ideal, as these are the only ones with which we shall need to deal.) Furthermore the first ideal contains the second. However, in general, the inclusion need not be proper, so that the quotient algebra may be the zero-dimensional C^*-algebra. This will happen if E is too small to carry the restriction to H of any representation of G. How this can easily happen when H is normal will become clear immediately after Proposition 2.2. Anyway, if we admit the zero-dimensional representation as a unitary representation of a group, and as the only Hermitian module over the zero-dimensional C^*-algebra, then these arguments give the proof of:

PROPOSITION 2.1. *Let H be a subgroup of a locally compact group G and let E be a locally closed subset of* $\mathrm{Prim}(H)$. *Let $\bar{E}$ denote the closure of E and let $E' = \bar{E} - E$. Then the category of unitary G-modules whose restriction to H lives on E is isomorphic to the category of Hermitian modules over the C^*-algebra*

$$[C^*(G)\ker(E')C^*(G)]/[C^*(G)\ker(\bar{E})C^*(G)]$$

(*which may be zero-dimensional*).

If N is a normal subgroup of G, if $J \in \mathrm{Prim}(N)$, and if the hypotheses of Theorem 1.1 are met, then it is the corresponding C^*-algebras defined as in Proposition 2.1 for J and $C^*(G_J)$ and for GJ and $C^*(G)$ which we shall show have an imprimitivity bimodule.

We pause now to obtain a bit of information concerning what kinds of sets can carry the restriction of a representation to a normal subgroup. This will clarify the remark preceding Proposition 2.1. If N is a normal subgroup of the locally compact group G, then, as mentioned, G acts as a group of automorphisms of $C^*(N)$. For $d \in C^*(N)$ and $y \in G$ we will let d^y denote the result of applying to d the automorphism corresponding to y. If I is an ideal in $C^*(N)$ and if $y \in G$, we let I^y be the ideal $\{d^y : d \in I\}$. If $I^y = I$ for all $y \in G$, we say that I is a *G-invariant ideal.* It is clear that I is a G-invariant ideal if and only if hull(I) is a G-invariant subset of $\mathrm{Prim}(N)$.

PROPOSITION 2.2. *Let N be a normal subgroup of a locally compact group G, let W be a unitary G-module, and let W_N denote the restriction of W to N. Then the kernel K of (the representation corresponding to) W_N is G-invariant. Thus the smallest closed subset of* $\mathrm{Prim}(N)$ *which carries W_N, namely the hull of K, is G-invariant.*

Proof. For any $d \in C^*(N)$, $y \in G$, and $w \in W$, it is easily seen that

$$d^y(w) = y(d(y^{-1}w)).$$

Now if $d \in K$ and if $y \in G$, $w \in W$, it follows that $d^y(w) = 0$, so that $d^y \in K$. Thus K is G-invariant. Q.E.D.

It follows that if C is a closed subset of $\mathrm{Prim}(N)$ that contains no non-empty G-invariant subset, then the restriction of no unitary G-module to C will live on C, and so the algebra of Proposition 2.1 will be zero-dimensional.

As mentioned following Proposition 2.1, our aim is to show that there is an imprimitivity bimodule between the C^*-algebra for J and $C^*(G_J)$ and that for GJ and $C^*(G)$. (We shall be more precise shortly.) To show this we shall need to use facts that depend on the normality of N. The main such fact relates the right and left actions of $C^*(N)$ on the various objects involved. This does not depend on G_J being a stability subgroup, and so for a while now we will assume only that we have a locally compact group G, a normal subgroup N, and a subgroup H containing N. Then, according to results in Section 4 of [36], particularly Proposition 4.10, $C_c(G)$ can be viewed as a Hermitian $C_c(H)$-rigged module over both $C_c(G)$ and $C_c(N)$.

Since we need to study ideals in the group C^*-algebras, and these are not in general contained in their dense subalgebras of continuous functions of compact support, we need to extend the preceding situation to the completions. Now if A and B are pre-C^*-algebras and if X is a Hermitian B-rigged A-module, then it is natural to equip X with the norm

$$\|x\|_X = \|\langle x, x\rangle_B\|^{1/2}$$

for $x \in X$. For a proof that this is in fact a norm see Proposition 2.10 of [36]. Since the action of the pre-C^*-algebra $C_c(H)$ on the right of $C_c(G)$ and the $C_c(H)$-valued inner product on $C_c(G)$ are easily seen to be continuous for this norm, as are the left actions of the pre-C^*-algebras $C_c(G)$ and $C_c(N)$ by Proposition 4.10 of [36], these actions all extend to the completions. We thus obtain:

PROPOSITION 2.3. *If X denotes the completion of $C_c(G)$ for the norm just defined, then X is a Hermitian $C^*(H)$-rigged module over both $C^*(G)$ and $C^*(N)$.*

It is easily seen that the inductive limit topology on $C_c(G)$ is finer than the topology from the norm defined on X. (The argument is given in the proof of Proposition 4.11 of [36].)

Now X is an essential right $C^*(H)$-module, and so, by applying the right-handed versions of either Proposition 3.9 of [36] or the proof of Theorem 21 of [29], X becomes an essential right module over $M(C^*(H))$. Since $C^*(N)$ is mapped into $M(C^*(H))$ by Proposition 4.1 of [36], X becomes a right $C^*(N)$-module, where the action of $\varphi \in C_c(N)$ on any element of X of the form $f \cdot \psi$ for $f \in C_c(G)$ and $\psi \in C_c(H)$ (where $\cdot$ is defined in 4.5 of [36]) is, by definition, given by

$$(f \cdot \psi)\varphi = f \cdot (\psi * \varphi).$$

Since N is normal in both G and H, the modular functions, Δ_G and Δ_H, of G and H, respectively, must coincide on N with the modular function of N (15.23 of [27]) and so $\gamma(s) = 1$ for all $s \in N$, where $\gamma(r) = (\Delta_G(r)/\Delta_H(r))^{1/2}$ for $r \in H$, as in 4.2 of [36]. From this fact it is easily seen that $\gamma(\psi * \varphi) = (\gamma\psi) * \varphi$

so that $(f \cdot \psi)\varphi = (f \cdot \psi) * \varphi$. It follows that the right action of $C^*(N)$ on X is just given by ordinary right convolution of $C_c(N)$ on $C_c(G)$.

As we saw previously, $C^*(N)$ also acts on the left as bounded operators on the space X, this action just coming from left convolution of $C_c(N)$ on $C_c(G)$. We wish to show that the left and right actions of $C^*(N)$ on X are related in terms of the action of G as automorphisms of N, and so of $C^*(N)$. To do this we now examine more specifically how these automorphisms are defined at the level of functions. Since G acts as a group of automorphisms of N, it will act as a group of automorphisms of the measure algebra of N. For any finite complex measure m on N and any $y \in G$ we will let m^y denote the measure obtained by applying to m the automorphism corresponding to y. If $\varphi \in C_c(N)$, we can view φ as a measure on N, as mentioned, and so φ^y is defined, as a measure. A simple computation shows that actually φ^y is defined by a density from $C_c(N)$, which we also denote by φ^y, and that in fact $\varphi^y(t) = \Delta_y \varphi(y^{-1}ty)$ for $t \in N$ and $y \in G$, where Δ_y is the factor by which the automorphism corresponding to y changes the left Haar measure on N (see the bottom of page 1103 of [6]).

The relation between the right and left actions of $C^*(N)$ on X is suggested by the easily verified relation

$$f * \varphi(y) = (\varphi\Delta)^y * f(y)$$

for $f \in C_c(G)$, $\varphi \in C_c(N)$, and $y \in G$, although this relation can have no meaning at the level of the group C^*-algebras. This relation is related to Lemma 2 of [6] or Lemma 16.1 of [23], as is the next lemma. Rather than stating the full relation between the left and right actions immediately, the next lemma is stated as a preliminary result (although it is the crux of the matter), because we will need this result in exactly this form later (Proposition 5.2 and Lemma 6.3).

LEMMA 2.1 (Main Lemma). *Let G be a locally compact group, N a normal subgroup of G, and H a subgroup of G containing N. Let X be the completion of $C_c(G)$ as a $C_c(H)$-rigged space, so that X is a Hermitian $C^*(H)$-rigged module over $C^*(N)$, on which $C^*(N)$ also acts on the right. Let $d \in C^*(N)$ and $f \in C_c(G)$, with f viewed as an element of X. Then*

$$fd \in [d^y : y \in \text{support}(f)]X, \qquad df \in X[d^y : y^{-1} \in \text{support}(f)],$$

(where the sets on the right are, as usual, taken to be closed linear spans in X).

Proof. For any $z \in G$ let δ_z denote the positive measure of mass one at z. Then for $\varphi \in C_c(N)$ it is easily seen that, as measures on G,

$$\delta_z * \varphi = \varphi^z * \delta_z,$$

in analogy with the relation previously mentioned. Let $h \in C_c(G)$. We begin by establishing the formulas

$$f * \varphi * h = \int_G (\varphi^z * \delta_z * h) f(z)\,dz, \qquad h * \varphi * f = \int_G (h * \delta_z * \varphi^{z^{-1}}) f(z)\,dz,$$

for the inductive limit topology on $C_c(G)$. Both integrands are easily seen to be continuous functions of compact support from G into $C_c(G)$ for the inductive limit topology, so, in particular, with values in $C(K)$ where K is an appropriate compact set in G depending on f, φ, and h. The integrals are taken as Bochner integrals for the supremum norm in $C(K)$. Now point evaluations are continuous linear functionals on $C(K)$, and the collection of point evaluations separates $C(K)$. But if either formula is evaluated at points, the two sides are seen to be equal. Thus the formulas are established.

Since the inductive limit topology on $C_c(G)$ is finer than the norm topology from X, so that the injection of $C(K)$ into X is continuous, it follows that the two previous formulas are also valid when the integrals are taken as Bochner integrals for the norm on X.

We would now like to replace φ with d in the formulas. Clearly the function $z \mapsto \delta_z * h$ is continuous for the inductive limit topology, and also for the norm topology from X. Since the action of G on $C^*(N)$ is strongly continuous, it follows that the function $z \mapsto [d^z(\delta_z * h)] f(z)$ is a continuous function of compact support with values in X. A similar statement holds for the other integrand. But routine estimates show that if φ is chosen close to d, then the corresponding integrals are close. In this way we establish the formulas

$$f(dh) = \int [d^z(\delta_z * h)] f(z)\,dz, \qquad h(df) = \int [(h * \delta_z) d^{z^{-1}}] f(z)\,dz,$$

in which $\delta_z * h$ and $h * \delta_z$ are regarded as elements of X.

Let $\epsilon > 0$ be given, and choose a partition of unity $\{k_i\}$ in $C_c(G)$ fine enough (p. 65 of [8]) so that if $y, z \in \text{support}(fk_i)$ for some i, then

$$\|d^y - d^z\|_{C^*(N)} \leqslant \epsilon/(\|f\|_1 M),$$

where $M = \|h\|_X$ for the first formula, but

$$M = \sup\{\|h * \delta_z\|_X : z \in \text{support}(f)\}$$

for the second. Now only a finite number of the k_i will be nonzero at some point of the support of f. For each of these choose $z^i \in \text{support}(fk_i)$ and discard the rest of the k_i. Then, from the formulas previously established and the fact that $\sum fk_i = f$ (where all such sums will be finite sums over the

i for which k_i has not been discarded), we have

$$f(dh) = \sum \int [d^z(\delta_z * h)] f(z) k_i(z)\, dz.$$

Then from this we have

$$\|f(dh) - \sum \int [d^{z^i}(\delta_z * h)] f(z) k_i(z)\, dz\|_X$$
$$\leqslant \sum \int \|d^z - d^{z^i}\|_{C^*(N)} \|\delta_z * h\|_X |f(z)| k_i(z)\, dz \leqslant \epsilon.$$

But it is easily seen that

$$\int [d^{z^i}(\delta_z * h)] f(z) k_i(z)\, dz = d^{z^i}(f k_i * h).$$

Since ϵ is arbitrary, it follows that

$$f(dh) \in [d^y : y \in \text{support}(f)]X.$$

In a similar way it is established that

$$h(df) \in X[d^y : y^{-1} \in \text{support}(f)].$$

Now let h run through an approximate identity consisting of positive elements of $C_c(G)$ whose supports shrink to the identity of G, with $\|h\|_1 = 1$. From the properties of the action of $C^*(G)$ on the left of X it is clear that $h(df)$ converges to df in X, and so we have completed the proof of the first relation of the lemma. However, $C^*(G)$ does not act on the right on X, and so we must argue directly that $f(dh)$ converges to fd. But

$$\|fd - f(dh)\|_X = \left\| fd \int h(z)\, dz - \int [(f * \delta_z) d^{z^{-1}}] h(z)\, dz \right\|_X$$
$$\leqslant \int \|fd - (f * \delta_z) d^{z^{-1}}\|_X h(z)\, dz,$$

which can be made as small as desired by choosing the support of h sufficiently small, since the rest of the integrand is a continuous function with value zero at the identity element of G. Q.E.D.

A simple approximation argument lifts this result to all elements of X, although we can no longer say anything about supports.

COROLLARY 2.1. *Let $d \in C^*(N)$ and $x \in X$. Then*

$$xd \in [d^G]X \qquad \text{and} \qquad dx \in X[d^G],$$

where $[d^G]$ denotes the set of elements in $C^*(N)$ of form d^y for $y \in G$.

If we let $G = H$, then it is easily seen that $X = C^*(H)$. We thus obtain:

COROLLARY 2.2. *Let N be a normal subgroup of the locally compact group H, let $a \in C^*(H)$, and let $d \in C^*(N)$. Then*

$$da \in C^*(H)[d^H] \qquad \text{and} \qquad ad \in [d^H]C^*(H).$$

COROLLARY 2.3. *Let N be a normal subgroup of the locally compact group H and let I be an ideal in $C^*(N)$ which is H-invariant. Then*

$$IC^*(H) = C^*(H)I.$$

COROLLARY 2.4. *Let N be a normal subgroup of the locally compact group H, and let C be a closed H-invariant subset of* Prim(N). *Then* ker(C) *is a H-invariant ideal, so that*

$$\ker(C)C^*(H) = C^*(H)\ker(C).$$

Suppose now that N is a normal subgroup of G and that $J \in \text{Prim}(N)$ with $\{J\}$ locally closed. Let G_J be the stability subgroup of J. Let $\{J\}^c$ denote the closure of $\{J\}$, and note that $\{J\}^c$ is a G_J-invariant closed subset of Prim(N) whose kernel is just the G_J-invariant ideal J itself, and that $(\{J\}^c - \{J\})$ is also a closed G_J-invariant subset. Let $J' = \ker(\{J\}^c - \{J\})$, which is equal to $C^*(N)$ if $\{J\}$ is already closed. Then J' is a G_J-invariant ideal which contains J. It follows from Corollary 2.4 that

$$C^*(G_J)J'C^*(G_J) = C^*(G_J)J', \qquad C^*(G_J)JC^*(G_J) = C^*(G_J)J.$$

Applying Proposition 2.1, we see that the category of unitary G_J-modules whose restriction to N lives on $\{J\}$ is naturally isomorphic to the category of Hermitian modules over the C^*-algebra $C^*(G_J)J'/C^*(G_J)J$. Furthermore, let GJ be the orbit of J under G, and assume that GJ is locally closed. Let

$$I = \ker((GJ)^c), \qquad I' = \ker((GJ)^c - GJ).$$

Then in the same way

$$C^*(G)I'C^*(G) = C^*(G)I', \qquad C^*(G)IC^*(G) = C^*(G)I$$

and the category of unitary G-modules whose restriction to N lives on GJ is naturally isomorphic to the category of Hermitian modules over the C^*-algebra $C^*(G)I'/C^*(G)I$. It is this pair of C^*-algebras that we shall show have an imprimitivity bimodule.

Now it follows from Section 7 of [36] that the process of inducing representations from G_J up to G establishes an equivalence between the category of Hermitian $C_c(G_J)$-modules and the category of Hermitian modules over the imprimitivity algebra $C_c(G \times G/G_J)$, with $C_c(G)$ serving as an imprimitivity bimodule for these two algebras. We need to study how this imprimitivity bimodule is related to the ideals in $C^*(G_J)$ and $C^*(G)$ defined previously, and to the corresponding quotient algebras.

3. Imprimitivity Bimodules for Ideals and Quotients

Let B and E be pre-C^*-algebras and let X be an E–B-imprimitivity bimodule (Definition 6.10 of [36]). We wish to relate X to ideals in the completions of B and E and to corresponding quotient algebras. For this reason we will need to consider the completion of X. Now both the B-valued and the E-valued inner products give norms with respect to which we can complete X. Thus it is reasuring that these norms agree.

Proposition 3.1. *Let E and B be pre-C^*-algebras and let X be an E–B-imprimitivity bimodule. Then for any $x \in X$,*

$$\|\langle x, x\rangle_E\| = \|\langle x, x\rangle_B\|.$$

Proof

$$\begin{aligned}\|\langle x, x\rangle_B\|^2 &= \|\langle x, x\rangle_B\langle x, x\rangle_B\| = \|\langle x, x\langle x, x\rangle_B\rangle_B\| = \|\langle x, \langle x, x\rangle_E x\rangle_B\| \\ &\leqslant \|\langle x, x\rangle_B\|^{1/2}\,\|\langle\langle x, x\rangle_E x, \langle x, x\rangle_E x\rangle_B\|^{1/2} \leqslant \|\langle x, x\rangle_B\|\,\|\langle x, x\rangle_E\|,\end{aligned}$$

where the first inequality is an application of the generalized Cauchy–Schwartz inequality in Proposition 2.9 of [36] and the second inequality is an application of condition 2 in the definition of an imprimitivity bimodule (6.10 of [36]). Dividing by $\|\langle x, x\rangle_B\|$ we obtain inequality in one direction. The opposite inequality is shown by a similar calculation. Q.E.D.

It follows that the completions of X with respect to the two norms are the same, and that we can simultaneously extend the actions of E and B to actions of their completions, $\bar{E}$ and $\bar{B}$, on the completion $\bar{X}$, so that $\bar{X}$ become an $\bar{E}$–$\bar{B}$-imprimitivity bimodule. Consequently we will usually, from now on, work with C^*-algebras rather than pre-C^*-algebras, although it is not usually important that X be complete as long as actions of the (complete) C^*-algebras can be defined on it.

The following result is the analog for C^*-algebras of part of Theorem 3.5(5) on p. 65 of [4] for algebras and Proposition 8.4 of [37] for W^*-algebras.

Theorem 3.1. *Let E and B be C^*-algebras and let X be an E–B-imprimitivity bimodule. Then there are natural isomorphisms among*

(1) *the lattice of (closed two-sided) ideals of B,*
(2) *the lattice of closed E–B-submodules of X, and*
(3) *the lattice of ideals of E,*

where ideals and submodules are ordered by inclusion. If J is an ideal in B, then the E–B-subspace of X corresponding to J under the first isomorphism is

$$X_J = \{y \in X : \langle x, y\rangle_B \in J \text{ for all } x \in X\}.$$

If Y is a closed E–B-subspace of X, then the ideal of B corresponding to Y under the isomorphism is

$$I_Y = \text{closed span of } \{\langle x, y\rangle_B : x \in X, y \in Y\}.$$

Similar statements hold for E.

LEMMA 3.1. *With the previously defined notation we have*

$$XJ = \{y \in X : \langle x, y\rangle_B \in J \text{ for all } x \in X\} = \{y \in X : \langle y, y\rangle_B \in J\}.$$

Proof. It is clear that the first space is contained in the second, which in turn is contained in the third, so we need to show that the third is contained in the first. Let $x \in X$ with $\langle x, x\rangle_B \in J$. Let $\{i_k\}$ denote a self-adjoint approximate identity for J (1.7.2 of [14]). Then, expanding $\langle x - xi_k, x - xi_k\rangle_B$, we obtain a sum of products of i_k with $\langle x, x\rangle_B$ which converges to zero. It follows that xi_k converges to x in the norm of X, so that $x \in XJ$. Q.E.D.

Proof of Theorem 3.1. Let J be an ideal in B. It is clear that J contains I_{X_J}. But by Lemma 3.1, $X_J = XJ$, and so I_{X_J} contains $\langle XJ, XJ\rangle_B = J\langle X, X\rangle_B J$, which spans a dense subset of J since $\langle X, X\rangle_B$ is assumed to span a dense subset of B. Thus $I_{X_J} = J$.

Conversely, let Y be a closed E–B-submodule of X. Then X_{I_Y} clearly contains Y. Now by Lemma 3.1, $X_{I_Y} = XI_Y$. Let $x, z \in X$ and $y \in Y$. Then

$$x\langle z, y\rangle_B = \langle x, z\rangle_E y,$$

which is in Y since Y is assumed to be E-invariant. But every element of XI_Y is a limit of linear combinations of elements of the form $x\langle z, y\rangle_B$ for $x, z \in X$, $y \in Y$. It follows that X_{I_Y} is contained in Y.

The fact that this correspondence between ideals and submodules preserves inclusion is clear. Q.E.D.

COROLLARY 3.1. *Let E and B be C^*-algebras and let X be an E–B-imprimitivity bimodule. Let J be an ideal in B and let K be the ideal in E which corresponds to X_J according to Theorem* 3.1. *Then both $\langle X, X_J\rangle_E$ and $\langle X_J, X_J\rangle_E$ are contained in and generate K. In particular, X_J is a K–J-imprimitivity bimodule.*

COROLLARY 3.2. *Let E and B be C^*-algebras and let X be an E–B-imprimitivity bimodule. Let I and J be ideals in B with $I \supset J$ and let $K(I)$ and $K(J)$ denote the ideals in E which correspond to I and J according to Theorem 3.1. Then the B-valued inner product on X drops to an (I/J)-valued inner product on X_I/X_J, and the E-valued inner product drops to a $(K(I)/K(J))$-valued inner product on X_I/X_J, so that X_I/X_J becomes a $(K(I)/K(J)))$–I/J-imprimitivity bimodule.*

We now show that the process previously described of passing to ideals and quotients behaves well with respect to the inducing process. Since for the inducing process we do not need an imprimitivity bimodule, but only a Hermitian rigged module, we work with these instead.

PROPOSITION 3.2. *Let A and B be C^*-algebras, let X be a Hermitian B-rigged A-module, and let I and J be ideals in B with $I \supset J$. Then XI and XJ are Hermitian I- and J-rigged A-modules, and XI/XJ is a Hermitian (I/J)-rigged A-module, which we will denote by Z. Let F_Z denote the functor consisting of inducing Hermitian (I/J)-modules up to A. Let $D = \mathrm{hull}(J) - \mathrm{hull}(I)$, so that the category of Hermitian (I/J)-modules is isomorphic to the category of those Hermitian B-modules which live on the locally closed set D. Let F_X denote the functor of inducing Hermitian B-modules up to A via X. Then F_Z is naturally unitarily equivalent to the restriction of F_X to the category of Hermitian B-modules which live on D (viewed as Hermitian (I/J)-modules).*

Proof. The assertions concerning XI, XJ, and X are easily verified. We must show that the indicated functors are unitarily equivalent. Let V be a Hermitian (I/J)-module, which we can view as a Hermitian B-module. Then the algebraic span IV (now not closed) is a dense submodule of V, and, as algebraic tensor products,

$$x \otimes_B IV = XI \otimes_B V.$$

Define a map t_V from $XI \times V$ into $Z \otimes_{I/J} V$ by

$$t_V(x, v) = (x + XJ) \otimes v$$

for $x \in XI$ and $v \in V$. It is easily verified that t_V is B-balanced and bilinear, so that it lifts to a linear map, also denoted by t_V, of $XI \otimes_B V$ into $Z \otimes_{I/J} V$. It is clear that t_V is in fact surjective, and it is easily seen that t_V is an A-module homomorphism. Furthermore, a simple calculation shows that t_V is an isometry with respect to the pre-inner products on these spaces (where that on $XI \otimes_B V$ comes from $X \otimes_B V$). But the image of $X \otimes_B IV$ is dense in $F_X(V)$, and so t_V extends to an isometry of $F_X(V)$ into $F_Z(V)$, which is surjective since the image of $Z \otimes_{I/J} V$ is dense in $F_Z(V)$. The collection of the unitary maps t_V gives the desired unitary natural equivalence of the restriction of F_X with F_Z. Q.E.D.

We conclude this section by showing that the kernels of representations induced by an imprimitivity bimodule behave well with respect to the bijection of Theorem 3.1, so that this bijection preserves primitive ideals.

PROPOSITION 3.3 *Let E and B be C*-algebras and let X be an E–B-imprimitivity bimodule. Let V be a Hermitian B-module and let J be the kernel of (the representation corresponding to) V. Let K be the ideal in E corresponding to J according to Theorem* 3.1 *and let* ${}^EV$ *be the Hermitian E-module obtained by inducing V to E via X. Then the kernel of (the representation corresponding to)* ${}^EV$ *is K.*

Proof. We use the notation of Theorem 3.1. Let $x, y \in X_J$, so that, in particular, $\langle y, z\rangle_B \in J$ for all $z \in X$. Let $z \in X$, $v \in V$. Then

$$\langle x, y\rangle_E(z \otimes v) = x\langle y, z\rangle_B \otimes v = x \otimes \langle y, z\rangle_B v = 0,$$

so that $\langle x, y\rangle_E$ is in the kernel of ${}^EV$. Since the elements of this form span K, it follows that K is contained in the kernel of ${}^EV$. Suppose conversely that e is an element of the kernel of ${}^EV$. Then for every $x \in X$, $v \in V$ we have $e(x \otimes v) = 0$, and so for every $y \in X$, $w \in V$ we have

$$0 = \langle e(x \otimes v), y \otimes w\rangle = \langle\langle y, ex\rangle_B v, w\rangle.$$

It follows that $\langle y, ex\rangle_B$ is in J for all $x, y \in X$, so that $ex \in X_J$ for all $x \in X$. Now $\langle X, X\rangle_E$ is assumed to span a dense subset of E, and E has an approximate identity, so e can be approximated by linear combinations of elements of the form

$$e\langle x, y\rangle_E = \langle ex, y\rangle_E.$$

But by definition each such element is in K, since $ex \in X_J$. Thus $e \in K$. Q.E.D.

COROLLARY 3.3. *Let E and B be C*-algebras and let X be an E–B-imprimitivity bimodule. Then the bijection of Theorem* 3.1 *between ideals of B and ideals of E restricts to a bijection between the primitive ideals of B and the primitive ideals of E. In particular,* Prim(B) *and* Prim(E) *are homeomorphic.*

Proof. By corollary 6.25 of [36], ${}^EV$ is irreducible if and only if V is. Q.E.D.

From Proposition 3.3 we obtain the following generalization of Theorem 6.1 of [26].

COROLLARY 3.4. *Let A and B be C^*-algebras and let X be a Hermitian B-rigged A-module. Let V_i, $i = 1, 2$, be Hermitian B-modules, let $J_i = \ker(V_i)$, and let $L_i = \ker({}^A V_i)$. If $J_1 \supseteq J_2$, then $L_1 \supseteq L_2$.*

Proof. Let $K_i = \ker({}^E V_i)$, so that $K_1 \supseteq K_2$ by Proposition 3.3. Then $L_1 \supseteq L_2$ follows from the fact that $L_i = \{a \in A : aE \subseteq K_i\}$. This is a special case of the easily verified fact that if C is any C^*-algebra, if E is any two-sided ideal of C, if A is a subalgebra of C, if W is a Hermitian C-module which is non degenerate as an E-module, and if K is the kernel of W as an E-module (so K is an ideal in C also), then the kernel of W as an A-module is $\{a \in A : aE \subseteq K\}$. Q.E.D.

4. THE HERMITIAN RIGGED MODULE OF AN ORBIT

In this section we apply the results of the last section to the situation of Theorem 1.1. Actually, for the first part of this section we still do not need to know that G_J is a stability subgroup, and so we begin by assuming only that we have a locally compact group G, a normal subgroup N, and a subgroup H of G which contains N. As in Section 2 we let X denote the completion of $C_c(G)$ as a $C_c(H)$-rigged space. Then from Corollary 2.1 we immediately conclude, in analogy with Corollary 2.3, the following:

COROLLARY 4.1. *If I is an ideal in $C^*(N)$ which is G-invariant, then $IX = XI$.*

Setting $H = N$, we are now in a position to prove that for a normal subgroup N of a locally compact group G, the $*$-homomorphism of $C^*(N)$ into $M(C^*(G))$ is injective.

PROPOSITION 4.1. *Let N be a normal subgroup of the locally compact group G and let V be a unitary N-module. Let $({}^G V)_N$ denote the restriction to N of the unitary G-module ${}^G V$ obtained by inducing V to G. Then $({}^G V)_N$ weakly contains V. In particular, the homomorphism of $C^*(N)$ into $M(C^*(G))$ is injective. Furthermore, if E is any locally closed subset of* Prim(N) *that carries V, then $({}^G V)_N$ is carried on any locally closed G-invariant subset of* Prim(N) *that contains E.*

Proof. Let X be as previously defined, but for $H = N$. Let K denote the kernel of (the representation on) $({}^G V)_N$. Then K is G-invariant according to Proposition 2.2, and so $KX = XK$ by Corollary 4.1. We need to show that $KV = \{0\}$. Working symbolically, with $B = C^*(N)$, we have

$$\{0\} = K(({}^G V)_N) = K(X \otimes_B V) = KX \otimes_B V = XK \otimes_B V = X \otimes_B KV.$$

This says that if the Hermitian B-module KV is induced to G, one obtains the zero-dimensional G-module. But this can only happen if $KV = \{0\}$ (as can be seen from Theorem 5.1 of [36]).

Now the kernel of the map of $C^*(N)$ into $M(C^*(G))$ must be contained in the kernel of the restriction to N of every unitary representation of G, and, in particular, in the kernel of $({}^GV)_N$ for every unitary N-module V. But according to the previous paragraph this kernel must then be in the kernel of every unitary representation of N, and so is zero.

Finally, suppose V lives on the locally closed subset E of $\mathrm{Prim}(N)$, and let D be a G-invariant locally closed subset of $\mathrm{Prim}(N)$ containing E, so that V also lives on D. Let $D' = \bar{D} - D$ and let $I = \ker(\bar{D})$, $I' = \ker(D')$, so that V is a Hermitian I'/I-module. Now I and I' are G-invariant, so that Corollary 4.1 is applicable. Working symbolically, we have

$$I(({}^GV)_N) = IX \otimes_B V = XI \otimes_B V = X \otimes_B IV = \{0\},$$
$$I'(({}^GV)_N) = I'X \otimes_B V = XI' \otimes_B V = X \otimes_B I'V = X \otimes_B V = ({}^GV)_N.$$

Thus $({}^GV)_N$ is a Hermitian I'/I-module, and so lives on D. Q.E.D.

We remark that the first part of Proposition 4.1 is Theorem 4.5 of [21] or Proposition 5.3 of [22], but without separability assumptions. J. M. G. Fell had shown (personal communication) that the separability assumption could be dropped, but this proof is somewhat different from his.

COROLLARY 4.2. *Let N be a normal subgroup of the locally compact group G and let E be a nonempty locally closed subset of* $\mathrm{Prim}(N)$ *which is G-invariant. Let $I = \ker(\bar{E})$ and $I' = \ker(\bar{E} - E)$, so that I and I' are G-invariant ideals. Then the algebra*

$$C^*(G)I'/C^*(G)I$$

is not the zero-dimensional algebra.

Proof. It is clear that I'/I cannot be the zero-dimensional algebra, and so there is at least one unitary N-module V which lives on E (use 2.10.4 of [14]). Then $({}^GV)_N$ will also live on E by Proposition 4.1. But ${}^GV$ is then a nonzero Hermitian module over $C^*(G)I'/C^*(G)I$ by Proposition 2.1, so this algebra cannot be the zero algebra. Q.E.D.

Suppose now that, as in Theorem 1.1, $J \in \mathrm{Prim}(N)$ and $\{J\}$ is locally closed. Let GJ be the orbit of J, and suppose that GJ is locally closed. Define the ideals J', I, and I' as in the paragraph after Corollary 2.4. Then from Corollary 4.2 the algebras $C^*(G_J)J'/C^*(G_J)J$ and $C^*(G)I'/C^*(G)I$ are nonzero. In particular, $C^*(G_J)J' \neq C^*(G_J)J$. It follows from Theorem 3.2 that

$XC^*(G_J)J' \neq XC^*(G_J)J$. It is clear that $XC^*(G_J)J' = XJ'$, and $XC^*(G_J)J = XJ$. Thus $XJ' \neq XJ$. Furthermore, $I \subseteq J$, so $IX = XI \subseteq XJ$. Combining this observation with the results at the end of Section 2 and in Section 3, we obtain:

COROLLARY 4.3. *The space XJ'/XJ is a nonzero Hermitian $C^*(G_J)J'/C^*(G_J)J$-rigged module over both $C^*(N)/I$ and $C^*(G)/C^*(G)I$. The inducing functor defined by this Hermitian rigged module is equivalent to the restriction to the category of unitary G_J-modules supported on $\{J\}$ of the ordinary inducing functor from unitary G_J-modules to unitary G-modules.*

PROPOSITION 4.2. *As an I'/I-module or a $C^*(G)I'/C^*(G)I$-module XJ'/XJ is still nondegenerate, and so is a Hermitian rigged module.*

Proof. We need to show that

$$I'(XJ'/XJ) = XJ'/XJ$$

or, since I' is G-invariant so that $I'X = XI'$, that

$$XI'J' + XJ = XJ'.$$

To show this it sufficies to show that

$$I'J' + J = J'.$$

Now $I'J' = I' \cap J'$ by 1.9.12 of [14], and so we need to show that

$$(I' \cap J') + J = J'.$$

Working with the definitions of these ideals, one finds that this relation is equivalent to

$$[((GJ)^c - GJ) \cup (\{J\}^c - \{J\})] \cap \{J\}^c = \{J\}^c - \{J\}.$$

But this relation is easily seen to be true. Q.E.D.

PROPOSITION 4.3. *The representation of I'/I on XJ'/XJ is faithful.*

Proof. Let K be the kernel of the representation of I' on XJ'/XJ, so that K consists of those elements k of I' for which $kXJ' \subseteq XJ$. If k is such an element and if $y \in G$, then

$$k^y XJ' = y(k(y^{-1}XJ')) \subseteq XJ.$$

Thus we see that K is G-invariant. Let C be the hull of K. Since $I' \supseteq K \supseteq I$, it follows that

$$(GJ)^c - GJ \subseteq C \subseteq (GJ)^c.$$

Since C is closed and G-invariant, it follows that either $C = (GJ)^c - GJ$ or $C = (GJ)^c$. In the former case we would have $K = I'$, so that every element of I' would act as the zero operator on XJ'/XJ. But this is impossible since XJ'/XJ is a nonzero Hermitian I'/I-module. It follows that $K = I$ and the representation is faithful. Q.E.D.

We will call XJ'/XJ the Hermitian rigged module associated with the orbit GJ.

5. The System of Imprimitivity

We keep the notation used at the end of the last section. Our aim is to show that $C^*(G)I'/C^*(G)I$ is the imprimitivity algebra (Definition 6.4 of [36]) of the $C^*(G_J)J'/C^*(G_J)J$-rigged space XJ'/XJ. Now from the definition, the imprimitivity algebra of XJ'/XJ is clearly the algebra of operators on XJ'/XJ spanned by the images of the operators $\langle xc, yd\rangle_E$ on X, for $x, y \in X$ and $c, d \in J'$, where E denotes the imprimitivity algebra of X. As shown in Corollary 7.13 of [36], E is normdense (for the norm on operators on X) in $C_c(G \times G/G_J)$, which is the transformation group algebra for the action of G on G/G_J (see 7.5 of [36] or 3.11 of [18]). But $C_c(G \times G/G_J) = C_c(G, C_c(G/G_J))$, and one of the ingredients of its action on X is the pointwise multiplication of elements of $C_c(G)$ by elements of $C_c(G/G_J)$ viewed as functions on G constant on cosets. In fact, E or $C_c(G \times G/G_J)$ can be viewed as a semidirect product of $C_c(G)$ with $C_c(G/G_J)$, in the sense of covariance algebras [40], and the algebra $C_c(G/G_J)$ together with its action of pointwise multiplication on $C_c(G)$ is the system of imprimitivity for the situation. Since $C_c(G/G_J)$ defines an algebra of bounded operators on X (Proposition 7.2 of [36]), it will also define an algebra of operators on XJ'/XJ. In this section we will study this action of $C_c(G/G_J)$ on XJ'/XJ. For much of what we do it is actually slightly more convenient to study the larger algebra $C(G/G_J)$ of bounded continuous functions on G/G_J, which also acts as an algebra of bounded operators on X (Proposition 7.2 of [36]), and so on XJ'/XJ.

Proposition 5.1. *The action of $C(G/G_J)$ on $C_c(G)$ commutes with the left action of $C_c(N)$. Thus the action of $C(G/G_J)$ on X commutes with the left action of $C^*(N)$.*

The proof consists of a routine computation.

Now one of the hypotheses of Theorem 1.1 is that the canonical map from G/G_J to GJ is a homeomorphism. As a result of this, GJ is Hausdorff, and we can identify $C(G/G_J)$ with $C(GJ)$. But GJ is (identified with) Prim(I'/I), and so $C(G/G_J)$ becomes identified with the C^*-algebra of bounded continuous functions on $\mathrm{Prim}(I'/I)$.

Now if A is any C^*-algebra, Dixmier [15] has shown that the algebra of bounded continuous functions on $\mathrm{Prim}(A)$ can be identified with the center of $M(A)$. Dixmier does not state his results in terms of double centralizers, but it is an easy matter to restate them in this way, as remarked by Busby on p. 371 of [11]. Dixmier based his proof on a theorem of Dauns and Hoffmann, but he also gave an alternate proof based on a result of Stormer. This result of Stormer has more recently been given a very short proof by Bunce [9], and consequently a quite direct path to Dixmier's theorem is via Stormer's result with Bunce's proof. Another short proof has just appeared in [19].

It follows from these considerations that $C(G/G_J)$ can be identified with the center of $M(I'/I)$. Since XJ'/XJ is a Hermitian rigged I'/I-module, it is also a Hermitian rigged module over $M(I'/I)$ (by Proposition 3.9 of [36]), and so is also a Hermitian rigged module over the center of $M(I'/I)$, which we have just identified with $C(G/G_J)$. We thus have defined two actions of $C(G/G_J)$ on XJ'/XJ, one coming from the pointwise action of $C(G/G_J)$ on $C_c(G)$ and the other from the identification of $C(G/G_J)$ with the center of $M(I'/I)$. The main result of this section is that these two actions coincide.

To distinguish between the two actions in the proof, we will, for $F \in C(G/G_J)$, denote by P_F the operator on XJ'/XJ obtained from pointwise multiplication on $C_c(G)$ by F, while we will denote by T_F the operator on XJ'/XJ obtained by identifying F with an element of the center of $M(I'/I)$.

PROPOSITION 5.2. *For any* $F \in C(G/G_J)$, $P_F = T_F$.

Proof. Let $d \in C^*(N)$ and let $f \in C_c(G)$, viewed as an element of X. Then according to Lemma 2.1,

$$d(P_F f) = d(Ff) \in X[d^Y : y^{-1} \in \mathrm{support}(F)],$$

since $\mathrm{support}(Ff) \subseteq \mathrm{support}(F)$. It follows that for any $x \in X$ we have

$$d(P_F x) \in X[d^Y : y^{-1} \in \mathrm{support}(F)].$$

Now view F as a function on GJ, and suppose it happens that $d \in \ker(\mathrm{support}(F))$. This means that $d \in J^y$ for each $y \in \mathrm{support}(F)$, where now F is viewed as a function on G. But then $d^y \in J$ for each $y^{-1} \in \mathrm{support}(F)$, and so $[d^y : y^{-1} \in \mathrm{support}(F)] \subseteq J$. Thus $dP_F X \subseteq XJ$, so that dP_F acts as the zero operator on XJ'/XJ.

Let $L(XJ'/XJ)$ denote the algebra of bounded operators on the rigged space XJ'/XJ (Definition 2.3 of [36]). Since the representation of I'/I on XJ'/XJ is faithful (Proposition 4.3), we can view I'/I as a subalgebra D

of the C^*-algebra $L(XJ'/XJ)$. Since XJ'/XJ is nondegenerate as an I'/I-module, it suffices to show that $T_F d = P_F d$ for all $d \in D$ in order to conclude that $T_F = P_F$. (Note that $P_F d = dP_F$ by Proposition 5.1.) Thus the proof will be complete once we have proven the following proposition (in which the primitive ideal space need not be assumed to be Hausdorff).

PROPOSITION 5.3. *Let A be a C^*-algebra with identity and let D be a C^*-subalgebra of A (possibly without identity). For $F \in C(\mathrm{Prim}(D))$ (the algebra of bounded continuous functions on* Prim(D)), *let T_F denote the corresponding double centralizer of D. Let P be a $*$-homomorphism of* $C(\mathrm{Prim}(D))$ *into A such that*

(1) *P_1 is the identity element of A.*
(2) *$P_F d = dP_F$ for all $F \in C(\mathrm{Prim}(D))$ and $d \in D$.*
(3) *If $d \in D$ and $F \in C(\mathrm{Prim}(D))$ and if $d \in \ker(\mathrm{support}(F))$, then $P_F d = 0$.*

Then $P_F d = T_F d$ for all $F \in C(\mathrm{Prim}\,(D))$ and $d \in D$.

Proof. Let $d \in D$ and $F \in C(\mathrm{Prim}(D))$. Suppose first that for some complex number m and some $\epsilon > 0$, we have $d \in \ker(R)$ where

$$R = \{J \in \mathrm{Prim}(D) : |F(J) - m| \geqslant \epsilon\}.$$

Let

$$S = \{J \in \mathrm{Prim}(D) : |F(J) - m| \geqslant 2\epsilon\},$$

so that S is contained in R. Let h be a continuous function from the complex plane to the interval $[0, 1]$ such that, for any complex number z, $h(z) = 1$ if $|z - m| \geqslant 2\epsilon$, and $h(z) = 0$ if $|z - m| \leqslant \epsilon$. Let $F_0 = h \circ F$, so that F_0 is a continuous function on Prim(D) with values in $[0, 1]$ and $F_0(J) = 1$ for $J \in S$, while $F_0(J) = 0$ for $J \notin R$. If we let $G = FF_0 + m(1 - F_0)$, then $F - G = (F - m)(1 - F_0)$, so that $\|F - G\|_\infty \leqslant 2\epsilon$. Furthermore $G - m$ is zero off of R, so that $d \in \ker(\mathrm{support}(G - m))$. Then by hypotheses (1) and (3) we see that $P_{(G-m)} d = 0$, so that $P_G d = md$. It follows that

$$\|P_F d - md\| = \|P_{(F-G)} d\| \leqslant 2\epsilon \|d\|.$$

Now let $d \in D$ and $F \in C(\mathrm{Prim}(A))$, and assume that $d = d^*$ and that F is real-valued. Let B be the C^*-subalgebra of A generated by the $T_H d$ for all $H \in C(\mathrm{Prim}(D))$ together with the image of all of $C(\mathrm{Prim}(A))$ under P. Note that B is commutative by hypothesis (2). Let $\epsilon > 0$ be given, and let $\{U_i\}$ be a finite covering of the closure of the range of F by open intervals of length strictly less than ϵ, and such that no U_i meets more than two other of the U_is. For each i choose a point m_i in U_i and let $V_i = F^{-1}(U_i)$. Then the V_i form a finite open covering of Prim(D), and $|F(J) - m_i| < \epsilon$ for $J \in V_i$.

Let $\{h_i\}$ be a partition of unity on the real line which is finer than the U_i together with the complement of the closure of the range of F. Discard the functions supported in the complement of the range of F, and for each remaining i set $G_i = h_i \circ F$. Then each G_i is a continuous function on Prim(D) having range in $[0, 1]$, and value 0 off of V_i. Furthermore $\sum G_i \equiv 1$ on Prim(D). In particular, $\sum T_{G_i} d = d$.

Now the relation defining how a double centralizer T_{G_i} is associated to the function G_i is

$$T_{G_i} d - G_i(J)d \in J$$

for every $J \in \text{Prim}(D)$. Since G_i vanishes off of V_i, it follows that $T_{G_i} d \in J$ for each $J \notin V_i$. This means that

$$T_{G_i} d \in \ker\{J \in \text{Prim}(D) : |F(J) - m_i| \geqslant \epsilon\},$$

so that by the first paragraph of the proof

$$\|P_{(F-m_i)} T_{G_i} d\| = \|P_F T_{G_i} d - m_i T_{G_i} d\| < 2\epsilon \|d\|.$$

Now for any i there are at most two j's different from i such that $h_i h_j \neq 0$, such that $T_{G_i} T_{G_j} \neq 0$, and such that

$$(P_{(F-m_i)} T_{G_i} d)(P_{(F-m_j)} T_{G_j} d) \neq 0.$$

Since all of this is happening in the commutative C^*-algebra B, it is easily seen that

$$\|\sum P_{(F-m_i)} T_{G_i} d\| < 6\epsilon \|d\|.$$

Since $P_F d = \sum P_F T_{G_i} d$, it follows that

$$\|P_F d - \sum m_i T_{G_i} d\| < 6\epsilon \|d\|.$$

However, $\|F - \sum m_i G_i\|_\infty < \epsilon$, so

$$\|T_F d - \sum m_i T_{G_i} d\| < \epsilon \|d\|.$$

Consequently

$$\|P_F d - T_F d\| \leqslant 7\epsilon \|d\|.$$

Since ϵ is arbitrary, $M_F d = T_F d$. This relation extends by linearity to arbitrary F and d. Q.E.D.

COROLLARY 5.1. *The homomorphism P of $C(G/G_J)$ into the algebra of bounded operators on XJ'/XJ coming from the action on X corresponding to pointwise multiplication on $C_c(G)$, is injective, and for every $d \in I'/I$ and $F \in C(G/G_J)$, we have $P_F d \in I'/I$ (viewed as a subalgebra of the algebra of bounded operators on XJ'/XJ).*

Proof. According to Proposition 4.3 the representation of I'/I on XJ'/XJ is faithful. If $P_F = 0$, then $T_F d = 0$ for all $d \in I'/I$, and it follows that $T_F = 0$, so that $F = 0$. Furthermore, for any F, $P_F d = T_F d \in I'/I$.

Q.E.D.

6. The Proof of the Main Theorem

We have not yet shown that the representation of $C^*(G)I'/C^*(G)I$ on XJ'/XJ is injective, and in fact we will leave the proof of this fact to the end. Consequently, we will here denote by C the image of $C^*(G)I'/C^*(G)I$ as operators on XJ'/XJ.

Lemma 6.1. *The imprimitivity algebra of XJ'/XJ is contained in C.*

Proof. According to the definition of the imprimitivity algebra of XJ'/XJ, it is the image of the subalgebra of E generated by $\langle XJ', XJ'\rangle_E$. Now it is easily seen that $I' + J \supseteq J'$. Furthermore, if x or y is in XJ, then $\langle x, y\rangle_E z$ $(= x\langle y, z\rangle_{C^*(G_J)})$ is in XJ for $z \in X$ (since J is G_J-invariant), so that $\langle x, y\rangle_E$ as an operator on XJ'/XJ is the zero operator. Thus the imprimitivity algebra for XJ'/XJ will be contained in the image of $\langle XI', XI'\rangle_E$. But $XI' = I'X$. Furthermore, since E is an ideal in the algebra of bounded operators on X, and elements of E come out of the E-valued inner product, elements of I' acting on the left of X will also come out of the E-valued inner product. Thus the imprimitivity algebra for XJ'/XJ will be contained in the image of $I'\langle X, X\rangle_E I'$, or $I'EI'$, and so it suffices to show that the image of this latter algebra is contained in C.

Now the transformation group algebra $C_c(G \times G/G_J)$ forms a dense subalgebra of E, and elements of the form $\Phi(y) = g(y)F$ for $g \in C_c(G)$, $F \in C_c(G/G_J)$ span a dense subspace of $C_c(G \times G/G_J)$. Furthermore, a simple calculation shows that the action of such a Φ on an $x \in X$ is given by $\Phi x = P_F(gx)$ (where P_F was defined just before Proposition 5.2). Thus it suffices to show that for $d, d' \in I'$, the operator on XJ'/XJ corresponding to $dP_F gd'$ is in C. But according to Proposition 5.2 the operator P_F acts like the double centralizer T_F, and so the above operator is the same as the operator $T_F(d)gd'$ (viewing d as an element of I'/I), and this operator is clearly in the image of $I'C^*(G)I'$, and so in C. Thus the imprimitivity algebra for XJ'/XJ is contained in C. Q.E.D.

Lemma 6.2. *The imprimitivity algebra of XJ'/XJ is all of C.*

Proof. Since the orbit GJ is Hausdorff, it is easily seen that $J' = I' + J$. Combining this fact with the discussion in the proof of Lemma 6.1, it follows that the imprimitivity algebra of XJ'/XJ will be all of the image of $I'EI'$, and so, as above, of the algebra generated by the $T_F(d)gd'$. But the elements of I'/I, which are of the form $T_F(d)$ as d ranges over I'/I and F ranges over $C_c(G/G_J)$, will span a dense ideal of I'/I, as is easily seen from Proposition 3.3.7 of [14]. Thus the image of $I'EI'$ contains a dense part of the image of $C^*(G)I'$, and so the imprimitivity algebra of XJ'/XJ is all of C. Q.E.D.

It follows that the functor consisting of inducing to G unitary G_J-modules whose restrictions to N live on $\{J\}$ is a full embedding of the category of such modules into the category of unitary G-modules whose restrictions to N live on GJ. We have thus proven what may be considered our version of Theorem 1 of [6].

To complete the proof of our Main Theorem (Theorem 1.1), which also includes our version of Theorem 2 of [6], we must show that the representation of $C^*(G)I'/C^*(G)I$ on XJ'/XJ is faithful. That the proof of this fact should still be somewhat delicate is perhaps not too surprising in view of the fact that the representation of $C^*(G)$ on X need not be faithful. It will be convenient to change our notation and hereafter let $C = C^*(G)I'/C^*(G)I$.

Now $C^*(G)I'$ is an ideal in $C^*(G)$, and so an ideal in $M(C^*(G))$ by Proposition 1.8.5 of [14]. This gives a natural homomorphism of $M(C^*(G))$ into $M(C^*(G)I')$. By Proposition 3.8 of [10] there is a natural homomorphism of $M(C^*(G)I')$ into $M(C)$ (which we do not know is surjective—see the comments after Theorem 4.2 of [1]). Thus we have a natural homomorphism of $M(C^*(G))$ into $M(C)$. Now I' and I can both be viewed as subalgebras of $M(C^*(G))$ by Proposition 4.1, and it is clear that the elements of I act as the zero operator on C. Thus we have a natural homomorphism of I'/I into $M(C)$. It is clear that C is nondegenerate as a right or left I'/I-module, and so an application of the left- and right-handed versions of Proposition 3.9 of [36] gives a natural homomorphism of $M(I'/I)$ into $M(C)$. Since we have identified $C(G/G_J)$ with the center of $M(I'/I)$, it follows that we have a homomorphism from $C(G/G_J)$ into $M(C)$, which is defined by the relation

$$F(dc) = (T_F(d))c$$

for $F \in C(G/G_J)$, $d \in I'/I$, $c \in C$, and similarly for the right action, where T_F is defined as in Proposition 5.2.

Each element of G also defines an element of $M(C^*(G))$, and so of $M(C)$. In addition, G acts by left translation on $C(G/G_J)$. We show that these two actions are related. Now T_F is defined by the fact that for each $z \in G$,

$$T_F(d) - F(J^z)d \in J^z$$

for every $d \in I'/I$. If we let yF denote the left translate of F by y, then a simple calculation shows that

$$(T_F(d))^y - yF(J^z)d^y \in J^z$$

for every $z \in G$. But this is the relation that defines $T_{yF}(d^y)$. It follows that

$$(T_F(d))^y = T_{yF}(d^y).$$

Then a simple calculation shows that for $c \in C$ we have

$$y(F(dc)) = (yF)(y(dc)).$$

By continuity we obtain:

LEMMA 6.3. *For $y \in G$, $F \in C(G/G_J)$, and $c \in C$, we have*

$$y(F(c)) = (yF)(yc).$$

Thus we see that $C(G/G_J)$ acts like a system of imprimitivity (as defined in Theorem 7.18 of [36]) for the action of G on C. Furthermore, in the proof of Lemma 6.2 we saw that I'/I is nondegenerate as a $C(G/G_J)$-module, and it follows that the same is true of C.

Let E denote the (norm-closed) imprimitivity algebra of the $C^*(G_J)$-rigged space X. In the proof of Theorem 7.18 of [36] it was shown that every unitary G-module on which $C_\infty(G/G_J)$ acts as a system of imprimitivity becomes an a natural way a Hermitian E-module. We would like to show here that in the same way the maps of G and $C_\infty(G/G_J)$ into $M(C)$ give a homomorphism of E into $M(C)$. Now from Proposition 7.11 of [36] it follows that $C_c(G \times G/G_J)$ is dense in E. Then for $c \in C$ and $\Phi \in C_c(G \times G/G_J)$, with Φ viewed as an element of $C_c(G, C_\infty(G/G_J))$, we can define, in analogy with 7.20 of [36],

$$\Phi c = \int \Phi(y)(yc)\,dy, \qquad c\Phi = \int ((c\Phi(y))y)\,dy. \tag{6.1}$$

Then it is easily seen that these actions define a $*$-homomorphism of $C_c(G \times G/G_J)$ into $M(C)$ (see the proof of Proposition 7.6 of [36]). We need to show that this $*$-homomorphism is continuous, so that it extends to all of E. Now for this purpose the most convenient information about the norm of E that we have is that if W is a unitary G-module on which $C_\infty(G/G_J)$ acts as a system of imprimitivity, then the corresponding $*$-representation of $C_c(G \times G/G_J)$ is continuous (see the proof of Theorem 7.18 of [36]). But suppose that W is a faithful Hermitian C-module, so that C can be identified with a subalgebra of the algebra $L(W)$ of the bounded operators on W. Then, according to Theorem 3.9 of [10], $M(C)$ can be identified with the idealizer of C in $L(W)$. The homomorphisms of G and $C_\infty(G/G_J)$ into $M(C)$ become homomorphisms into $L(W)$, so that W becomes a unitary G-module on which $C_\infty(G/G_J)$ acts as a system of imprimitivity. The homomorphism of $C_c(G \times G/G_J)$ into $M(C)$ becomes the homomorphism into $L(W)$ corresponding to this system of imprimitivity, and so is continuous for the reason previously indicated. Thus we do obtain a homomorphism of E into $M(C)$. Since we saw that C is nondegenerate as a $C_\infty(G/G_J)$-module, it follows that C is nondegenerate as an E-module. We thus obtain:

LEMMA 6.4. *The actions defined in* (6.1) *extend by continuity to give a homomorphism of E into $M(C)$ under which C, as a left or right E-module, is nondegenerate.*

The proof of the following lemma, which is the crux of the proof of the faithfulness of the representation of C on XJ'/XJ, was motivated by that part of the proof of Theorem 2 of [6] which appears on p. 1108.

LEMMA 6.5. *Let $a \in C^*(G)I'$, and suppose that $a(XJ') \subseteq XJ$. Then $a \in C^*(G)I$.*

Proof. If we let b denote the image of a in C, then it will suffice to show that $b^* = 0$. Now $b^* \in a^*C$, and C is nondegenerate as a left I'-module, and so $b^* \in I'a^*C$. From Lemma 6.4, C is nondegenerate as an E-module, and so $b^* \in \langle X, X\rangle_E I'a^*C$ (where we always take closed linear spans). Thus it suffices to show that $\langle y, x\rangle_E d^*a^*c = 0$ for all $x, y \in X$, $d \in I'$, and $c \in C$. Now E is an ideal in $L(X)$, so by application of Proposition 3.9 of [36], we obtain a homomorphism of $L(X)$ into $M(C)$. Then if $\Phi \in C_c(G \times G/G_J)$ and $f \in C_c(G)$, it is easily seen that, for any $c \in C$,

$$\Phi(f(c)) = (\Phi * f)(c),$$

where $\Phi * f$ is the element of $C_c(G \times G/G_J)$ [corresponding to composition of operators in $L(X)$], defined [in $C_c(G, C_c(G/G_J))$] by

$$(\Phi * f)(z) = \int \Phi(zy) f(y^{-1})\, dy.$$

By continuity it follows that

$$e(a'(c)) = (ea')c$$

for any $e \in E$, $a' \in C^*(G)$, and $c \in C$. However, for any element S of $L(X)$, it is easily seen that

$$\langle y, x\rangle_E S^* = \langle y, Sx\rangle_E$$

(see Proposition 6.3 of [36]). From all this it follows that

$$\langle y, x\rangle_E d^*a^*c = \langle y, adx\rangle_E c,$$

so it suffices to show that the right-hand side is 0.

Now $adx \in a(I'X) = a(XI') \subseteq a(XJ')$, which by hypothesis is contained in XJ. Thus it suffices to show that

$$\langle y, xd \rangle_E c = 0, \tag{6.2}$$

where now $d \in J$, while $x, y \in X$, $c \in C$.

It is easily seen that if U is a compact neighborhood of the identity element e_G of G, then $U\{J\}^c$ is a closed subset of $\mathrm{Prim}(N)$, and $\{J\}^c$ is the intersection of the $U\{J\}^c$ as the U shrink down to e_G. It follows that $d \in J$ can be approximated in norm by elements from the various $\ker(U\{J\}^c)$. Thus it suffices to show that (6.2) holds for every $d \in \ker(U\{J\}^c)$ for every compact neighborhood U of e_G.

Let us now fix a compact neighborhood U of e_G and a $d \in \ker(U\{J\}^c)$. It suffices to show that

$$\langle y, xd^* \rangle_E c = 0$$

for all $x, y \in X$, $c \in C$. Now for any $g \in C_c(G)$, viewed as an element of X, we have

$$\langle y, g \rangle_E \langle g, xd^* \rangle_E c = \langle y\langle g, g \rangle_B, xd^* \rangle_E c, \tag{6.3}$$

where $B = C^*(G_J)$. But as g ranges over positive elements of $C_c(G)$ whose supports shrink to e_G, it is easily seen that $\langle g, g \rangle_B$ ranges over positive elements of $C_c(G_J)$ whose supports shrink to e_G. Thus if the g are suitably normalized, the $\langle g, g \rangle_B$ will form an approximate identity for B, so that the right-hand side of (6.3) will converge to $\langle y, xd^* \rangle_E c$. It thus suffices to show that

$$\langle g, xd^* \rangle_E c = 0 \tag{6.4}$$

for all $x \in X$, $c \in C$, and all $g \in C_c(G)$ which are supported in a suitably small neighborhood of e_G.

In preparation for this we need to relate the action of $C(G/G_J)$ on C with its action on $C_c(G)$ by pointwise multiplication. For $F \in C(G/G_J)$ we will let P_F denote the operator on $C_c(G)$ consisting of pointwise multiplication by F. Then for $f, g \in C_c(G)$ and $y \in G$ it is easily verified that

$$F(\langle f, g \rangle_E(y)) = \langle P_F f, g \rangle_E(y),$$

where we identify $\langle f, g \rangle_E$ with its integral kernel defined in 7.8 of [36], and view it as an element of $C_c(G, C_c(G/G_J))$. Then a straightforward calculation using the preceding identity shows that, for $c \in C$,

$$F(\langle f, g \rangle_E c) = \langle P_F f, g \rangle_E c.$$

We return now to the situation in which we have a compact neighborhood U of e_G, $d \in \ker(U\{J\}^c)$, $x \in X$, and $c \in C$. We wish to show that (6.4) holds for all $g \in C_c(G)$ which are supported in a suitably small neighborhood of e_G. As that suitably small neighborhood we choose a compact symmetric neighborhood V of e_G such that $V^3 \subseteq U$, and assume from now on that g is supported in V. Choose $F \in C_c(CJ)$ such that $F \equiv 1$ on VJ while $F \equiv 0$ outside V^2J. Viewing F as a function in $C_c(G/G_J)$ and on G, we see that $P_F g = g$. Then

$$\langle g, xd^* \rangle_E c = \langle P_F g, xd^* \rangle_E c = F(\langle gd, x \rangle_E c). \tag{6.5}$$

According to the Main Lemma (Lemma 2.1),

$$gd \in [d^Y : y \in V]X.$$

Furthermore, for the reasons given in the proof of Lemma 6.1, the d^Y will come outside the E-valued inner product. But any element of the support of F is of the form J^z with $z \in V^2$, and if $y \in V$, then $y^{-1}z \in V^3 \subseteq U$ so that $d \in J^{y^{-1}z}$ and $d^y \in J^z$. It follows that $Fd^y = 0$ for $y \in V$, and so (6.5) is equal to 0. Q.E.D.

The fact that the representation of $C^*(G)I'/C^*(G)I$ on XJ'/XJ is faithful is an immediate consequence of Lemma 6.5. Thus we have concluded the proof of the Main Theorem, once we have remarked that the statement about weak containment follows easily from Proposition 6.26 of [36].

We remark that examples showing how the Main Theorem can fail when its hypotheses are weakened can be found in [6, 32].

7. Alternative Hypotheses

In this section we collect various conditions which imply that the hypotheses of the Main Theorem are satisfied. These conditions were suggested in part by results appearing in [17] and [3], and there are certainly additional conditions which can be concocted from the results appearing in these papers. We recall now several definitions.

Definition 7.1. A topological space is said to be *almost Hausdorff* if every nonempty closed subset contains a nonempty open Hausdorff subset, or equivalently, if every closed subset contains an open dense Hausdorff subset.

The equivalence of the two formulations in this definition is shown by a simple argument (found on p. 125 of [24]) using Zorn's lemma. It is easily seen that any subset of an almost Hausdorff space is almost Hausdorff, and that any subset consisting of only one element of an almost Hausdorff space will be open in its closure.

DEFINITION 7.2. Let N be a normal subgroup of the locally compact group G, so that G acts on Prim(N). Then N is said to be *regularly embedded* in G if the orbit space Prim(N)/G with the quotient topology is almost Hausdorff.

This definition was introduced by Blattner [6]. If N is regularly embedded in G, then from the fact that points of Prim(N)/G will be open in their closures, it follows that the orbits in Prim(N) will be open in their closures, that is, locally closed.

We recall further that if N is a normal subgroup of the locally compact group G, then the Borel structure on the orbit space Prim(N)/G is defined to be the quotient of the Borel structure on Prim(N) generated by the topology of Prim(N), and not the Borel structure generated by the quotient topology on Prim(N)/G. A Borel structure is said to be *countably separated* if it contains a countable collection of subsets which separate the points of the space. Finally, a subset of a topological space is a G_δ if it is the intersection of a countable number of open sets.

ALTERNATE HYPOTHESES 7.1. *In the statement of the Main Theorem the following alternate sets of hypotheses can be used.*

A. *Hypothesis* 2 *can be replaced by*:
  (2) *GJ is locally closed in* Prim(N) *and is almost Hausdorff, and G/G_J is σ-compact.*

B. *Hypotheses* 1 *and* 2 *can be replaced by*:
  (1) Prim(N) *is almost Hausdorff.*
  (2) *N is regularly embedded in G and G/G_J is σ-compact.*

C. *Hypothesis* 2 *can be replaced by*:
  (2) *$\overline{GJ}$ is almost Hausdorff (which will be true if* Prim(N) *is almost Hausdorff), GJ is a G_δ-subset of $\overline{GJ}$, and G/G_J is σ-compact.*

D. *Hypotheses* 1 *and* 2 *can be replaced by*:
  (1) Prim(N) *is almost Hausdorff and second countable.*
  (2) Prim(N)/G *is a T_0 topological space and G/G_J is σ-compact.*

E. *Hypotheses* 1 *and* 2 *can be replaced by*:
  (1) *G is second countable and* Prim(N) *is almost Hausdorff.*
  (2) Prim(N)/G *has countably separated Borel structure.*

We now give the proofs that these alternate hypotheses imply that the hypotheses of the Main Theorem are satisfied.

Proof for A. The primitive ideal space of any C^*-algebra is a T_0 locally quasi-compact space (see 3.3.8 of [14]). Furthermore it is easily seen that any subset of a locally quasi-compact space which is locally closed is itself locally quasi-compact, so that GJ is locally quasi-compact. But a simple argument shows that any almost Hausdorff locally quasi-compact space is a Baire space. (For facts about Baire spaces see p. 109 of [8]. The condition of almost Hausdorffness cannot be dropped, as can be seen by considering a countable set whose closed subsets are the finite subsets.) Thus GJ is a Baire space. The proof now follows from the following proposition.

PROPOSITION 7.1. *Let G be a locally compact group which acts as a transformation group on the T_0 topological space M. Let $m \in M$, and let G_m denote the stability subgroup of m. Suppose that the orbit Gm is a Baire space and almost Hausdorff in the relative topology, and that G/G_m is σ-compact. Then the canonical mapping q of G/G_m onto Gm is a homeomorphism.*

Proof. Note that G_m is closed by Lemma 1 of [6], so that G/G_m is Hausdorff. We begin by showing that the orbit Gm is Hausdorff. Since Gm is assumed to be almost Hausdorff, we can find a dense relatively open Hausdorff subset H of Gm. Then the preimage of H under q will be an open subset of G/G_m and since G/G_m is assumed σ-compact, a countable number of translates of this preimage will cover G/G_m. It follows that a countable number of translates of H will cover Gm, but each translate of H is again open and dense, and so, since Gm is assumed to be a Baire space, the intersection of such a countable number of translates will be nonempty. But a point in this intersection can be separated by open neighborhoods from any other point in Gm. Since Gm is a homogeneous space, it follows that Gm is Hausdorff.

The rest of the proof follows by a quite standard argument using the fact that Gm is a Baire space (see the proof on p. 65 of [33]). Q.E.D.

We remark that this argument provides a direct proof of "(1) implies (6)" in Theorem 1 of [24]. Phil Green has pointed out (personal communication) that in Proposition 7.1 and in Alternate Hypothesis A we can replace "almost Hausdorff" by "almost T_1".

Proof for B. As mentioned, any subset consisting of a single point in an almost Hausdorff space will be locally closed. (In particular, the space will be T_0.) Thus hypothesis 1 of the Main Theorem will be satisfied when Prim(N) is almost Hausdorff. We also mentioned before that if N is regularly embedded, then orbits in Prim(N) will be locally closed. Thus the hypotheses of alternate hypotheses A are satisfied.

Proof for C. We recall from exercise 12 on p. 115 of [8] that a topological space is said to be *totally inexhaustible* (*totalement inepuisable*) if no nonempty closed subset of it is meager in itself. It is then easily verified that a locally quasi-compact almost Hausdorff space will be totally inexhaustible. According to part c of the exercise 6 in [8], any G_δ subset of a totally inexhaustible space is itself totally inexhaustible, and hence Baire. Thus GJ will be a Baire subset of Prim(N). Thus we can apply Proposition 7.1 to conclude that the map from G/G_J to GJ is a homeomorphism. In particular, GJ will be Hausdorff and locally compact in the relative topology. The proof will then be completed once we have shown:

PROPOSITION 7.2. *Let G be a group that acts as a transformation group on an almost Hausdorff topological space M. Let $m \in M$ and suppose that the orbit Gm of m is Hausdorff and locally compact in the relative topology. Then Gm is locally closed in M.*

Proof. We may assume that M is in fact the closure of Gm. Let H be an open dense Hausdorff subset of M. Then H must contain some point, say n, of Gm. Then $H \cap Gm$ is a neighborhood of n in Gm and so, since Gm is locally compact, there must be a compact neighborhood V of n in Gm contained in $H \cap Gm$. Let W be an open subset of H whose intersection with Gm is the interior of V. Then $W \cap Gm$ must be closed in W, for any net of elements in $W \cap Gm$ which converges to an element of W must have a subnet which also converges to an element of V, since V is compact. But H is Hausdorff so that limits in H are unique, so the limit of such a net must be in $W \cap V$ which is contained in $W \cap Gm$. Thus $W \cap Gm$ is closed. We can now translate this situation to any other point of Gm, and we find in this way that Gm is locally closed. Q.E.D.

We remark that Lemma 17.2 of [23] shows that the conditions of Proposition 7.2 are always satisfied if G is compact.

Proof for D. If a group G acts as a transformation group on a space M which is second countable, then it is easily seen that the orbit space is second countable. (See Lemma 2.3 of [17]—Hausdorffness of M and the topology of G are not used here.) In a second countable T_0 space, every point is a G_δ in its closure, for by T_0-ness it must be the intersection of the intersections with its closure of those members of a countable base for the topology which contain it. Thus GJ is a G_δ-subset of its closure, and so alternate hypotheses C are satisfied.

We remark that this argument provides a direct proof of "(2) implies (1) and (6)" in Theorem 1 of [24].

Proof for E. This is essentially the statement "(3) implies (1) and (6)" in Theorem 1 of [24]. We have found no direct proof of this implication, and so the proof seems to depend on the very deep arguments of [24].

8. G-Stable Representations

When N is a normal subgroup of a locally compact group G which is regularly embedded in G, the Main Theorem provides a major step in trying to describe the unitary representations of G in terms of those of N and of the G_J/N. In this section we will, under suitable hypotheses, complete that description. Our treatment has the feature that it involves neither cocycles nor measure theory. However, almost all of the results of this section already appear in one form or another in the literature, especially in [6] and [23], and so we will allow our discussion to be quite sketchy in places.

Suppose now that N is a regularly embedded normal subgroup of the locally compact group G, so that orbits in Prim(N) are locally closed. Then according to Lemma 9 of [6] the restriction to N of any primary (i.e., factor) representation of G will live on an orbit of Prim(N). We indicate here a proof of this fact which does not use Glimm's projection-valued measure.

Proposition 8.1. *Let N be a regularly embedded normal subgroup of the locally compact group G and let W be a primary unitary G-module. Then there is a G-orbit in* Prim(N) *(necessarily locally closed) on which W_N lives.*

Proof. Let D be any closed G-invariant subset of Prim(N) and let I be the kernel of D. Then I is a G-invariant ideal, so IW is a G-invariant subspace of W. But it is easily seen that IW is also invariant under the commutant of the action of G on W. Since W is assumed to be primary, it follows that IW is either all of W or $\{0\}$.

Let C be the intersection of all the closed G-invariant subsets of Prim(N) on which W_N lives. Then it is easily verified that W_N lives on C, so that C is the smallest closed G-invariant set on which W_N lives. In particular, if D is a proper closed G-invariant subset of C, then $(\ker(D))W = W$.

Since C is closed and G-invariant, its image in Prim$(N)/G$ is closed and, by assumption, contains a relatively open dense Hausdorff subset. Suppose this subset contains more than one point. Then, since it is Hausdorff, it must contain two disjoint nonempty relatively open subsets. Let the preimages in Prim(N) of these two subsets be denoted by E_1 and E_2, so that E_1 and E_2 are relatively open disjoint nonempty G-invariant subsets of C. Then $C - E_1$ and $C - E_2$ are proper G-invariant closed subsets of C whose union is C.

By the previous paragraph, $(\ker(C - E_i))W = W$ for $i = 1, 2$. Since

$$\ker(C) = (\ker(C - E_1))(\ker(C - E_2)),$$

it follows that $(\ker(C))W = W$, which contradicts the fact that W_N lives on C. Thus C contains a dense relatively open orbit, say GJ. Since $C - GJ$ is a proper closed G-invariant subset of C,

$$(\ker(C - GJ))W = W,$$

so that W lives on GJ as desired. Q.E.D.

Thus, in the case in which N is regularly embedded in G and $\mathrm{Prim}(N)$ is almost Hausdorff, the Main Theorem says that in order to classify the irreducible representations of G it suffices to look at each $J \in \mathrm{Prim}(N)$ and classify the representations of G_J whose restrictions to N live on $\{J\}$, at least as long as the G/G_J are σ-compact. Thus, letting $G = G_J$, we need to analyze the case in which J is G-invariant and find all representations of G whose restriction to N lives on $\{J\}$.

We recall from [6] that an irreducible representation of a C^*-algebra is called *semicompact* if its range contains at least one nonzero compact operator, and hence all compact operators, and that a primitive ideal is called semicompact if it is the kernel of a semicompact representation. If J is a semicompact primitive ideal which is the kernel of the semicompact representation R, then it is easily seen that the preimage under R of the ideal of all compact operators will be the smallest ideal properly containing J, and so will be the kernel of $\{J\}^c - \{J\}$. In particular, $\{J\}$ will be open in its closure. Then, from standard facts about the algebra of compact operators (4.1 of [14]), one immediately obtains the following special case of Lemma 8 of [6].

PROPOSITION 8.2. *Let J be a semicompact ideal of a C^*-algebra. Then any representation which lives on $\{J\}$ is a direct sum of copies of the unique (within unitary equivalence) irreducible representation whose kernel is J.*

Note that a primitive ideal can easily be open in its closure without being semicompact. For example, this will be true of the zero ideal in any simple C^*-algebra, such as the Calkin algebra. In fact, the requirement that every primitive ideal of a C^*-algebra be semicompact is equivalent to the requirement that the algebra be GCR (i.e., postliminaire; see 4.3.1 of [14], and so type I by [38]), in which case the primitive ideal space is automatically almost Hausdorff by 4.4.5 of [14].

Anyway, we see from Proposition 8.2 that if J is a semicompact G-invariant element of $\mathrm{Prim}(N)$, then the restriction to N of an irreducible unitary G-module will live on $\{J\}$ if and only if this restriction is a direct sum of copies of the irreducible unitary N-module whose kernel is J. We now discuss how to classify the latter.

Let J be a semicompact G-invariant primitive ideal and let V be the corresponding irreducible unitary N-module. Since J is G-invariant, V^x also must have J as kernel for every $x \in G$, so, since J is semicompact, V^x must be equivalent to V. Thus for each $x \in G$ we can find a unitary operator P_x on V such that

$$(xsx^{-1})v = P_x s P_x^{-1} v \tag{8.1}$$

for all $s \in N$, $v \in V$. Since V is irreducible, P_x must be unique up to a scalar multiple of modulus one. Let $U(V)$ denote the group of unitary operators on the Hilbert space V and let $PU(V)$ denote the corresponding projective unitary group, which is the quotient of $U(V)$ by the subgroup consisting of the scalar multiples of the identity operator. On $U(V)$ the strong and weak operator topologies agree, and with them $U(V)$ is a topological group. We always consider $PU(V)$ to have the corresponding quotient topology. For each $x \in G$ let P_x also denote its image in $PU(V)$, which now is uniquely determined. Then it is easily verified that P is a group homomorphism from G into $PU(V)$, but in general we do not know whether it is continuous. However, if G is second countable, then one can use Polish topology to show that P is continuous, as was done, for example, in Proposition 17.2 of [23]. Alternatively, if V is semicompact as in the case we are considering, then P will again be continuous, as shown in Proposition 17.1 of [23]. We include the argument here, since we can give a somewhat simpler proof of the lemma on which it depends (Lemma 17.1 of [23]).

LEMMA 8.1. *Let V be a Hilbert space, let $P \in U(V)$, and let $\{P_i\}$ be a net of elements of $U(V)$. Then the (images of the) P_i converge in $PU(V)$ to P if and only if $P_i T P_i^{-1}$ converges strongly to PTP^{-1} for every rank one operator T on V.*

Proof. Routine arguments show that the direct implication actually holds for all bounded operators T. We show the converse. Choose a vector v_0 of unit length in V and let T_0 denote the projection on the one-dimensional subspace spanned by v_0. Then $P_i T_0 P_i^{-1} P v_0$ converges to $PT_0P^{-1}Pv_0 = Pv_0$. Let v_i denote $P_i T_0 P_i^{-1} P v_0$ divided by its length when its length is nonzero, and eliminate the terms of length zero, so that $\{v_i\}$ is a net in V that converges to Pv_0. Then each v_i is a unit vector in the range of the rank one operator $P_i T_0$. It follows that there is a complex number of modulus one c_i such that $v_i = c_i P_i v_0$. Let $Q_i = c_i P_i$. Then $Q_i v_0 = c_i P_i v_0 = v_i$, which converges to Pv_0.

We show now that the Q_i (which have the same image in $PU(V)$ as the P_i), converge in $U(V)$ to P. Note that

$$Q_i T Q_i^{-1} = P_i T P_i^{-1},$$

so that $Q_i T Q_i^{-1}$ converges to PTP^{-1} for any rank one operator T. Let v be any element of V, and chose as T any rank one operator such that $Tv_0 = v$. Then

$$Q_i v = Q_i T v_0 = Q_i T Q_i^{-1} Q_i v_0 = Q_i T Q_i^{-1}(Q_i v_0 - P v_0) + Q_i T Q_i P v_0.$$

The first term converges to zero, while the second term converges to $PTP^{-1}Pv_0 = Pv$ as desired. Q.E.D.

Let us return to the notation used in the paragraph preceding the lemma. Then relation (8.1) lifts to elements of $C^*(N)$, so if V is semicompact, the hypotheses of Lemma 8.1 will be satisfied, and the continuity of P follows. If V is a Hilbert space and G is a topological group, then a continuous homomorphism of G into $PU(V)$ will be called a *projective representation* of G on V.

Let V be a Hilbert space, G a locally compact group, and P a projective representation of G on V. Let p denote the canonical projection of $U(V)$ onto $PU(V)$, and form the pullback of P and p, that is,

$$\tilde{G} = \{(x, U) \in G \times U(V) : P(x) = p(U)\}.$$

Then it is easily verified that $\tilde{G}$ is a central topological group extension of G by the group T of complex numbers of modulus one. As such, it is locally compact (p. 52 of [33]). The isomorphism classes of central extensions of G by T form an Abelian group under Baer multiplication (see [28, 34] and p. 147 of [23]), which in the separable case is isomorphic with the second cohomology group (defined in terms of cocycles) of G with coefficients in T [31, 34, 35]. If each element (x, U) of $\tilde{G}$ is sent to the corresponding operator U on V, one obtains an ordinary representation of $\tilde{G}$ whose restriction to T is a multiple of the standard representation of T, that is, is of the form $t \mapsto tI_V$ for $t \in T$, where I_V is the identity operator on V. Conversely, every ordinary representation of a central extension of G by T whose restriction to T is a multiple of the standard representation defines in an obvious way a projective representation of G. A projective representation of G lifts to an ordinary representation of G [that is, factors through $p: U(V) \to PU(V)$)] if and only if the central extension $\tilde{G}$ as previously constructed from it splits. But even then it need not come from a unique ordinary representation. Two ordinary irreducible representations of G define the same projective representation if and only if each is the inner tensor product of the other with some one-dimensional representation (character) of G.

Let us now return to the situation in which we have a normal subgroup N of G and a semicompact G-invariant primitive ideal J of $C^*(N)$, with corresponding representation R on V. Let P be the projective representation of G which implements the equivalence of V^x with V for $x \in G$, as defined earlier. Then for $s \in N$ the equivalence of V^s with V can be implemented by R_s, so for $s \in N$ we can assume that $P_s = R_s$, so that P is an extension of the ordinary representation of N to a projective representation of G. This result can be summarized as follows.

PROPOSITION 8.3. *Let N be a normal subgroup of the locally compact group G. Then any irreducible semicompact G-invariant ordinary representation of N can be extended to a projective representation of G.*

With the previous notation, R can be viewed as a projective representation of N (coinciding with the restriction of P to N) and we can form the corresponding central extension $\tilde{N}$ of N. This extension splits via the map $s \mapsto (s, R_s)$, but $\tilde{N}$ can be viewed as a normal subgroup of $\tilde{G}$. We then obtain the following diagram of exact sequences, which is commutative in rows and columns, but not with its diagonal maps.

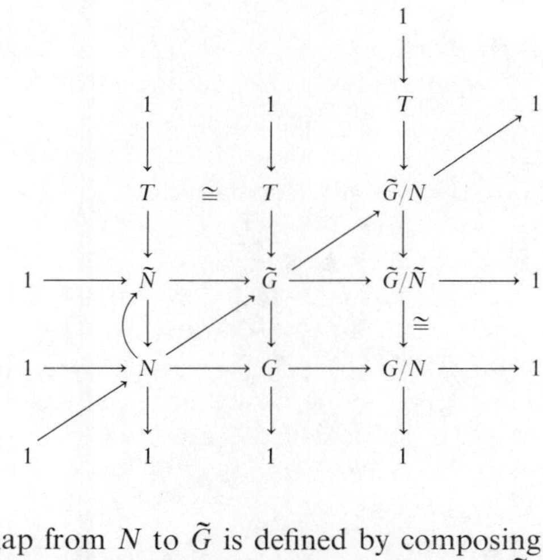

The diagonal map from N to $\tilde{G}$ is defined by composing the map from $\tilde{N}$ to $\tilde{G}$ with the splitting map defined previously from N to $\tilde{N}$. The isomorphism class of the extension $\tilde{G}/N$ of G/N is called the *Mackey obstruction* of R or J.

We are looking for the ordinary representations of G whose restriction to N is a multiple of R. Now any such representation of G clearly defines (see the diagram) an ordinary representation of $\tilde{G}$ whose restriction to N is R. Recall that R extends to an ordinary representation, namely P, of $\tilde{G}$. Then we can find the desired representations of G by the two-step process of first finding all ordinary representations of $\tilde{G}$ whose restriction to N is R, and then determining which of these drop to G. The first step is accomplished by the following proposition, in which H plays the role of $\tilde{G}$.

PROPOSITION 8.4. *Let N be a normal subgroup of the locally compact group H and let R be an ordinary irreducible representation of N. Suppose that R extends to an ordinary representation P of H. For any ordinary representation Q of H/N, viewed as a representation of H, let $P \# Q$ denote the inner tensor product of P with Q. Then the functor $Q \mapsto P \# Q$ is an equivalence between the category of ordinary representations of H/N and the category of ordinary representations of H whose restriction to N is a multiple of R.*

Proof. It is clear that the indicated functor gives representations of the desired type. Suppose, conversely, that S is a representation of H whose restriction to N is a multiple of R. If we let V denote the space for R and P, then the space for S can be written in the form $V \otimes W$ where W is a Hilbert space of dimension equal to the multiplicity of R in the restriction of S to N. We can also assume that for each $s \in N$,

$$S_s = R_s \otimes I_W.$$

For each $x \in H$ let

$$T_x = (P_x^{-1} \otimes I_W) S_x.$$

Then a simple calculation shows that each T_x commutes with all operators of the form $R_x \otimes I_W$. Since R is irreducible, it follows from the double commutant theorem (p. 42 of [13]) that each T_x commutes with all operators of the form $M \otimes I_W$, where M is any bounded operator on V. It follows (p. 24 of [13]) that $T_x = I_V \otimes Q_x$ for some operator Q_x on W. It is then easily verified that Q is a unitary representation of H which is trivial on N, so that it gives a representation of H/N, and that $S = P \# Q$. Finally, arguments similar to these show that the previously defined functor is an isomorphism between the corresponding spaces of intertwining operators.

Q.E.D.

We now carry out the second step, namely, we determine which of the ordinary representations of $\tilde{G}$ of the form $P \# Q$ drop to G, where Q is an ordinary representation of $\tilde{G}/N$. Now from the diagram it is clear that these will be exactly the representations that have the subgroup T of $\tilde{G}$ in their kernel. But it is easily calculated that these will be exactly the representations in which the restriction of Q to T is the inverse of the standard representation, that is, $Q_t = t^{-1}I_W$. Finally, if we now form the extension of G/N by T which is inverse under Baer multiplication to the given extension, then it is easily verified that the preceding Q's are exactly those representations of this inverse extension whose restriction to T is a multiple of the standard extension. We summarize these results as follows.

PROPOSITION 8.5. *Let N be a normal subgroup of the locally compact group G and let J be a semicompact G-invariant ideal in* Prim(N). *Then the category of unitary G-modules whose restriction to N lives on $\{J\}$ is equivalent to the category of projective representations of G/N which belong to the inverse of the Mackey obstruction of J.*

We remark that Dixmier has shown (Proposition 6 of [16]) that this equivalence preserves weak containment, at least for irreducible representations when G is second countable and N is type I.

Note Added in Proof. Extensive generalizations of the results and techniques of this article, with applications, are contained in Philip Green, The local structure of twisted covariance algebras, *Acta Math.* **140** (1978), 191–250.

REFERENCES

1. C. A. AKEMANN, G. T. PEDERSEN, AND J. TOMIYAMA, Multipliers of C^*-algebras, *J. Functional Anal.* **13** (1973), 277–301.
2. L. BAGGETT, A description of the topology on the dual spaces of certain locally compact groups, *Trans. Amer. Math. Soc.* **132** (1968), 175–215.
3. L. BAGGETT AND A. KLEPPNER, Multiplier representations of Abelian groups, *J. Functional Anal.* **14** (1973), 299–324.
4. H. BASS, "Algebraic K-theory," Benjamin, New York, 1968.
5. R. J. BLATTNER, On induced representations, *Amer. J. Math.* **83** (1961), 79–98.
6. R. J. BLATTNER, Group extension representations and the structure space, *Pacific J. Math.* **15** (1965), 1101–1113.
7. N. BOURBAKI, "Topologie générale," 3rd ed., Chapter 1, Act. Sci. Ind. No. 1142, Hermann, Paris, 1961.
8. N. BOURBAKI, "Topologie générale," Chapter 9, Act. Sci. Ind. No. 1045, Hermann, Paris, 1958.
9. J. BUNCE, A note on two-sided ideals in C^*-algebras, *Proc. Amer. Math. Soc.* **28** (1971), 635–636.
10. R. C. BUSBY, Double centralizers and extensions of C^*-algebras, *Trans. Amer. Math. Soc.* **132** (1968), 79–99.
11. R. C. BUSBY, On structure spaces and extensions of C^*-algebras, *J. Functional Anal.* **1** (1967), 370–377.

12. A. H. Clifford, Representations induced in an invariant subgroup, *Ann. of Math.* **38** (1937), 533–550.
13. J. Dixmier, "Les algèbres d'opérateurs dans l'espace hilbertien," 2nd ed., Gauthier-Villars, Paris, 1969.
14. J. Dixmier, "Les C^*-algèbres et leurs représentations," 2nd ed., Gauthier-Villars, Paris, 1969.
15. J. Dixmier, Ideal center of a C^*-algebra, *Duke Math. J.* **35** (1968), 375–382.
16. J. Dixmier, Bicontinuité dans la méthode du petit groupe de Mackey, *Bull. Sci. Math.* **97** (1973), 233–240.
17. E. G. Effros, Transformation groups and C^*-algebras, *Ann. Math.* **81** (1965), 38–55.
18. E. G. Effros and F. Hahn, Locally compact transformation groups and C^*-algebras, *Mem. Amer. Math. Soc.* **75** (1967).
19. G. A. Elliott and D. Olesen, A simple proof of the Dauns–Hofmann theorem, *Math. Scand.* **34** (1974), 231–234.
20. J. M. G. Fell, A new proof that nilpotent groups are CCR, *Proc. Amer. Math. Soc.* **13** (1962), 93–99.
21. J. M. G. Fell, Weak containment and induced representations of groups, *Can. J. Math.* **14** (1962), 237–268.
22. J. M. G. Fell, Weak containment and induced representations of groups, II, *Trans. Amer. Math. Soc.* **110** (1964), 424–447.
23. J. M. G. Fell, An extension of Mackey's method to Banach *-algebraic bundles, *Mem. Amer. Math. Soc.* **90** (1969).
24. J. Glimm, Locally compact transformation groups, *Trans. Amer. Math. Soc.* **101** (1961), 124–138.
25. J. Glimm, Families of induced representations, *Pacific J. Math.* **12** (1962), 885–911.
26. E. C. Gootman, Primitive ideals of C^*-algebras associated with transformation groups, *Trans. Amer. Math. Soc.* **170** (1972), 97–108.
27. E. Hewitt and K. A. Ross, "Abstract Harmonic Analysis," Springer-Verlag Berlin and New York, 1963.
28. G. Hochschild, Group extensions of Lie groups, I, *Ann. Math.* **54** (1951), 96–109.
29. B. E. Johnson, An introduction to the theory of centralizers, *Proc. London Math. Soc.* **14** (1964), 299–320.
30. G. W. Mackey, Induced representations of locally compact groups, I, *Ann. Math.* **55** (1952), 101–139.
31. G. W. Mackey, Les ensembles Borélien et les extensions des groupes, *J. Math. Pures Appl.* **36** (1957), 171–178.
32. G. W. Mackey, Unitary representations of group extensions, I, *Acta Math.* **99** (1958), 265–311.
33. D. Montgomery and L. Zippen, "Topological Transformation Groups," Wiley (Interscience), New York, 1955.
34. C. C. Moore, Extensions and low dimensional cohomology theory of locally compact groups, I. *Trans. Amer. Math. Soc.* **113** (1964), 40–63.
35. M. A. Rieffel, On extensions of locally compact groups, *Amer. J. Math.* **88** (1966), 871–880.
36. M. A. Rieffel, Induced representations of C^*-algebras, *Advan. Math.* **13** (1974), 176–257.
37. M. A. Rieffel, Morita equivalence for C^*-algebras and W^*-algebras, *J. Pure Appl. Algebra* **5** (1974), 51–96.

38. S. Sakai, On a characterization of type I C^*-algebras, *Bull. Amer. Math. Soc.* **72** (1966), 508–512.
39. I. Schochetman, The Dual Topology of Certain Group Extensions, preprint.
40. M. Takesaki, Covariant representations of C^*-algebras and their locally compact automorphism groups, *Acta Math.* **119** (1967), 273–303.

STUDIES IN ANALYSIS
ADVANCES IN MATHEMATICS SUPPLEMENTARY STUDIES, VOL. 4

On the Moduli of Linear Dynamical Systems†

CHRISTOPHER I. BYRNES‡

Department of Mathematics
University of Utah
Salt Lake City, Utah

AND

NORMAN E. HURT§

Department of Mathematics
Cleveland State University
Cleveland, Ohio

Contents

INTRODUCTION

The construction of various canonical forms for finite-dimensional, linear dynamical systems has long been recognized as an important tool in the design and simulation of such systems. In Section 1, we shall recall for the reader the definitions and main results pertaining to this area of systems theory, systems synthesis, phrasing such problems within the scope of what may be properly referred to as the discrete inverse scattering problem. As is well known (cf., Section 1), the set of all systems with m inputs, p outputs, and n degrees of freedom may be represented as k^N, for k a fixed field of scalars and $N = n(n + m + p)$. For fixed n, m, there corresponds two open

† Partially supported by National Science Foundation Grant GJ-35759, through the generosity of Professors M. A. Arbib and E. G. Manes.

‡ Present address: Harvard University, Cambridge, Massachusetts 02138.

§ Present address: Mechanics Research, Inc., McLean, Virginia.

ISBN 0-12-599150-9

subspaces, $\tilde{X}_p$ and M_p, of k^N to each $p \geqslant 0$, which arise naturally in systems theory. Now each system admits $\mathscr{G} = \mathscr{G}\ell(n, k)$ as a group of symmetries in such a way that $\mathscr{G}$ acts naturally on both $\tilde{X}_p$ and M_p. For the sake of completeness, we remark that the genesis of this chapter is the solution of

PROBLEM A. Find canonical forms for systems in M_p, $p \geqslant 0$, for this action of $\mathscr{G}$.

In this connection, we refer the reader to the special volume [14] set aside by the IEEE in its *Transactions on Automatic Control*. This issue discusses the problems of system identification and systems parameterization, emphasizing the importance of such problems in a wide variety of disciplines (aside from systems synthesis, these included econometrics, seismology, and statistics). In addition to the historical remarks, this volume also contains the perspective that a more theoretical, global analysis was needed for the solution of various problems related to Problem A. Now, for n arbitrary and $m = p = 1$, the solution is classical and is referred to as the "phase canonical form." The general solution we give here is due entirely to Kalman [16] and is local in nature; i.e., M_0 may be covered by open sets V_α, on each of which a canonical form may be defined. This is an analogy with the formation of projective space $\mathbf{IP}_l$, where the open set is $k^{l+1} - \{0\}$ and the group is $\mathscr{G}\ell(1, k) = k^*$. Here the canonical forms, viz., the homeogeneous coordinates of the lines in k^{l+1}, also serve to coordinatize the orbit space $\mathbf{IP}_l = (k^{l+1} - \{0\})/\mathscr{G}$. Thus Problem A may be considerably refined as

PROBLEM B. Construct the orbit space $\Sigma_p = M_p/\mathscr{G}$ as explicitly as possible. In addition, what are its (algebro-) geometric properties; e.g., nonsingularity, irreducibility. Compute its invariants; e.g., its dimension, its picard group $\mathscr{P}ic(\Sigma_p)$, or its integral cohomology ring.

Aside from the intrinsic interest of such questions in the global analysis of linear systems, these questions have a deep interpretation in the solution of Problem A. For example, we have been able to show that a complete obstruction to finding global algebraic canonical forms lies in $\mathscr{P}ic(\Sigma_p)$. In fact, the obstruction for systems in M_0 generates $\mathscr{P}ic(\Sigma_0)$. Such a result comes, in part, from interpreting canonical forms as sections of the quotient map $\varphi_p: M_p \to \Sigma_p$. An analysis of this map is made possible by Mumford's geometric invariant theory, a modernization of the classical invariant theory typically used to obtain partial solutions to Problem A. Now, the map φ_p gives an algebraic parameterization of the equivalence classes of systems and, here again, the techniques in GIT (i.e., [21]) enable us to answer some questions, raised by Kalman in [17], regarding the global analysis of linear dynamical systems, viz.,

PROBLEM C. How do certain types of systems behave under perturbation?

Partial solutions to Problem B were given by Kalman [16], [17], Hermann [11], Brockett (unpublished), and later by Kalman in collaboration with Hazewinkel [10]—all with an eye toward Problem C. Kalman sketched a proof that Σ_0 is nonsingular of dimension nm by defining local canonical forms, coordinatizing Σ_0, and producing homogeneous equations, supposedly defining Σ_0 as a quasi-projective variety in $\mathbf{P}_\mathbf{N}$, for $\mathbf{N} \gg 0$, in [16] and [17]. Later, he and Hazewinkel gave another set of equations and asserted (as in [17]) that $\Sigma_p \cong \Sigma_0 \times k^{pn}$. As we show in Section 3, neither set of equations actually cut out Σ_0. Moreover, we have shown that, although $\Sigma_p \to \Sigma_0$ $(p \geqslant 1)$ may be regarded as a vector bundle of rank pn, an obstruction to the triviality of this bundle, which lies in $\mathscr{Pic}(\Sigma_0)$, is nonvanishing if $m > 1$. In fact, this is also equivalent to the nonexistence of global algebraic canonical forms.

Although we do not give explicit equations defining Σ_0 in some projective space, we are still able to show, via GIT, that Σ_0 is quasi-projective. In fact, these techniques also provide a single criterion for the solution of Problem C, which we exploit in Section 5. We should remark here that the major notions in systems theory arise naturally in the setting of geometric invariant theory, a fact which itself may be as important as the solutions to these problems. We find this development to be the most surprising contribution. As we indicate in Section 5, we can prove, via invariantive methods, the Fundamental theorem of linear realization theory and the Minimal realization theorem (cf., Section 1), which are undoubtedly among the most important theorems in the field. As a final touch, we solve the following.

PROBLEM D. Give an explicit solution to the discrete inverse scattering problem.

It is our intention to make this article as self-contained, at least on the conceptual level, as possible. Accordingly, Section 1 contains a brief summary of the basic definitions and results that we will need from linear systems theory. We also include a discussion of Problem A, together with a pertinent illustration of what is now known as the *moduli problem*. This is a discussion of the Grassmann variety, which Kalman has used in solution of Problems A and B, and serves as a good motivation for what follows. We have also included this in order to set some notation (e.g., to fix which bundle we call *universal*) and to introduce some concepts in a familiar context.

Section 2 contains the computations needed to make the general results we derive in Section 4 more explicit. These computations are on the set-theoretic level, but we have assumed that the ground field supports some analytic structure (e.g., **R**, **C** or $\mathbf{Q}_p$), since this is technically the most convenient setting for applications. The major results are that $M_0 \to \Sigma_0$ is an analytic principal fiber bundle with group $\mathscr{G}$ and base Σ_0, an analytic manifold, and that the analytic vector bundles, $\Sigma_p \to \Sigma_0$, satisfy the equations $\Sigma_p = \bigoplus^p \Sigma_1$. These results are predicted by geometric invariant theory but it seems more natural for the exposition to make these calculations beforehand. We also include a proof of Kalman's theorem, that Σ_0 is nonsingular of dimension nm, and his construction of canonical forms.

In Section 3, we explicitly compute some examples. The moduli space Σ_0 for "prime machines" (Example 3.3) gives a counterexample to Kalman's assertion that $\Sigma_p = \Sigma_0 \times k^{pn}$ and to a conjecture of Hazewinkel's regarding the projectivity of Σ_0 for $m = 3$. These computations make essential use of two projective imbeddings of Σ_0, which we show are imbeddings by sections of the line bundle $\det \Sigma_1 \to \Sigma_0$. The first is the restriction of a suitable Plucker imbedding and the second, which we call the *Kalman imbedding*, was originally defined in [16]. We also use these imbeddings to show that the relations given by Kalman [16] and by Hazewinkel and Kalman [10] do not define Σ_0 as was claimed (Examples 3.3 and 3.4).

Our main theorems, answering parts of Problem B, are contained in Section 4. These assert that M_0 arises as the principal orbit type of the action on k^{n^2+nm} and may also be represented as the set of properly stable points relative to the standard character $\mathscr{G}$-linearization of the trivial line bundle. Since the usual numeric criterion, the Hilbert–Mumford criterion, is not available for affine varieties of positive dimension, our proof starts "from scratch," relying only on a few well-known classical facts. As a corollary, we obtain that Σ_0 is an irreducible quasi-projective variety of dimension nm which is nonsingular, by Section 2. Furthermore, each quotient $M_p \to \Sigma_p$ is geometric and all bundles are algebraic. We are able to compute $\mathscr{P}ic(\Sigma_p)$, showing that this group is, canonically, **Z** when $m > 1$, so that the relevant obstructions do not vanish. Finally, having filled in the gaps in the Hazewinkel–Kalman machine, we state their theorem that the moduli functor for families of systems is representable.

In Section 5, these results are applied to the global analysis of linear dynamical systems, viz., to solve Problems C and D. It is found that various system-theoretic properties, primarily those related to control and observation, are generic. A particular example, the existence of a quotient for $\tilde{X}_1$, generalizes Hermann's theorem. Conspicuous in its absence, however, is the property of asymptotic stability which we show is not generic for the Zariski topology. Finally, we sketch a solution of the discrete inverse scattering problem for scalar input–output systems, obtaining a new algorithm for systems design as well as new proofs of some fundamental theorems. The general case will appear in detail, in [3].

1. Linear Sequential Machines, Their Classification and Canonical Forms

In this section we shall follow the terminology in Eilenberg's ALM [5], except that we shall compute our maps "on the left." Now, viewing a dynamical system in terms of the usual "blackbox" picture, one may observe that such a system is dependent on three parameters, viz., input, state, and output vectors. Frequently, the evolution of a system with time is assumed to be linear in these parameters. In this case, the only relevant possibilities depend upon

(i) a ring k;
(ii) three left k-modules, A, Q, B (the input, state, and output spaces);
(iii) four maps, $F \in \mathscr{H}om_k(Q, Q)$, $G \in \mathscr{H}om_k(A, Q)$, $H \in \mathscr{H}om_k(Q, B)$, and $J \in \mathscr{H}om_k(A, B)$.

The evolution, over discrete time intervals, of such a system is given by the equations

$$x(t + 1) = Fx(t) + Gu(t), \qquad y(t) = Hx(t) + Ju(t). \tag{1.1}$$

Thus a (discrete-time) linear sequential machine is given by data (i)–(iii) with (1.1) understood implicitly. We shall always take our scalars to lie in a field k and we make the further assumption that $\dim_k(Q) = n$, $\dim_k(A) = m$, and $\dim_k(B) = p$ are finite; i.e., we ask that the linear sequential machine be finite-dimensional. For the sake of brevity, such objects will be referred to as "machines" (over k). Fixing bases of A, Q, and B, (1.1) is replaced by a system of matrix equations. The resulting object is classically referred to as a (discrete-time) *finite-dimensional, linear dynamical system* and shall be referred to here as a "machine," also. Returning to the blackbox picture, one may observe that by numbering the input and output terminals, it is possible to naturally choose bases for A and B. However, one cannot choose a canonical basis for the state space Q and, thus, a natural group of symmetries for machines, of a fixed dimension (n, m, p), is $\mathscr{G} = \mathscr{G}\ell(n, k)$. In accordance with (1.1), two machines $M_i = (F_i, G_i, H_i, J_i)$ are equivalent mod $\mathscr{G}$, provided that for at least one $g \in \mathscr{G}$, $gM_1 = (gF_1g^{-1}, gG_1, H_1g^{-1}, J_1) = (F_2, G_2, H_2, J_2) = M_2$.

The orbit $\mathscr{O}(M)$ of a machine under $\mathscr{G}$ is its equivalence class; i.e., the set $\mathscr{O}(M) = \{M : gM = M \text{ for some } g \in \mathscr{G}\}$. In general, equivalence mod $\mathscr{G}$ refines the standard relation of systems equivalence, viz., the equivalence of input–output pairs. Explicitly, an input sequence $(a_0, a_1, \ldots, a_l, \ldots)$ may be canonically regarded as an element of $A[[z]]$. Since the same is true for state and output sequences, (1.1) yields for each machine M a map $f_M : A[[z]] \to B[[z]]$, which is known to be a map of $k[[z]]$-modules, hence a map of

$k[z]$-modules. The map f_M is called the result of the machine M, and the system-theoretic equivalence of M_1 and M_2 is defined by $f_{M_1} = f_{M_2}$. Of course, the inverse scattering problem in this context is to construct M from f_M. Now, as a $k[[z]]$-module map, f_M may be developed in a power series expansion

$$f_M = \sum_{j=0}^{\infty} f_j z^j, \tag{1.2}$$

with $f_j \in \mathscr{H}om_k(A, B)$. Furthermore, from this coordinate-free definition of f_M, it follows that $f_j \in \mathscr{H}om_k(A, B)^{\mathscr{G}}$. In fact, in ALM (Chapter 16) it is shown that these invariants are

$$f_0 = J, \qquad f_j = HF^{j-1}G \qquad \text{for} \quad j > 0. \tag{1.3}$$

In particular, two systems that are equivalent mod $\mathscr{G}$ give rise to the same result function, as we previously stated. The result is related to another invariant, defined for continuous-time machines. The evolution of such a machine is given by the differential equations

$$\dot{x}(t) = Fx(t) + Gu(t), \qquad y(t) = Hx(t) + Ju(t), \tag{1.4}$$

where we assume, for the moment, that $k = \mathbf{R}$ or $\mathbf{C}$. By Laplace transform methods, the solution of the evolution equations may be calculated as

$$y(t) = \int_0^t H \exp\{F(t-s)\}Gu(s)\,ds + H\exp\{F(t)\}x(0) + Ju(0). \tag{1.5}$$

Now, the function $I: k \to \mathscr{H}om_k(A, B)$ defined by

$$I(t) = H\exp\{Ft\}G \tag{1.6}$$

is known as the *impulse response function*, while its Laplace transform

$$\Phi(s) = H(F - s)^{-1}G \tag{1.7}$$

is known as the *frequency response function*. The main problem in system synthesis is the realization problem, which asks for a machine (F, G, H) with a prescribed frequency response function. Since (1.6) may be developed as

$$I(t) = HG + \sum_{n=1}^{\infty} f_n t^n/n!, \tag{1.8}$$

the realization problem (in the time domain) is exactly the inverse scattering problem alluded to previously. In particular, the class of machines for which the realization problem has a unique solution is contained in the class for which equivalence mod $\mathscr{G}$ and system-theoretic equivalence coincide, viz., the class for which the assignment $\mathscr{O}(M) \to f_M$ is bijective.

Some definitions are in order. First of all, it is important to know when a machine may be controlled (e.g., moved to any state) by a finite sequence of inputs. Following Eilenberg, we consider the map $\mathbf{G}_M: A[z] \to Q$ defined by $\mathbf{G}_M(az^n) = F^nGa$. Treating Q as a $k[z]$-module in the standard way, one may regard $\mathbf{G}_M$ as a map of $k[z]$-modules. The condition on controllability is as follows.

DEFINITION 1.1. The machine M is *accessible* if $\mathbf{G}_M$ is surjective.

For each p, the set of accessible machines is denoted by M_p. In particular, M_0 is the set of accessible machines for which the only output is the change of state. Note that $M_p = M_0 \times k^{pm}$, since the property is independent of H. Now, we have defined a map $\mathbf{G}: k^N \to \mathscr{H}om_k(A[z], Q)$, where $N = (n(n+m)$, which is well known to be injective on the set M_0. In fact, fixing bases in A and Q and then choosing the standard basis in $A[z]$, one may express $\mathbf{G}_M$ as the matrix of blocks

$$\mathbf{G}_M = (G, FG, F^2G, \ldots, F^nG, \ldots).$$

By the Cayley–Hamilton theorem, rank $\mathbf{G}_M =$ rank $R(F, G)$, where $R(F, G) = (G, FG, \ldots, F^nG)$. Thus, M_p is Zariski open, for each $p \geqslant 0$.

The dual notion to control is that of observation. More precisely, by observing enough successive outputs one would hope to be able to distinguish between the various states of the machine. This has a mapping interpretation as well, viz., consider the map

$$\mathbf{H}_M: Q \to B[[z]], \qquad \text{defined by} \quad x \to \sum_{j=0}^{\infty} (HF^jx)z^j.$$

DEFINITION 1.2. M is *reduced* provided $\mathbf{H}_M$ is injective. In case M is accessible and reduced, M is said to be *minimal*.

Note that one has an honest duality here, M is reduced if and only if the machine ${}^tM = ({}^tF, {}^tG, {}^tH, {}^tJ)$ is accessible, and vice versa. In particular, the point sets $M_p{}^*$ are also Zariski open. As the terminology suggests, these classes of machines have a pretty categorical interpretation (due to Arbib and Manes; see [1], for example), involving the notion of simulation. Briefly, fix A, B and (the direct part) J, allowing the other objects and maps to vary. For accessible machines M, we say M_1 is a *simulation* of M_2 just in case there is a map $\phi \in \mathscr{H}om_k(Q_1, Q_2)$ making the diagram

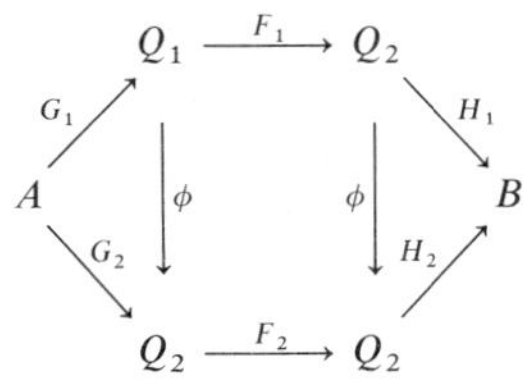

(1.9)

commute. Using (1.3), it is easy to see that $f_{M_1} = f_{M_2}$ if M_1 simulates M_2. In fact, one can show that a minimal realization of a result f is just a terminal object in the category of accessible machines M with $f_M = f$ (and A, B, J fixed, as before). In particular, we have what Kalman calls in [18], the "fundamental theorem of linear realization theory."

PROPOSITION 1.1. *Two minimal machines M_1 and M_2 are equivalent* mod $\mathcal{G}$ *if and only if the corresponding results f_{M_1} and f_{M_2} are equal.*

Proof. See ALM, Theorem 5.6, Chapter 6. In fact, the statement in ALM is a bit sharper, implying, in particular, that if $f_{M_1} = f_{M_2}$, then there is a unique $g \in \mathcal{G}$ for which $gM_1 = M_2$. Of course, this is also inherent in the definition of a terminal object.

This proposition follows from the deeper

THEOREM 1.1 (Minimal realization theorem). *If $f: A[[z]] \to B[[z]]$ is rational, then $f = f_M$ for a unique minimal machine M.*

The definition of rationality can be found in Chapter 16, Section 9 of ALM, while a statement and proof of this theorem can be found as Theorem 10.2 in that treatise. The proof and relevant concepts rely on the interpretation of the various objects and morphisms within the category of finitely generated $k[z]$-modules. In particular, the structure theory of such modules over a principal ideal domain (P.I.D.) serves to linearize the constructs of parallel and series composition, which play an important role in Section 4 of this chapter. Kalman first gave a proof of the existence of a solution to this inverse scattering problem, based on the module theory over P.I.D.s. In ALM, Eilenberg has given a similar theorem for more arbitrary scalar rings, while Arbib and Manes have given a development of these ideas in a categorical setting, showing that the proof of Theorem 1.1 relies only on categorical algebra.

Of more practical importance, however, is the development of an algorithm for producing a machine from a given rational f. For the scalar input–output case ($m = p = 1$), an algorithm is known that is based on the existence of the "phase canonical form" and the a priori existence of a realization (see [23]). More generally, Ho has given an algorithm (see [18]) which is cumbersome relying on the invariant factor algorithm and the consideration of generalized Hankel matrices. In Section 5, we shall give another algorithm for a generic case, sketching the general case (see [3]). This derivation comes from an investigation of the geometry involved in Problem B, which we are now in a position to examine.

The earlier interest among the engineering community was in classifying pairs (F, G) with some imposed rank conditions, e.g., accessibility. Of course, one extreme—the case $G = 0$—is just the similarity problem. Notice that, if $m = 1$, the condition of accessibility is just that G be a cyclic vector for F. In this case, the rational canonical form for (F, G) classifies the pair. More precisely, there is a unique $g \in \mathcal{G}$, such that $g(F, G) = (F_1, G_1)$, where F_1 is the rational canonical form for F and $G_1 = e_1$. Thus the orbit space Σ_0 is canonically k^n. In fact, the quotient map φ_0 has as its components the coefficients $\chi(F)_i$ of the characteristic polynomial

$$\chi(F, s) = (-1)^n s_n{}^n + \sum_{i=1}^{n-1} \chi_i(F) s^{n-i}$$

(see Example 3.1 or [2, section 4]). The form (F_1, G_1) is a special case ($p = 0$) of the phase canonical form and is referred to in [18] as the "control canonical form."

Before Kalman's solution of Problem A, the best result along these lines, for $m > 1$, was obtained by Popov ([24]), who obtained canonical forms for accessible machines (F, G) with rank $G = m$. Necessarily, $m \leqslant n$, and thus Popov generalized the scalar input case by showing that (F, G) was classified, mod $\mathcal{G}$, by a sequence of integers $(\kappa_1, \ldots, \kappa_m)$ which he referred to as "Kronecker invariants" and "k-valued algebraic functions" (α_{ijk}). We now know that the α_{ijk} arise as the coefficients of another set of well-known forms, the local coordinates of the Grassman variety. As we have seen previously, the assignment $M \to \mathbf{G}_M$ induces a map R,

$$R: (F, G) \to (G, FG, \ldots, F^n G),$$

which is bijective for accessible machines (F, G). One of Kalman's fundamental contributions to the solution of Problem B was to exploit this observation as follows: R maps M_0 into $\mathcal{M}_*(n, (n+1)m)$, the set of n by $(n+1)m$ matrices of (full) rank n. Moreover, R intertwines the actions of $\mathcal{G}$ on systems with the action of left multiplication by $\mathcal{G}$ on $\mathcal{M}_*(n, (n+1)m)$—an action for which canonical forms are well known. Notice that this approach is a rather clever generalization of the scalar input case. Here, $R: M_0 \to \mathcal{M}_*(n, n+1)$, which may be replaced by $\mathcal{M}_*(1, n+1) = k^{n+1} - \{0\}$. The coordinates $\chi_i(F)$ arise in this context as the homogeneous coordinates on a certain subspace of the quotient $\mathbf{P}_n$ as we point out in Example 3.1. In the general case, he tried to use the local coordinates (i.e., the canonical forms) on the quotient $\mathcal{M}_*(n, (n+1)m)/\mathcal{G}$ to explicitly describe Σ_0. This shall be our starting point in Section 2, but we would like to review some basic facts regarding this quotient, in the setting of a moduli problem.

Now the connection between classical invariant theory and the orbit space problem, familiar in geometry and dynamics, is a part of geometric folklore and is one of the starting points for Mumford's GIT. In this treatise, Mumford divides the general moduli problem into two subproblems: the first is that of finding "canonical forms" for the classification; i.e., finding a space X and a group G in the same category such that the desired equivalence classes are exactly the orbits of G on X—and the second is to parameterize the orbits by means of a geometric object; i.e., to represent the moduli space as a quotient object in this category. For example, the Grassmann variety of r-planes in s-space, $\mathbf{Gr(r,s)}$ may be constructed in this way. Explicitly, two matrices in $\mathcal{M}_*(r,s)$ generate the same row space in k^s if and only if they differ by a multiple in $\mathcal{G}\ell(r,k)$. Thus, an orbit of the left action

$$\mathcal{G}\ell(r,k) \times \mathcal{M}_*(r,s) \to \mathcal{M}_*(r,s)$$

is just an r-plane in k^s; i.e., the problem of classifying such subspace of k^s has been "brought into canonical form." In addition, the quotient $\mathbf{Gr(r,s)}$ is well known to be smooth and locally affine. In fact, local coordinates are given in terms of the canonical forms for this action (cf., Section 2), which are also sections of the parameterization map

$$\pi: \mathcal{M}_*(r,s) \to \mathbf{Gr(r,s)} \tag{1.10}$$

Moreover, it is well known and easy to check that the local sections of π exhibit (1.10) as a principal $\mathcal{G}\ell(r,k)$-bundle on $\mathbf{Gr(r,s)}$ (see Hirzebruch [12]). Now in the usual way, the left action of $\mathcal{G}\ell(r,k)$ on k^r gives rise to a vector bundle (of rank r) on $\mathbf{Gr(r,s)}$ associated to the principle bundle (1.10), say $V \to \mathbf{Gr(r,s)}$. For algebrogeometric reasons, we prefer to use the dual bundle V^*, denoted by $\mathbf{U_{r,s}}$, which may be referred to as a *universal bundle* ([8, see especially Chapter 5]). This terminology is tied up with the existence of the moduli functor; i.e., in order to classify analytic families of r-spaces consider an analytic vector bundle V of rank r on a compact, complex manifold X such that $\dim H^0(X, V) = s$, where $H^0(X, V)$ is the vector space of (global) sections of the map $V \to X$. Then, essentially by evaluation, one can construct an analytic map

$$f: X \to \mathbf{Gr(r,s)}$$

classifying V; i.e., such that $V \simeq f^*\mathbf{U_{r,s}}$. Explicitly, if V is a line bundle and if $\varphi_1^*, \ldots, \varphi_s^*$ generate $H^0(X, V)^*$ then $f(p)$ is defined, in homogeneous coordinates, by $[(\varphi_1 * \sigma(p), \ldots, \varphi_s * \sigma(p))]$, where σ is any section of V not vanishing at p. Evidently, this construction only requires that $\varphi_1(p), \ldots, \varphi_s(p)$ generate each fiber (i.e., some φ_i does not vanish at each p) V_p, so that compactness is not essential. Taking into account the natural duality between spaces and their structure sheaves, this may be viewed as a projective analog of the imbedding of a linear space into its second dual. In the general case, $r > 1$, $\varphi_1 * \sigma(p)$ must be replaced by the expansion of $\sigma(p)$ in a basis of cardinality r, giving rise to an $r \times s$ matrix, which determines a class in $\mathbf{Gr(r, s)}$. Categorically speaking, the moduli functor is the functor $\mathscr{F}$ which assigns to each space X the set of isomorphism classes of such vector bundles on X, while the universal object

$$\mathbf{U_{r,s}} \to \mathbf{Gr(r, s)}$$

represents the functor $\mathscr{F}$, viz.,

$$\mathscr{F}(X) = \mathscr{H}om(X, \mathbf{Gr(r, s)}),$$

naturally. Thus, a more sophisticated version of the moduli problem asks for the representability of a suitable functor $\mathscr{F}$, corresponding to families of the objects under consideration.

Notice that if $r = 1$, one has a universal line bundle

$$\mathbf{U_{1,s}} \to \mathbf{P_{s-1}(C)}$$

for which

$$\mathscr{F}(X) = \mathscr{H}om(X, P_{s-1}).$$

For example, if $X = \mathbf{Gr(r, t)}$, the line bundle $\mathbf{L_1} = \Lambda^{\mathbf{r}}\mathbf{U_{r,t}}$ (usually referred to as the determinant line bundle, $\det \mathbf{U_{r,t}}$) determines an element of $\mathscr{F}(\mathbf{Gr(r, t)})$ and thus a map

$$p: \mathbf{Gr(r, t)} \to \mathbf{P^N} \qquad \text{for} \quad \mathbf{N} \gg \mathbf{0},$$

for which $p^*\mathbf{U_{1,N}} = \mathbf{L_1}$. In fact, p is an imbedding—the Plucker imbedding—exhibiting $\mathbf{Gr(r, t)}$ as a projective variety (cf., Section 3). Moreover, p is the minimal imbedding of $\mathbf{Gr(r, t)}$ in a projective space. To see this, notice that the set of (isomorphism classes of) line bundles on any X forms a group under tensor product, the picard group $\mathscr{P}ic(X)$. Now $\mathscr{P}ic(\mathbf{Gr(r, t)}) \simeq \mathbf{Z}$, canonically; in fact, under this correspondence c, $c(\mathbf{L_1}) = 1$. From the fact that $\mathbf{L_1}$ generates the picard group, one can deduce that, if classifying $\mathbf{L_1}$ imbeds $\mathbf{Gr(r, t)}$, then this imbedding is minimal. After all, one may use the projectivization of the dual space, $H^0(\mathbf{Gr(r, t)}, \mathbf{L_p})$, for the projective space and this dimension goes up when $\mathbf{L_1}$ is replace by $\mathbf{L_p} = \otimes^p \mathbf{L_1}$ for $p \geqslant 2$. Now there are various ways to show that $\mathscr{P}ic(\mathbf{Gr(r, t)}) \simeq Z$; for example, one can first show that any line bundle on $\mathbf{Gr(r, t)}$ is associated to the fiber bundle

$$\mathscr{M}_*(r, t) \to \mathbf{Gr(r, s)}$$

by a character of $\mathscr{G}\ell(r, k)$. However, each of these has the form $\gamma_i(g) = \det(g)^i$, for $i \in \mathbf{Z}$ (see [4, Theorem, p. 22]). For instance, it may be directly shown that the character corresponding to $\mathbf{L_1}$ is γ_{-1}. These statements can be verified by the same techniques used in Section 4 to compute $\mathscr{P}ic(\Sigma_0)$. Moreover, the proof given there extends to show that the map

$$\widehat{\mathscr{G}\ell}(r, k) \to \mathscr{P}ic(\mathbf{Gr(r, t)}),$$

defined by $\gamma_i \to -i$, is an isomorphism.

2. The Analytic Structure of the Moduli Spaces

In this section k denotes a topological field, complete with respect to a nontrivial absolute value (see Serre [25]). Following the traditional approach, we shall first describe the orbit space of pairs (F, G) modulo the action of $\mathscr{G}$. Recall that we have an injective map R which is an intertwining operator, i.e., the diagram

$$\begin{array}{ccc} M_0 & \xrightarrow{R} & \mathscr{M}_*(n, (n+1)m) \\ \downarrow g & & \downarrow g \\ M_0 & \xrightarrow{R} & \mathscr{M}_*(n, (n+1)m) \end{array} \tag{2.1}$$

commutes, for all $g \in \mathscr{G}$. We shall label the columns of a matrix $M \in \mathscr{M}_*(n, (n+1)m)$ as follows: $(01, \ldots, 0m, 11, \ldots, nm)$. In this notation a multi-index $\alpha = (\alpha_1, \ldots, \alpha_n)$, with $\alpha_1 \leqslant \cdots \leqslant \alpha_n$ has as its components $\alpha_i = j_i k_i$. For each M there corresponds a submatrix $M_\alpha \in \mathscr{M}_*(n, n) = \mathscr{G}$ for at least one α and thus the canonical form for M, $M_\alpha^{-1}M$, which is unique up to choice of α. This forms a local coordinate chart for a rational atlas on $\mathbf{Gr(n, (n+1)m)}$, as is well known. In this way, one may obtain canonical forms for systems, viz.,

$$(F, G) \to (R^{-1}(R(F, G)_\alpha^{-1} R(F, G))$$

corresponding to appropriate multi-indices α. Now R descends to a map r making the diagram

$$\begin{array}{ccc} M_0 & \xrightarrow{R} & \mathscr{M}_*(n, (n+1)m) \\ \downarrow \varphi_0 & & \downarrow \pi \\ \Sigma_0 & \xrightarrow{r} & \mathbf{Gr(n, (n+1)m)} \end{array} \tag{2.2}$$

commute, and one might expect that the canonical forms exhibit Σ_0 as a subvariety $r\Sigma_0$ of $\mathbf{Gr(n, (n+1)m)}$. A key observation, made by Kalman, is that the choice of α can be made considerably sharper in case $M = R(F, G)$.

Kalman [16] calls a multi-index α *nice*, just in case whenever $jk \in \alpha$, then $ik \in \alpha$, for $0 \leqslant i \leqslant j$. The interest in this concept stems from the following two results and Theorem 3.1.

LEMMA 2.1 (Kalman). $M_0 = \{(F, G) : R(F, G)_\alpha$ *is regular for some nice* $\alpha\}$.

Proof. The proof is clear, but we would like to set some notation. Starting at any column $0i_1$ add the columns $1i_1$, etc., until a dependence relation occurs, the last independent column being denoted by $\kappa_{i1}i_1$. Since M has the form $R(F, G)$, we can choose enough i_ns so that

$$\sum_{l=1}^{j} (\kappa_{il} - 1) = n. \qquad \text{Q.E.D.}$$

The integers κ_{ij} are referred to in [24] as Kronecker invariants for (F, G). Notice that each sequence (κ_{ij}) determines a nice multi-index, and conversely. We may also relate another sequence to a nice multi-index. Explicitly, for $l = 0, \ldots, n-1$, set μ_1 is equal to the cardinality of the $\{li_j \in \alpha, l \leqslant \kappa_{ij}\}$. Thus, (κ_{il}) and (μ_l) are the marginal cardinalities of α, viewed as a rectangular $j \times n$ array. Each $\mu_l \geqslant 0$ and

$$\sum_{l=0}^{n-1} \mu_l = n, \qquad \sum_{l=1}^{j} \kappa_{il} = n + j. \tag{2.3}$$

The μ_l's have also been referred to as "Kronecker indices" (see [15]).

If U_α is the invariant open subset of $\mathscr{M}_*(n, (n+1)m)$ consisting of matrices M for which M_α is regular, then $V_\alpha = \pi(U_\alpha)$ is open in $\mathbf{Gr(n, (n+1)m)}$. Thus the family (V_α) is an open cover. Lemma 2.1 says that

$$r\Sigma_0 \subset U_{\alpha\,\text{nice}} V_\alpha.$$

Locally, we have the following theorem.

THEOREM 2.1 (Kalman, 15). *The canonical form* $R(F, G)_\alpha^{-1}\, R(F, G)$ *for* α *nice induces a regular bijection of* $W_\alpha = r\Sigma_0 \cap V_\alpha$ *with* k^{nm}.

Proof. A detailed proof has been given by Hazewinkel and Kalman [10]. We shall give a simple alternate proof, based on "Kronecker indices". The point $\pi(R(F, G)) \in W_\alpha$ is canonically represented by the form $R = R(F, G)_\alpha^{-1} R(F, G)$. Let B_i, for $i = 0, \ldots, n$, denote the block of R consisting of the columns $(i1, \ldots, im)$. Suppose α corresponds to the total Kronecker invariant $(\kappa_{i1}, \ldots, \kappa_{ij})$, which latter determines the sequence (μ_l). Then, in B_0, there are $M - \mu_0$ free columns, while the remaining columns $\alpha \cap B_0$ consist of the unit vectors $e_1, \ldots, e_{\mu 0}$. Let $L_0 \subset \{1, \ldots, m\}$ be the subset of those l for which $0l \in B_0 - \alpha \cap B_0$. Now, assuming the columns in $B_0 - \alpha \cap B_0$ are fixed, the free columns in B_1 are determined as follows. Let $L' \subset \{1, \ldots, m\}$ be the subset of those l for which $1l \in \alpha \cap B_1$. If $L_1 = \{1, \ldots, m\} - (L' \cup L_0)$, then $\operatorname{card}(L_1) = m - (\mu_1 + \mu_0) = \mu_0 - \mu_1$. Moreover, supposing F were known, the only free columns in B_1 are those in L_1. Assuming this for the moment, it is easy to induce on this formula; viz., if L_i indexes the free columns in B_i, $0 \leqslant i \leqslant n - 1$, then $\operatorname{card}(L_i) = \mu_{i-1} - \mu_i$. Thus if L is the number of free columns in R, $L = \bigcup_{i=0}^{n} L_i$ and, in particular,

$$\operatorname{card}(L) = (m - \mu_n) + \mu_n = m, \tag{2.4}$$

since μ_n is the number of free columns in B_n, assuming F is fixed by the data indexed by L. Now, to know F on $e_1, \ldots, e_n$ is to know F. However, the Fe_i are either unit vectors or are column vectors indexed by L. Of course regularity of the chart is clear. Q.E.D.

Based on this result and some computations, Hazewinkel and Kalman have concluded that Σ_0 is a quasi-projective variety of dimension nm, and that $\Sigma_p = \Sigma_0 \times k^{pn}$ is then also quasi-projective. Unfortunately, their computations are in error and, more seriously, Σ_p is not a global product with factor Σ_0, unless $m = 1$. Actually, the situation is much more interesting, as we can see by pushing forward in the analytic case.

Now it follows from Kalman's theorem that Σ_0 is a topological k-manifold, and it is easily checked that this coordinate cover provides $r\Sigma_0$ with a rational atlas. Our next theorem shows that this analytic structure on Σ_0 is related to the natural structure on M_0 in a fairly conventional way. We shall give a constructive proof of this fact, since we shall need the computations later. As a matter of fact, these computations imply Kalman's theorem.

THEOREM 2.2. *Σ_0 has a natural analytic structure for which $\varphi_0 : M_0 \to \Sigma_0$ is a k-analytic left principal $\mathscr{G}$-bundle.*

Proof. First of all, M_0 is open in k^{n^2+nm}, and may be regarded as an analytic manifold. In this context, it is clear that the action $\alpha : \mathscr{G} \times M_0 \to M_0$ is analytic. We claim that R is an analytic imbedding. Assuming the claim, we may regard M_0 as a subspace of $\mathscr{M}_*(n, (n+1)m)$, having the metric space Σ_0 as a topological quotient. On the topological level, we have for nice α.

LEMMA 2.2. *$\varphi_0^{-1}(W_\alpha) \simeq W_\alpha \times \mathscr{G}$, canonically, as $\mathscr{G}$-spaces.*

Proof. The idea is to define a section of φ_0,

$$\psi_\alpha : W_\alpha \to \varphi_0^{-1}(W_\alpha)$$

by $\psi_\alpha([R(F,G)]) = R(F,G)_\alpha^{-1}R(F,G)$, which is clearly well defined. Since φ_0 is a quotient map, continuity follows immediately from the equation

$$\psi_\alpha \circ \varphi_0(R(F,G)) = R(F,G)_\alpha^{-1}R(F,G)$$

and the continuity of σ. To obtain the lemma, define $\varphi_\alpha : \varphi_0^{-1}(W_\alpha) \to W_\alpha \times \mathscr{G}$, by

$$\varphi_\alpha(R(F,G)) = ([R(F,G)], R(F,G)_\alpha). \tag{2.5}$$

Thus φ_α has an inverse, viz.,

$$([R(F,G)], g) \to g\psi_\alpha([R(F,G)]).$$

That φ_α preserves the appropriate $\mathscr{G}$-actions is a straightforward computation.

Of course, since (W_α) is a cover of Σ_0, $M_0 \to \Sigma_0$ is a principal (left) $\mathscr{G}$-bundle. As for the claim, notice that R is even a polynomial map. To see that R is an immersion, factor R as follows: First,

$$\mu : \mathscr{H}om_k(Q,Q) \times \mathscr{H}om_k(A,Q) \to \left(\bigoplus^n \mathscr{H}om_k(Q,Q)\right) \times \mathscr{H}om_k(A,Q)$$

defined by $(F,G) \to ((I, F, \ldots, F^n), G)$, followed by the map

$$\left(\bigoplus^n \mathscr{H}om_k(Q,Q)\right) \times \mathscr{H}om_k(A,Q) \xrightarrow{\gamma} \bigoplus^n \mathscr{H}om_k(A,Q) \simeq \mathscr{H}om_k\left(\bigoplus^n A, Q\right),$$

defined by composition. The only stage at which the rank could decrease is the last, but the Jacobian of γ on the image of μ is easily expressible in term of $R(F,G)$ and, in fact, surjectivity of $R(F,G)$ implies that the Jacobian has maximal rank. Finally, to say R^{-1} is continuous is to say that whenever $R(F_i, G_i) \to R(F,G)$, $(F_i, G_i) \to (F,G)$. Suppose we have such a sequence, $G_i \to G$. To recover the statement $F_i \to F$ we pass first to canonical forms. These converge, by continuity of σ and by continuity of inversion in $\mathscr{G}$. As we have seen in the proof of Kalman's theorem, F can be read directly from $R(F,G)_\alpha^{-1}R(F,G)$. Thus convergence of the canonical forms implies $\tilde{F}_i \to \tilde{F}$, where $(\tilde{F}_i, \tilde{G}_i) = R^{-1}(R(F_1,G_1)_\alpha^{-1}R(F_1,G_1))$. Again, since σ is continuous, (2.1) with $g_i = R(F_i,G_i)_\alpha$ yields

$$F_i = R(F_i,G_i)_\alpha \tilde{F}_i R(F_i,G_i)_\alpha^{-1} \to R(F,G)_\alpha \tilde{F} R(F,G)_\alpha^{-1} = F,$$

and the claim is proved. Returning to the proof of the theorem, since M_0 is an analytic submanifold of $\mathscr{M}_*(n,(n+1)m)$, the orbits of $\mathscr{G}$ are closed and immersed while Lemma 2.2 shows that the action is free. By standard theorems concerning analytic actions, we may deduce the existence of an analytic structure on Σ_0 for which $\varphi_0 : M_0 \to \Sigma_0$ is a $\mathscr{G}$-bundle as stated. The analytic structure on Σ_0 is uniquely determined by the condition that φ_0 be a submersion, thus an atlas for this manifold is given by the canonical forms described in the proof of Kalman's theorem. Q.E.D.

COROLLARY 2.1. *If (F, G) and (F', G') are accesible machines, congruent mod $\mathscr{G}$, then there is a unique $g \in \mathscr{G}$ such that $(gFg^{-1}, gG) = (F', G')$. Moreover, $g = R(F', G')_\alpha R(F, G)_\alpha^{-1}$ for any nice α such that $[R(F, G)] \in W_\alpha$.*

COROLLARY 2.2. *Let $k \subset \Omega$ be a subfield and let $(F, G), (F', G')$ be accessible machines defined over k. If $(F, G) \equiv (F', G') \bmod \mathscr{G}\ell(n, \Omega)$, then $(F, G) \equiv (F', G')$ mod $\mathscr{G}\ell(n, k)$.*

We remark that Hans Fischer has shown us a proof of this theorem in a coordinate-free setting, without reference to canonical forms, in the context of analytic transformation groups.

We will now determine the class ξ of the fiber bundle M_0 in $H^1(\Sigma_0, \mathscr{G})$. The 1-cocycles for ξ, relative to the cover (W_α), are

$$\theta_{\alpha\beta}(p) = R(F, G)_\beta^{-1} R(F, G, \ )_\alpha, \tag{2.6}$$

going from the α-trivialization to the β-trivialization. Here, $p = [R(F, G)]$. Either from the cocycles or the construction, it is clear that $\xi^* = \eta_1$, where η_1 is the class of the restricted universal bundle $\mathbf{U}_{\mathbf{n},(\mathbf{n}+1)\mathbf{m}|\Sigma_0}$. Notice the nontriviality of η_1 is equivalent to the nonexistence of global canonical forms for machines. Now η_1 also arises as the class determined by the bundle $\Sigma_1 \to \Sigma_0$. One has a commutative diagram of spaces and continuous maps

$$\begin{array}{ccc} M_p & \xrightarrow{\text{proj}_1} & M_0 \\ \downarrow{\scriptstyle \varphi_p} & & \downarrow{\scriptstyle \varphi_0} \\ \Sigma_p & \xrightarrow{\tau_p} & \Sigma_0 \end{array} \tag{2.7}$$

where τ_p is the map $\varphi_0 \circ \text{proj}_1$ descended to Σ_p. Of course, proj_1 is the projection $M_p = M_0 \times k^{pn} \to M_0$. In particular, τ_p has k^{pn} as its fibers and the local product structure descends, by the existence of local sections of φ_0. Explicitly, we have the following lemma.

LEMMA 2.3. *$\tau_p^{-1}(W_\alpha) \simeq W_\alpha \times k^{pn}$, via the map $([R(F, G)], H) \to (\psi_\alpha(R(F, G), HR(F, G)_\alpha)$.*

In fact, the corresponding "twisting data" defines $\Sigma_p \to \Sigma_0$ as a vector bundle of rank pn. Denote the class of this bundle in $H^1(\Sigma_0, \mathscr{G}\ell(pn, k))$ by $\bar{\xi}_p$. Pushing this one step further, notice that

$$\varphi_p^{-1}(\tau_p^{-1}(W_\alpha)) \simeq \psi_\alpha^{-1}(W_\alpha) \times k^{pn} \simeq W_\alpha \times \mathscr{G} \times k^{pn}, \tag{2.8}$$

canonically, given by

$$(R(F,G),H)\to(R(F,G)_\alpha^{-1}R(F,G),R(F,G)_\alpha,HR(F,G)_\alpha). \tag{2.9}$$

Of course, Lemma 2.3 and equation (2.8) give canonical forms for accessible machines (F,G,H). However, they also serve to relate the vector bundle $\Sigma_p\to\Sigma_0$ to the principal bundle $M_0\to\Sigma_0$ in a standard way. Formally, the content of (2.9) is that the map R_p, defined by

$$R_p:M_p\to\mathcal{M}_*(n,(n+1)m)\times k^{pn},$$

is an intertwining map (the action on k^{pn} is just $H\to Hg^{-1}$). As before, R_p descends to give an imbedding r_p of the quotients. Moreover, one has a commutative diagram

$$\begin{array}{ccc}\Sigma_p & \xrightarrow{r_p} & \mathcal{M}_*(n,(n+1)m)\times k^{pn}/\mathcal{G}\\ \Big\downarrow{\scriptstyle\varphi_p} & & \Big\downarrow{\scriptstyle\pi_p}\\ \Sigma_0 & \xrightarrow{r} & \mathbf{Gr(n,(n+1)m)}\end{array} \tag{2.10}$$

since the action is a product action. The vertical diagram on the right is a vector bundle, viz., the vector bundle associated to $\mathcal{M}_*(n,(n+1)m)\to$ $\mathbf{Gr(n,(n+1)m)}$ by the representation $\theta(p):\mathcal{G}\ell(n,k)\to\mathcal{G}\ell(pn,k)$, defined by $\theta(p)(g)H=Hg^{-1}$.

THEOREM 2.3. *$\Sigma_p\to\Sigma_0$ is, canonically, an analytic vector bundle of rank pn.*

In fact, it is the vector bundle associated to $M_0\to\Sigma_0$ by $\theta(p)$. With this analytic structure, $M_p\to\Sigma_p$ is an analytic principal $\mathcal{G}$-bundle.

Proof. The precise statement of the first claim is that $\theta(p)_*(\xi)=\bar{\xi}_p$, where $\theta(p)_*$ is the map

$$\theta(p)_*:H^1(\Sigma_0,\mathcal{G}\ell(n,k))\to H^1(\Sigma_0,\mathcal{G}\ell(np,k))$$

of cohomology sets induced by the representation $\theta(p)$. However, the equality $\theta(p)_*(\xi)=\bar{\xi}_p$ follows, in a straightforward manner, from equations (2.6) and (2.9) and the definition of $\theta(p)_*$. As for the latter statement, recall that the associated bundle may be constructed as a quotient

$$\Sigma_p\simeq\Sigma_0\times k^{pn}/\mathcal{G}\simeq M_p/\mathcal{G}$$

by the intertwining map R_p. This determines M_p, up to bianalytic equivalence. Q.E.D.

COROLLARY 2.3. *$\Sigma_p\simeq\bigoplus^p\Sigma_1$, as bundles over Σ_0. Moreover, $\Sigma_1\simeq$ $\mathbf{U}_{\mathbf{n},(\mathbf{n}+1)\mathbf{m}|\Sigma_0}$, and thus Σ_1 is trivial if and only if $M_0\to\Sigma_0$ is trivial.*

Proof. Now $\theta(p)$ is just the adjoint representation $g \to {}^t g^{-1}$, followed by the standard product representation of $\mathcal{G}$ on $\bigoplus^p k^n$. This not only proves the first statement, but also points out that $\theta(1)$ is just the adjoint representation. In particular, $\theta(1)_*(\xi) = \xi^* = \eta_1$, thus the second statement follows. Q.E.D.

COROLLARY 2.4. *The imbedding* $r: \Sigma_0 \to \mathbf{Gr(n, (n + 1)m)}$ *induced from the map* R *arises from classifying the vector bundle* Σ_1.

If $\sigma_p = \det \Sigma_p = \Lambda^{np} \Sigma_p$, then from multilinear algebra, one has

COROLLARY 2.5. $\sigma_p = \otimes^p \sigma_1$. *Of course*, $\sigma_1 = \det \mathbf{U}_{\mathbf{n},(\mathbf{n}+1)\mathbf{m}|\Sigma_0}$.

This also follows, at least for k algebraically closed and of characteristic zero, by computing the cocycles relative to (W_α) for σ_p in terms of those for σ_1. Explicitly, notice that the composite map

$$\mathcal{G}\ell(n, k) \xrightarrow{\theta(p)} \mathcal{G}\ell(np, k) \xrightarrow{\det} k^*$$

is a rational character γ on $\mathcal{G}\ell(n, k)$. Choosing $\alpha \in k^*$ not to be a root of unity, one calculates

$$\alpha^{-np} = \det \theta(p)(\alpha I) = \det(\alpha I)^l = \alpha^{nl}.$$

Thus $\gamma(g) = \det(g)^{-p}$ and the cocycles defining σ_p are

$$(\varphi_{\alpha\beta}) = (\det R(F, G)_\alpha^{-p} \det R(F, G)_\beta{}^p). \tag{2.11}$$

One should also notice that the subgroup $(\sigma_1) \subset \mathcal{P}ic(\Sigma_0)$ is the homomorphic image of $\mathcal{P}ic(\mathbf{Gr(n, (n + 1)m)})$ under restriction. As it turns out, restriction is an isomorphism whenever $m > 1$. In the next section, we shall get a feel for this situation by considering some examples.

3. PROJECTIVE IMBEDDINGS. EXAMPLES AND COUNTEREXAMPLES

First, there is an obvious imbedding of Σ_0 in a high-dimensional projective space, given by restricting the Plucker imbedding of $\mathbf{Gr(n, (n + 1)m)}$. In order to set notation, we shall recall a few more details. Consider the polynomial ring $k[P_\alpha]$, where α runs over some fixed ordering of the multi-indices. The codomain of the Plucker map p is $\mathbf{Proj}\, k[P_\alpha] \simeq \mathbf{P_N}$, where $N = \binom{(n+1)m}{n} - 1$. In homogeneous coordinates, p can be expressed globally as

$$p: [M] \to [(\det M_\alpha)]. \tag{3.1}$$

Now, via p, $\mathbf{Gr(n, (n + 1)m)}$ is a projective variety in $\mathbf{P_N}$; i.e., it may be defined by homogeneous relations in $k[p_\alpha]$. One such set of relations is the well-known Plucker relations, which are homogeneous quadratics (cf. [8, especially Chapter 5]). For the purpose of defining Σ_0 as a quasi-projective variety, Kalman produced another set of relations for **Gr(r, s)**, which are of higher degree but are much easier to set down explicitly. We shall need these for a counterexample which we shall produce presently. However, these relations are of interest, for their simplicity, in their own right. Kalman's relations are

$$f_{\alpha\beta}(P) = p_\alpha \det(p_{\alpha-\alpha_i+j})_{\substack{\alpha_i\in\alpha \\ j\in\beta}} - p_\alpha{}^n p_\beta. \tag{3.2}$$

By $p_{\alpha-\alpha_i+j}$ is meant (sign σ) p_γ, where γ is the multi-index obtained from α by replacing α_i by j and permuting the indices by σ to obtain $\gamma_1 < \gamma_2 < \cdots < \gamma_n$. In the case $\gamma_i = \gamma_j$, for $i \neq j$, $p_{\alpha-\alpha_i+j}$ should be taken to be 0. The proof that the $f_{\alpha\beta}$ cut out the Grassman variety is straightforward, using only the canonical forms $M_\alpha^{-1}M$; see [16] or [17].

REMARK 3.1. Although the Plucker imbedding is defined relative to a fixed ordering of the αs, it is easy to see [say, by (3.2)] that image p is invariant under the natural permutation action by $\mathscr{S}_{N+1} \subset \mathscr{A}ut\,\mathbf{P_N} = \mathscr{PGL}(N)$. In particular, over $\mathbf{C}$, the Fubini–Study metric on $\mathbf{P_N}$ is invariant under $\mathscr{S}_{N+1}$ and thus restricts, canonically, to **Gr(n, (n + 1)m)**. Thus **Gr(n, (n + 1)m)** is a Kähler manifold (in fact, it is Hodge, as is well known) and so the same is true for Σ_0; i.e. Σ_0 is a Kähler manifold, with integral Kähler form, in a natural way (see [26]).

For the sake of applications, it is worthwhile to examine two important examples. First, we consider the case when the machine may be controlled through one input terminal, i.e., $m = 1$.

EXAMPLE 3.1 (Scalar input machines). In this case, the representation of $\Sigma_p \to \Sigma_0$ given by Kalman coincides with that given by Section 2. By the proof of Theorem 2.2, there is an immersion which, by duality, may be written as

$$\Sigma_0 \xrightarrow{r} \mathbf{Gr(n, n+1)} \simeq \mathbf{Gr(1, n+1)} \simeq \mathbf{P_n}.$$

Note that the composite map is just the Plucker imbedding, since under this duality $\det \mathbf{U_{n,n+1}} = \mathbf{U_{1,n+1}}$. Moreover, the imbedding of Σ_0 by sections of $\det \mathbf{U}_{\mathbf{n,n+1}|\Sigma_0}$ may be directly calculated to be

$$\Sigma_0 \simeq U_\alpha \subset \mathbf{P_n},$$

where U_α is the open affine submanifold corresponding to the multi-index $\alpha = (01, \ldots, (n-1)1)$. To see this, recall that to say (F, G) is accessible is to say that $k[F] \cdot G$ spans Q, or that G is a cyclic vector for F. Now, $R^{-1}(R(F,$

$G)_\alpha^{-1}R(F,G)) = (F_1, G_1)$, where $G = e$ and F_1 is similar to F. In particular, $\chi(F,s) = \chi(F_1,s)$, but since F_1 is the companion matrix to χ, F_1 is the rational canonical form for F. Recalling the map induced by duality, one has

$$(F,G) \to P = [(p_0, \ldots, p_\alpha, \ldots, p_\omega)],$$

where $p_\alpha = 1$ and the p_β are the coefficients of $\chi(F,s)$. Thus $\Sigma_0 \simeq k^n$ and, since α is the only nice multi-index, $M_0 \to \Sigma_0$ is trivial. In fact, the global section ψ_α takes (F,G) to (F_1,G_1)—the "control canonical form" for (F,G) (see [18, Chapter 2]). By Corollary 2.3, $\Sigma_p \simeq \Sigma_0 \times k^{pn}$, for all $p \geqslant 1$. Finally note that, trivially, Σ_0 is quasi-projective; i.e., it is open in its projective closure $\mathbf{P_n}$, which (in addition) is nonsingular.

The importance of the next example stems from the structure theory of machines. Briefly, each machine may be decomposed into the parallel sum of "primary machines" (ALM, Chapter 16, §8). Each of these may be represented as the composite in series of systems for which $n = 1$.

Example 3.2 (Prime machines). In this case, our techniques yield an imbedding

$$\Sigma_0 \to \mathbf{Gr(1, 2m)} = \mathbf{P_{2m-1}}.$$

Explicitly, in homogeneous coordinates a machine is expressible as $[(g_1, \ldots, g_m, \alpha g_1, \ldots, \alpha g_m)]$, where $\alpha \in k$ and, of course, some $g_i \neq 0$ since (α, g) is accessible. Thus Σ_0 may be biregularly identified with $k \times \mathbf{P_m}$. Moreover, the projective closure of Σ_0 in $\mathbf{P_{2m-1}}$ is nonsingular; in fact, it is the image of $\mathbf{P_1} \times \mathbf{P_m}$, imbedded in $\mathbf{P_{2m-1}}$ by the Segre map (with the multi-indices ordered lexicographically). In particular, Σ_0 is again quasi-projective. However, in contrast to Example 3.1, $M_0 \to \Sigma_0$ is nontrivial. Indeed, σ_1 restricted $\mathbf{P_m}$ is just $\mathbf{U_{1,m}}$. Thus, for each $p \geqslant 1$, σ_p is nontrivial. In fact, in this case $\mathcal{P}ic(\Sigma_0)$ is generated by σ_1, and thus

$$\mathcal{P}ic(\Sigma_0) \simeq \mathcal{P}ic(\mathbf{P_m}) \simeq H^2(\mathbf{P_m}, \mathbf{Z}) \simeq \mathbf{Z}.$$

Of course, this even implies that each Σ_p is algebraically stably nontrivial.

Remark 3.2. The same kind of situation holds whenever $n < m$, viz., there is a natural imbedding of $k \times \mathbf{Gr(n,m)}$ in Σ_0. Again one has that $\sigma_1|_{\mathbf{Gr(n,m)}}$ generates $\mathcal{P}ic(\mathbf{Gr(n,m)}) \simeq \mathbf{Z}$. Thus this family of examples shows that, in general, $M_0 \to \Sigma_0$ is nontrivial. Moreover, this shows that Σ_0 is not Stein, whenever $n < m$.

Remark 3.3. The question of whether or not the projective closure of Σ_0 is nonsingular is of definite interest. Hazewinkel, in his colloquium at

Udine, Italy (June 1975), conjectured that for n arbitrary, $m = 3$, Σ_0 is projective. Of course, our examples indicate that this conjecture is false for $n = 1$ or 2, but we are not prepared to conjecture either on this special case or on the nonsingularity of $\mathscr{C}\ell(\Sigma_0)$ in general.

Of course, this latter question is relative to the projective imbedding of Σ_0. It is known that the Segre imbedding, just as the Plucker imbedding, is minimal and, moreover, in Examples 3.1 and 3.2 it is clear that these maps arise as imbeddings of Σ_0 by sections of σ_1. More generally recall that a map of Σ_0 to a projective space $\mathbf{P_M}$ corresponds to a line bundle L on Σ_0, together with a choice of $M + 1$ global sections generating each fiber. For example, the Plucker imbedding corresponds to the line bundle σ_1 together with the restrictions of global sections $\gamma_1, \ldots, \gamma_{N+1}$ on $\mathbf{Gr(n, (n + 1)m)}$ of $\det \mathbf{U_{n,(n+1)m}}$. Explicitly, it is known (see [8, especially Chap. 5]) that

$$\dim H^0(\mathbf{Gr(n, (n + 1)m), L_1}) = \binom{(n+1)m}{n} = N + 1, \quad \text{where} \quad \mathbf{L_1} = \det U_{\mathbf{n,(n+1)m}},$$

as in Section 1 [compare (3.1)]. It is reasonable to expect this number to go down when we restrict to Σ_0, and it usually does. Explicitly, Kalman calls a multi-index β a successor of a nice α just in case β may be obtained from α as a permutation of the multi-index $\alpha - \alpha_i + j$, where $\alpha_i \in \alpha$ is arbitrary and $ab_j \in \alpha$ for $a \leqslant a_j$, where $j = a_j b_j$. For example, if $n = m = 2$, then the nice multi-indices are $\alpha = (01, 02)$, $\beta = (01, 11)$, and $\gamma = (02, 12)$, while the successor are $\beta_1 = (01, 21)$, $\beta_2 = (02, 11)$, $\beta_3 = (11, 21)$, $\gamma_1 = (01, 12)$, $\gamma_2 = (02, 22)$, and $\gamma_3 = (12, 22)$. One should also note that all the multi-indices that occur in Examples 3.1 and 3.2 are either nice or successors. Thus the proper generalization of the Plucker and Segre imbeddings, in this context, is Kalman's imbedding.

THEOREM 3.1 (Kalman). *The map*

$$\kappa: \Sigma_0 \to \mathbf{P_k}$$

defined in homogeneous coordinates by

$$\kappa[R(F, G)] = [(\det R(F, G)_\alpha)],$$

for α either a nice or a successor multi-index, is a birational imbedding.

Proof. See [16], [17], or [10] for a sketch. The problem is clearly to recover $R(F, G)$ from $\det R(F, G)_\alpha$. The appropriate candidate is the $n \times (n + 1)m$ matrix $(p_{\alpha - \alpha_i + j})$ and all that need be checked is that knowledge of the p_β, for β nice or a successor, suffices to calculate each p_γ. Assuming that the point $[(p_\alpha, p_\beta, \ldots, p_\omega)]$ comes from a matrix $R(F, G)$ in the Plucker fashion, this is certainly the case.

Now κ was originally defined in order to prove that Σ_0 is quasi-projective. In [16] and [17], Kalman produced a set of relations "defining" $\mathscr{C}\ell(\Sigma_0)$ in $\mathbf{P_k}$, claiming that Σ_0 is the open subset of $\mathscr{C}\ell(\Sigma_0)$ defined by the conditions $p_\alpha \neq 0$, for α nice. Actually these relations are not restrictive enough.

EXAMPLE 3.3. Again, consider $n = m = 2$. First of all, p imbeds Σ_0 in $\mathbf{P_{14}}$, whereas κ imbeds Σ_0 in $\mathbf{P_8}$. The claim in [16] is that Σ_0 is the open subset of the projectivized locus of the 27 polynomials $f_{\delta\varepsilon}$ [see (3.2)], where $\delta = \alpha, \beta, \gamma$ and ϵ is either nice or a successor, defined by the additional conditions $p_\delta \neq 0$. As a counterexample to this claim, consider the point P in $\mathbf{P_8}$

$$P = [(1, 1, 1, -1, 1, -1, 0, -1, 1)]$$

where we have fixed the ordering as

$$\mathbf{P_8} = \mathbf{Proj}\, k[p_\alpha, p_\beta, p_{\beta_1}, p_{\beta_2}, p_{\beta_3}, p_\gamma, p_{\gamma_1}, p_{\gamma_2}, p_{\gamma_3})].$$

One may show that P satisfies the relations $f_{\delta\varepsilon}$ together with the condition $p_\alpha \neq 0$. In fact, to avoid the computations, first note that, in general, κ factors as

$$\begin{array}{ccc} \mathbf{Gr(n, (n+1)m)} & \xrightarrow{p} & \mathbf{P_N} \\ \cup & & \downarrow \text{proj} \\ \Sigma_0 & \xrightarrow{\kappa} & \mathbf{P_k} \end{array}$$

Now notice that P is the image under $\text{proj}\, p$ of the matrix $[M]$ in $\mathbf{Gr(2, 6)}$, where

$$M = \begin{pmatrix} 1 & 0 & 1 & 1 & 0 & 1 \\ 0 & 1 & 1 & 0 & 1 & 1 \end{pmatrix}.$$

Now, if P represents a machine $[(R(F, G)]$ in Σ_0, then, since $p_\alpha \neq 0$, $R(F, G)$ may be taken in canonical form (I, F, F^2). Assuming P does represent a machine, all but two of the entries in the canonical form are determined by the p_α's; viz.,

$$R(F, G) = \begin{pmatrix} 1 & 0 & 1 & 1 & x & 1 \\ 0 & 1 & 1 & 0 & 1 & y \end{pmatrix},$$

as in the proof of Theorem 3.1. However, in canonical form, the only machine having these entries is given by

$$R(F, G) = \begin{pmatrix} 1 & 0 & 1 & 1 & 2 & 1 \\ 0 & 1 & 1 & 0 & 1 & 1 \end{pmatrix}.$$

All that remains to be calculated is that

$$\kappa([R(F,G)]) = [(1,1,1,-1,-1,-1,0,-1,1)],$$

which differs from P in the p_{β_3} coordinate.

EXAMPLE 3.4. In an expanded, more detailed presentation [10], a different set of relations was derived, defining Σ_0 in $\mathbf{P}_8$. We refer the reader to the list of equations on pp. 23–24 of [10]; the list on p. 23 gives the relations (3.2) defining $\mathbf{Gr}(2,6)$ in $\mathbf{P}_{14}$, while on p. 24 the equations defining Σ_0 as a subvariety of $\mathbf{Gr}(2,6)$ are given. For convenience, we shall translate the notation in [10] into that used here (and in [16]). According to Hazewinkel and Kalman, the equations for $\mathscr{C}\ell(\Sigma_0)$ in $\mathbf{P}_8$ are

$$\begin{aligned} p_\alpha p_{\gamma_2} &= p_\gamma p_{\beta_2} - p_{\beta_3} p_\gamma, & p_\alpha p_{\beta_1} &= p_\beta p_{\gamma_1} - p_\beta p_{\beta_2}, \\ p_\beta p_{\gamma_1} &= p_\alpha p_{\beta_1} - p_\beta p_{\beta_2}, & p_\gamma p_{\beta_2} &= p_\gamma p_{\gamma_1} - p_\alpha p_{\gamma_2} \end{aligned} \tag{3.3}$$

Unfortunately, (3.3) is too restrictive, for if $R(F,G)$ is the machine corresponding to $F = G = I$, then $\kappa([R(F,G)])$ fails to satisfy the first identity in (3.3) (this is the second equation on p. 24 of [10]).

Due mainly to its relation with the Plucker map and its dependence on fewer variables, the Kalman imbedding is a useful computational tool. There is strong evidence which suggests that κ is the minimal projective imbedding of Σ_0. In the next section, it shall be shown that $\mathscr{P}ic(\Sigma_0) = (\sigma_1)$, as in example 3.2. Thus the minimal imbedding of Σ_0 is given by sections of σ_1 and, since $\kappa^*\mathbf{U}_{1,\mathbf{k}} = \sigma_1$, one need only check that κ is minimal among such imbeddings.

4. THE GEOMETRIC INVARIANT THEORY OF LINEAR MACHINES

The construction of the moduli space in the two extreme cases leads us to expect that, in general, Σ_0 will carry a familiar geometric structure. In fact, Σ_0 is quasi-projective. One can show this, using the results of Section 2 directly by the techniques outlined in [9] (see especially Section 6) for the construction of quotient preschemes. Actually, we shall come at this result within the framework of modern invariant theory. On one hand, this gives us more information concerning the quotient $M_0 \to \Sigma_0$ and, in addition, one of the more interesting consequences of these techniques is that M_0 arises naturally from an analysis of the stability of orbits. This viewpoint, in the scalar input case, has been exploited in [2] and reposes on calculations of the dimension of the orbits $\mathcal{O}(F,G)$, the accessible orbits being those of maximal dimension.

In this section we shall adopt the notation and conventions used in GIT. In addition, we shall denote by $\mathscr{G}$ the algebraic group (prescheme) $\mathscr{G}\ell(n,k)$ over an algebraically closed field, of characteristic zero. The latter conditions are, at least partially, justified by Corollary 2.2 and entail no loss of generality. We shall use X to denote the affine variety

$$\mathbf{A}^{\mathbf{n}^2+\mathbf{nm}} = \mathscr{spec}(k[x_{ij}] \otimes_k k[y_{k1}]).$$

Since $\mathscr{G}$ is reductive, we may propose to study the algebraic action $\sigma\colon \mathscr{G} \times X \to X$ via the ring of invariants $R^{\mathscr{G}}$ in the coordinate ring $R = k[x_{ij}] \otimes_k k[y_{k1}]$. That is, by the Hilbert–Nagata theorem (cf. [4, Theorem, p. 42]), $R^{\mathscr{G}}$ is finitely generated over k. Now the inclusion $R^{\mathscr{G}} \to R$ induces a map of prevarieties

$$X = \mathscr{spec}(R) \xrightarrow{\pi} \mathscr{spec}(R^{\mathscr{G}}).$$

Mumford's theorem, in the affine case (see [16, Theorem 5.8, p. 159]), states that π is a categorical quotient. For the sake of illustration, let us just consider the geometric points. Note that $R^{\mathscr{G}} \to R$ induces a map, the restriction of π,

$$k^{n^2+nm} \simeq \mathscr{H}om_k(R,k) \xrightarrow{\pi} \mathscr{H}om_k(R^{\mathscr{G}},k),$$

where $\mathscr{H}om_k(R^{\mathscr{G}},k)$ may be interpreted as an affine algebraic variety by the *nullstellensatz*. Unfortunately, this variety is not the set-theoretic quotient of k^{n^2+nm} by $\mathscr{G}$. To see this, notice that there is a natural inclusion

$$k[x_{ij}]^{\mathscr{G}} \simeq k[x_{ij}]^{\mathscr{G}} \otimes_k k[y_{kl}]^{\mathscr{G}} \to (k[x_{ij}] \otimes_k k[y_{kl}])^{\mathscr{G}}.$$

Remarkably, this map is surjective as well; this is the content of Lemma 4.3 of [2]. Now $k[x_{ij}]^{\mathscr{G}}$ is generated by the coefficient functions $\chi_i(s)$ of the characteristic polynomial $\chi(F,s)$. Since these polynomials are algebraically independent, the *nullstellensatz* gives $\mathscr{H}om_k(R^{\mathscr{G}},k) \simeq k^n$, so that Mumford's theorem (for the geometric points) reduces to

PROPOSITION 4.1. *k^n is a concrete affine model for $R^{\mathscr{G}}$. In fact, $\chi(\cdot,s)\colon k^{n^2+nm} \to k^n$ is the quotient map.*

Now, to say that k^n is the orbit space for the action is to say that the orbits of $\mathscr{G}$ are exactly the fibers of $\chi(\cdot,s)$. However, taking $G=0$, it is classical that there are nonequivalent machines $(F,0)$, $(F',0)$ with $\chi(F,s) = \chi(F',s)$. It is also not hard to see that $\chi(\cdot,s)$, restricted to M_0, is surjective as well. In general, such a construction can be easily carried out using the Minimal realization theorem, where in (1.7) the term $(F-s)^{-1}$ is replaced by the usual formula involving $\chi(F,s)$. Since M_0 is an invariant open subspace of k^{n^2+nm}, one may compute, using Proposition 4.1, the invariants on M_0 as follows.

PROPOSITION 4.2. *The ring of invariants in* $H^0(M_0, \mathcal{O}_x)$ is $k[x_{ij}]\mathcal{G}$.

Proof. If $m = 1$, the proof of the proposition is contained in [2] (see Corollary 4.4), so we shall assume $m > 1$. In this case consider the nice multi-indices α_1, α_2, where $\alpha_i = (0i, li, \ldots, (n-1)i)$. If $f \in H^0(M_0, \mathcal{O}_x)$ is invariant, then the restriction $f \in H^0(\varphi_0^{-1}(W_{\alpha_i}), \mathcal{O}_x)$ is invariant as well. Now $\varphi_0^{-1}(W_\alpha) \simeq W_{\alpha_i} \times \mathcal{G}$, by Lemma 2.2, as $\mathcal{G}$-spaces and thus, to say f is invariant on $\varphi_0^{-1}(W_{\alpha_i})$ is to say that f is determined as $\varphi_0^{-1}(W_{\alpha_i})$ by its restriction to W_{α_i}. That is, for $(F, G) \in \varphi_0^{-1}(W_{\alpha_1})$, f is a polynomial in the entries of the canonical form (F', G'). Now the nm free variables (F', G') occur as follows: in G' all entries except the 01 column are free, while the first column is the unit vector e_1; on the other hand, in F' the free variables occur as the $\chi_i(F)$. As in the proof of Lemma 4.3 of [2], if λ is not a root of unity, then the equation

$$f(F', G') = f(\lambda F \lambda^{-1}, \lambda G') = f(F', \lambda G')$$

implies that f can depend only on the $\chi_i(F)$ and on the variables y_{kl} corresponding to the entries of e_1. Since the same statement holds for α_2, f depends only on F. Q.E.D.

Of course, for $m = 1$, this is another way to compute Example 3.1. In particular, by restricting to M_0, we have discarded enough orbits so that the fiber of $\chi(\cdot, s)$ contains only one orbit. Such a construction is quite general and is closely connected with the existence of nonclosed orbits, a pathology we shall mention only briefly. Since $\mathcal{G}$ is irreducible, the fibers of $\chi(\cdot, s)$ are irreducible, Zariski, closed subvarieties of k^{n^2+n}. Thus, an obvious obstruction to the existence of a quotient structure on $k^{n^2+n}/\mathcal{G}$ is the existence of of nonclosed orbits. By the Closed orbit lemma (see [13, Proposition, p. 60]), all the minimal dimensional orbits are closed, but in this case the minimal-dimensional orbits are actually fixed points, viz., the machines $(\lambda I, 0)$, by Schur's lemma. In fact, nonclosed orbits abound, even among the machines $(F, 0)$ (see [2] or [21], for a discussion of this pathology for the similarity problem). On the other hand, for prime machines Proposition 4.2 gives the categorical quotient $\mathbf{k}$ (in the category of affine algebraic varieties), while our computations in Example 3.2 give the quasi-projective variety $\mathbf{k} \times \mathbf{P}_\mathbf{m}$ as an orbit space. Thus, by staying affine, we are missing whole projective spaces of orbits in each fiber of $\chi(\cdot, s)$.

An essential concept in GIT is that of a $\mathcal{G}$-linearization of an invertible sheaf or, dually, of a line bundle on X. However, since $\mathcal{P}ic(X)$ is trivial, a

$\mathscr{G}$-linearization is just an algebraic, fiber-preserving action of $\mathscr{G}$ on $L_0 = X \times k$, covering the action of $\mathscr{G}$ on X; i.e., a function $\theta(x, g)$ in End(k) such that the diagram

$$\begin{array}{ccc} X \times k & \xrightarrow{g \times \theta(x,g)} & X \times k \\ \big\downarrow & & \big\downarrow \\ X & \xrightarrow{\quad g \quad} & X \end{array} \tag{4.1}$$

commutes. Thus by the rigidity lemma (see [13, Proposition, p. 106]), $\theta(x, g)$ is independent of x and so each $\mathscr{G}$-linearization of the trivial bundle $X \times k$ is given by a character $\gamma \in \widehat{\mathscr{G}\ell}(n, k)$. That is, the action is given, up to isomorphism, by

$$g(x, \lambda) = (gx, \gamma_l(g) \cdot \lambda) = (gx, \det(g)^l \cdot \lambda). \tag{4.2}$$

In particular, the group of $\mathscr{G}$-linearized line bundles, $\mathscr{P}ic^{\mathscr{G}}(X)$, is canonically isomorphic to Z. If L_i denotes the pair $(L_0, \gamma_i) \in \mathscr{P}ic^{\mathscr{G}}(x)$, then the invariant sections of $L_i, H^0(X, L_i)^{\mathscr{G}}$, are just the sections of $L_0 \to X$ (i.e., polynomials) making the diagram

$$\begin{array}{ccc} X \times k & \xrightarrow{g \times \gamma_i(g)} & X \times k \\ \big\uparrow {\scriptstyle f} & & \big\uparrow {\scriptstyle f} \\ X & \xrightarrow{\quad g \quad} & X \end{array} \tag{4.3}$$

commute. Thus $H^0(X, L_0)^{\mathscr{G}} = R^{\mathscr{G}}$, while elements of $H^0(X, L_i)^{\mathscr{G}}$ are relative invariants of weight γ_i.

Again since $\mathscr{P}ic(X) = (0)$, Mumford's definition of stability reduces to the following.

DEFINITION 4.1 (Mumford). Let x be a geometric point of X. Then

(i) x is prestable, just in case there is an invariant affine open subspace X_f containing x and for which the action of $\mathscr{G}$ on X_f is closed.

(ii) x is semistable (with respect to L_i), just in case there exists $f \in H^0(X, L_i)^{\mathscr{G}}$ such that $f(x) \neq 0$.

(iii) x is stable (with respect to L_i), just in case there exists $f \in H^0(X, L_i)^{\mathscr{G}}$ such that $f(x) \neq 0$ and the action of $\mathscr{G}$ on X_f is closed.

Here, X_f denotes the co-locus of f. The corresponding invariant open subspaces are denoted by $X^s(\text{Pre})$, $X^{ss}(L_i)$, and $X^S(L_i)$; note that

$$X^s(L_i) \subset X^s(\text{Pre}) \cap X^{ss}(L_i) \qquad \text{for all } i \in \mathbf{Z}. \tag{4.4}$$

There is another point set which arises naturally, the principal orbit type. Explicitly, denote by X_{reg} the invariant open subspace determined by those geometric points x for which the stabilizer $\mathcal{S}_x$ of x in $\mathcal{G}$ has minimal dimension. For example, in the scalar input case, $X^s(\text{Pre}) = X_{\text{reg}} = \Sigma_0$, while $X^{ss}(L_0) = \{(F, G)$: F is not nilpotent$\}$ and $X^s(L_0) = \varnothing$ (see [2, Theorem 4.6 and Proposition 4.8). The proofs of the latter two identities extend, verbatim, to the general case, reposing on the identity $R^{\mathcal{G}} = k[X_{ij}]^{\mathcal{G}}$. The first identity implies, by Proposition 1.9 of GIT, that $M_0 \to \Sigma_0$ is a geometric quotient (see Fogarty [6] or Mumford and Suominen [22] for a very readable account of the notions of categorical and geometric, or strict, quotients). However, we are after the structure of Σ_0, so Proposition 1.9 is not enough. Since the information on the quotients of the $X^s(L_i)$ is more explicit, this will be our program. Of course, $X^s(L_0) = \varnothing$ so the next obvious candidate is the standard character linearization L_1.

THEOREM 4.1. *$\varphi_0: M_0 \to \Sigma_0$ is a universal geometric quotient with Σ_0 a quasi-projective prevariety and φ_0 an affine map.*

Proof. As in GIT, $X^s_{(i)}(L_1)$ will denote the subspace of $X^s(L_1)$ determined by geometric points X for which $\dim \mathcal{S}_X = i$. For example, the set of properly stable points $X^s_{(0)}(L_1)$ consists of those points with $\dim \mathcal{S}_x = 0$ or, equivalently, with $\dim \mathcal{O}(X) = n^2$. The method of proof is to obtain the inclusions

$$M_0 \subset X^s_{(0)}(L_1) \subset X_{\text{reg}} \subset M_0 \tag{4.5}$$

and then to apply Mumford's theorem (in the general case, GIT, Theorem 1.19). The first inclusion is easy; explicitly, the polynomials $f_\alpha(F, G) = \det R(F, G)_\alpha$ are clearly in $H^0(X, L_1)^{\mathcal{G}}$, thus $X_\alpha = X_{f_\alpha}$ is invariant. Since $M_0 \subset \bigcup_{\alpha \text{ nice}} X_\alpha$, it is enough to show that the action on each X_α is closed. However, since the action is free, by Lemma 2.2, each orbit is of minimum dimension, viz., 0, hence closed by the Closed orbit lemma. For this reason, $M_0 \subset X_{\text{reg}}$, and, in particular, to say a geometric point x is in X_{reg} is to say $\dim \mathcal{O}(X) = n^2$. Trivially, one also has the inclusion $X^s_{(0)}(L_1) \subset X_{\text{reg}}$.

LEMMA 4.1. $X_{\text{reg}} \subset M_0$.

Proof. Suppose $X = (F, G) \in X_{\text{reg}}$. Decompose the machine (F, G) into its parallel summands

$$(F, G) = \bigoplus_{i=1}^{l} (F_i, G_i).$$

According to ALM this corresponds, in part, to the Jordan decomposition of F and Q. Let $d_i = \dim Q_i$, where $Q = \bigoplus_{i=1}^{l} Q_i$. Of course, the group of symmetries of a machine with state space Q_i is $\mathcal{G}\ell(d_i k)$; thus, Q admits, naturally, the action of

$$\bigoplus_{i=1}^{l} \mathscr{G}\ell(d_i, k) \subset \mathscr{G}\ell(n, k)$$

in the standard way. In particular, the stabilizer $\mathscr{S}_i$ of (F_i, G_i) in $\mathscr{G}\ell(d_i, k)$ is finite, since this inclusion restricts to $\bigoplus \mathscr{S}_i \to \mathscr{S}_x$. This observation yields

$$\dim \mathscr{O}(F_i, G_i) = d_i^2. \tag{4.6}$$

By Proposition 4.5 of [2], this integer may also be calculated by the formula

$$\dim \mathscr{O}(F_i, G_i) = \dim \mathscr{O}(F_i) + \dim(Z(F_i)G_i), \tag{4.7}$$

where $\mathscr{O}(F_i)$ is the orbit of F_i under conjugation by $\mathscr{G}\ell(d_i, k)$ and $Z(F_i)$ is the centralizer of F_i in $\mathscr{G}\ell(d_i, k)$.

Note that each Jordan block F_i of F is nonderogatory matrix, since the superdiagonal of F_i consists entirely of 1's. Thus $\dim \mathscr{O}(F_i) = d_i^2 - d_i$, by Proposition 2.5 of [2]. Intuitively, a rough form of the formula obtained in [2] for $\dim \mathscr{O}(F_i)$ is

$$\dim \mathscr{O}(F_i) = n^2 - \dim Z(F_i). \tag{4.8}$$

By a classical theorem of Frobenius, $Z'(F_i) = k[F_i]$ (where $Z'(F_i)$ is the centralizer of F_i in the $d_i \times d_i$ matrices) if and only if F_i is nonderogatory and thus, by the Cayley–Hamilton theorem, $\dim Z'(F_i)$ achieves its minimum value d_i when F_i is nonderogatory. Moreover, using this result again, with the inclusion $Z(F_i) \subset Z'(F_i)$, one computes $\dim(k[F_i]G_i) = d_i$, by substituting $\dim \mathscr{O}(F_i)$ in (4.7) and combining with (4.6). But, to say $\dim(k[F_i]G_i) = d_i$ is to say that the matrix $(G_i, F_iG_i, \ldots, F^{d_i}G_i)$ has rank d_i; i.e., (F_i, G_i) is accessible. Finally, assume that (F, G) is not accessible; i.e., by definition, the image Q' of $\mathbf{G}_M$ is a proper (F-invariant) subspace of Q. However, $Q = \bigoplus Q_i$ with each Q_i being F-invariant and, since (F_i, G_i) is accessible, Q_i is the image of $\mathbf{G}_i$. Thus one may define an F-invariant complementary subspace for Q', say Q'', in Q. In particular, the projection P of Q onto Q'' centralizes F, while $PG = 0$. In other words, the tangent space to $\mathscr{O}(F, G)$ at (F, G), i.e., the image of $T_{Id}(\mathscr{G})$ under $d\sigma$, has dimension $\leqslant n^2 - 1$, contradicting $\dim \mathscr{O}(F, G) = n^2$. This proves the lemma.

Now, Mumford's theorem states that a universal geometric quotient $\varphi: X^s(L_1) \to Y$ exists for the action of $\mathscr{G}$ on $X^s(L_1)$, and that φ is an affine map. Moreover, the theorem asserts the existence of an ample line bundle $\mathscr{L}$ on Y such that $\varphi^*\mathscr{L} = L_0$. That is, some tensor power of this line bundle imbeds Y in a projective space, as a quasi-projective subvariety. Now since $X^s_{(0)}(L_1)$ is an invariant, Zariski, open subspace of $X^s(L_1)$ and since the quotient is geometric, the same remarks apply to the restriction φ_0 of φ

to $X^s_{(0)}(L_1)$; i.e.,

$$\varphi_0 : M_0 \to \Sigma_0$$

is a universal geometric quotient with φ_0 an affine map. Q.E.D.

Notice that equalities (4.5) exhibit M_0 not only as the set of properly stable points for the character γ_1, but also as the principal orbit type of the action of $\mathscr{G}$ on machines. Since a geometric quotient is categorical we have the following corollary.

COROLLARY 4.1. *Σ_0 inherits the following properties of M_0: it is reduced and irreducible (hence connected). In addition, Σ_0 is nonsingular of dimension nm.*

Proof. See GIT, paragraph (2), p. 5.

COROLLARY 4.2. *M_0 is an algebraic principal fiber bundle over Σ_0 with group $\mathscr{G}$.*

Proof. This is just Proposition 0.9 of GIT.

From Section 2, follows.

COROLLARY 4.3. *Σ_p is canonically an algebraic vector bundle over Σ_0.*

It is quite interesting to examine Mumford's proof (GIT, pp. 38–39) in this context. He first obtains quotients $(\Sigma_\alpha, \varphi_\alpha)$ of the affine prevarieties X_α, which exist by Mumford's theorem, in the affine case. For each pair of (nice) multi-indices (α, β) consider the function $f_{\alpha\beta} = f_\alpha / f_\beta$ on X_β. Since all the f_α's are invariants of the same weight, $f_{\alpha\beta}$ is an invariant element in $H^0(X_\beta, \mathscr{O}_X)$ and thus descends to a function $\varphi_{\beta\alpha}$ on Σ_β. One can construct, from the $\varphi_{\beta\alpha}$ and the uniqueness of the quotients $(\Sigma_\alpha, \varphi_\alpha)$, patching data giving the quotient $\Sigma_0 = (\bigcup_{\alpha \text{ nice}} \Sigma_\alpha)/\sim$. Furthermore, the family $(\varphi_{\beta\alpha})$ gives a 1-cocycle relative to the cover (Σ_α) of Σ_0, and thus defines a line bundle $\mathscr{L}$ on Σ_0, which Mumford shows to be ample. Now, first of all, consider the scalar input case. Here one only needs to deal with one X_α, in which case $\Sigma_\alpha \simeq k^n$. More generally, the Σ_α are precisely the open sets W_α, for α nice, used in Section 2. Of course, the $\varphi_{\beta\alpha}$ are the cocycles $\det R(F, G)_\alpha \cdot \det R(F, G)_\beta^{-1}$, given with respect to the cover (W_α). Equation (2.11), with $p = 1$, shows that these are exactly the cocycles for σ_1; i.e., $\mathscr{L} = \sigma_1$. Thus the imbedding of Σ_0 by sections of $\mathscr{L}$ may be taken to be the Kalman imbedding. Notice that this shows that $\mathscr{L}$ is very ample, i.e., $\mathscr{L}$ itself imbeds Σ_0 in $\mathbf{P}_\mathbf{N}$, $\mathbf{N} \gg \mathbf{0}$. These observations allow us to compute $\mathscr{P}ic(\Sigma_p)$.

THEOREM 4.2. $\mathscr{P}ic(\Sigma_0) \simeq \mathscr{P}ic^{\mathscr{G}}(M_0) \simeq \mathbf{Z}$, provided $m > 1$. Of course, in the scalar input case, $\mathscr{P}ic(\Sigma_0) = \{0\}$.

Proof. We already have a map $\varphi_0{}^*: \mathscr{P}ic(\Sigma_0) \to \mathscr{P}ic^{\mathscr{G}}(M_0)$, for which there is a natural left inverse δ. In this case, the construction of δ is quite simple, since $\mathscr{P}ic^{\mathscr{G}}(M_0) \simeq \hat{\mathscr{G}}$ and since $M_0 \to \Sigma_0$ is a principal $\mathscr{G}$-bundle. Explicitly, L_i descends as follows: since L_i corresponds to the pair (L_0, γ_i), let $\delta(L_i)$ be the line bundle associated to the principal bundle $\mathscr{G} \to M_0 \to \Sigma_0$ by the character γ_i. Of course, $\delta(L_i)$ is just $\sigma_i \in \mathscr{P}ic(\Sigma_0)$. A straightforward computation shows that δ is a left inverse for $\varphi_0{}^*$, thus δ is surjective. For $m > 1$, δ is injective as well, i.e., σ_p is very ample, since σ is very ample. Now supposing σ_p is trivial, the sections of the trivial line bundle imbed Σ_0 in $\mathbf{P_N}$, $\mathbf{N} \gg \mathbf{0}$. In particular, since the global sections of $\mathscr{O}_{\Sigma_0}$ are just the invariants in $H^0(M_0, \mathscr{O}_X)$, the functions χ_i must separate inequivalent machines, by Proposition 4.2. By Theorem 4.6 and Corollary 4.7 of [2], this is certainly the case if $m = 1$, implying $\delta = 0$. On the other hand if $m > 1$, it is not hard to construct inequivalent accessible machines $(F_1, G_1), (F_2, G_2)$ with $\chi_i(F_1) = \chi_i(F_2)$. We have already seen that this is true for prime machines, while the general case can be done, for example, by using the Minimal realization theorem. Q.E.D.

COROLLARY 4.4. *In any case,* $\mathscr{P}ic(\Sigma_0) = (\sigma_1)$.

Recalling that σ_1 is the restriction of $\det \mathbf{U_{n,(n+1)m}}$ and that the Kalman imbedding is by sections of σ_1, we have the following, for $m > 1$.

COROLLARY 4.5. *Each algebraic line bundle extends uniquely to a line bundle on* $\mathbf{Gr(n, (n+1)m}$. *In fact, the restriction map induces an isomorphism of the Picard groups. Similarly,* $\kappa^*: \mathscr{P}ic(\mathbf{P_k}) \simeq \mathscr{P}ic(\Sigma_0)$ *and* $p^*: \mathscr{P}ic(\mathbf{P_N}) \simeq \mathscr{P}ic(\Sigma_0)$.

COROLLARY 4.6. *The algebraic vector bundles* $\Sigma_p \to \Sigma_0$ *are nontrivial.*

COROLLARY 4.7. *There do not exist global algebraic canonical forms for machines* (F, G, H), *when* $m > 1$.[†]

REMARKS. 1. This was proved in [10], by showing, at least in a somewhat representative case, that $\det \mathbf{U}_{\mathbf{n},(\mathbf{n}+1)\mathbf{m}|\Sigma_0}$ is nontrivial, interpreting the existence of canonical forms in another way. Ironically, the nontriviality of

[†] *Added in Proof*. The authors have recently shown that $\sigma_1(\sigma_1) \neq 0$ and, hence, there do not exist continuous canonical forms (this is all for $m \neq 1$). Recently, M. Hazewinkel has shown, over $\mathbf{R}$, that $\omega_1(\sigma_1) \neq 0$ in mod 2 cohomology, by constructing a nontrivial element in $\pi_1(\Sigma_0)$—thus giving the same conclusion over $\mathbf{R}$.

$\det \mathbf{U}_{\mathbf{n},(\mathbf{n}+1)\mathbf{m}|\Sigma_0}$ also gives Corollary 4.5, at least for $p = 1$, while in [10] it was claimed that $\Sigma_p \simeq \Sigma_0 \times k^{pn}$.

2. These computations employ, indirectly, the fact that $\mathscr{Pic}(M_0) = (0)$. Now, it can be shown that the restriction map $\mathscr{Pic}(X) \to \mathscr{Pic}(M_0)$ is surjective, while it is well known that $\mathscr{Pic}(X) = (0)$. Furthermore, since M_0 is $\mathscr{G}$-invariant, this induces a surjection, $\hat{\mathscr{G}} = \mathscr{Pic}^{\mathscr{G}}(X) \to \mathscr{Pic}^{\mathscr{G}}(M_0)$. In terms of group cohomology, this is just the map $\hat{\mathscr{G}} = H^1(\mathscr{G}, \mathbf{C}^*) \to H^1(\mathscr{G}, \mathscr{O}^*_{M_0}) = \mathscr{Pic}^{\mathscr{G}}(M_0)$, induced by the inclusion of $\mathscr{G}$-modules $\mathbf{C}^* \to \mathscr{O}^*_{M_0}$, the group of invertible functions on M_0. By the Rigidity lemma this map on cohomology is surjective, while the content of Proposition 4.2 is the statement that this map is injective when $m > 1$. Of course, when $m = 1$, $\mathscr{Pic}^{\mathscr{G}}(M_0) = (0)$ since the unit f_α^l (see the proof of Theorem 4.1) is a coboundary of the character χ_l, regarded as a cocycle with values in $\mathscr{O}^*_{M_0}$. In particular, one always has an isomorphism $\delta: \mathscr{Pic}^{\mathscr{G}}(M_0) \to \mathscr{Pic}(\Sigma_0)$ (compare GIT, p. 32).

COROLLARY 4.8. $\mathscr{Pic}(\Sigma_p) \simeq \mathbf{Z}$, provided $m > 1$. In the other case, the Picard group is trivial.

Proof. Just as before, one can show that $\tau_p^*: \mathscr{Pic}(\Sigma_p) \simeq \mathscr{Pic}^{\mathscr{G}}(M_p) \simeq \mathbf{Z}$ has a left inverse δ_p. Thus, the only thing to check is nontriviality of $\delta_p(i)$. If $m > 1$, then $\tau_p^*(\sigma_i) = \delta_p(i)$. However, if $\tau_p^*(\sigma_i)$ were trivial, then its restriction σ_i to the image of the 0-section of $\Sigma_p \to \Sigma_0$, viz., Σ_0, would also be trivial, contrary to fact.

There are several extensions of Theorem 4.1 that are useful for applications. The first concerns the quotient $M_p \to \Sigma_p$. By definition, the universal geometric quotient $M_0 \to \Sigma_0$ has the following property: If $\Sigma' \to \Sigma_0$ is a morphism and if $M' = M_0 \times_{\Sigma_0} \Sigma'$, then the map $\mathrm{proj}_2: M' \to \Sigma'$ is a geometric quotient. If $\Sigma' = \Sigma_p$, then $M' = M_p$ and, thus, one has

PROPOSITION 4.3. *$M_p \to \Sigma_p$ is a geometric quotient. In fact, as before, M_p is an algebraic principal fiber bundle over Σ_p with group $\mathscr{G}$.*

Since the action of $\mathscr{G}$ on the direct part J of a machine is trivial, the moduli space for machines with a direct part is the space $\Sigma_p \times k^{pm}$ and thus may also be naturally regarded as a vector bundle on Σ_0. If M_{pm} denotes the set of machines with a direct part, then $M_{pm} = M_p \times_{\Sigma_p} (\Sigma_p \times k^{pm})$ and, again by universality, one has a geometric quotient

$$M_{pm} \to \Sigma_p \times k^{pm}.$$

Now, by duality, we also have a similar theorem for reduced machines. In this case, Σ_0^* is quasi-projective, nonsingular, and irreducible of dimension

np. Each $\Sigma_m{}^* \to \Sigma_0{}^*$ is a vector bundle of rank nm, obtained by considering the inputs to the machines in $M_0{}^*$. Finally, a geometric quotient $M_{mp}^* \to \Sigma_m{}^* \times k^{pm}$ exists for the action of $\mathscr{G}$ on reduced machines with a direct part. As before the local coordinates on this moduli space give canonical forms for reduced machines. Of course, these may not be taken to be global unless $p = 1$, since $\mathscr{P}ic(M_{mp}^*) \simeq \mathbf{Z}$.

Now, Hazewinkel and Kalman have shown that the moduli functor for families of accessible machines is representable. To be precise, they define a family of machines to be a vector bundle E of rank n over a prevariety S, equipped with an endomorphism $F \in \mathscr{H}om(E, E)$ and sections $g_1, \ldots, g_m \in H^0(S, E)$. The family $(E; F, g_1, \ldots, g_m)$ is accessible just in case the machine $(F(p), G(p))$ is accessible at each point p. As in the Grassmann case, pullbacks of such families exist and thus we have defined a functor $\mathscr{F}$ from the category of prevarieties over k to the category of sets, viz., $\mathscr{F}(S)$ is the set of isomorphism classes of accessible families of machines on S

$$\mathscr{F} : \mathscr{P}re(k) \to \text{Sets}.$$

Of course, its possible to construct a canonical element F_0 in $\mathscr{F}(\Sigma_0)$, which is referred to as the *universal family of machines on* Σ_0. One should note that the underlying bundle of F_0 is exactly the restriction of $\mathbf{U}_{\mathbf{n},(\mathbf{n}+1)\mathbf{m}}$ to Σ_0, viz., Σ_1, while the machine $(F(x), G(x))$ is just the system $x = (F, G)$. From this description, one should expect that $F_0 \to \Sigma_0$ is universal for $\mathscr{F}$, that is that the following theorem holds.

THEOREM 4.3 (Hazewinkel–Kalman). *$\mathscr{F}$ is representable. Explicitly, there is an isomorphism of functors $\mathscr{F} \simeq \mathscr{H}om(\cdot, \Sigma_0)$.* In particular, for each family $(E, F, g_1, \ldots, g_m)$ on a prevariety S, there is a map $\varphi: S \to \Sigma_0$ with $\varphi^* F_0 = (E; F, g_1, \ldots, g_m)$.

Proof. The proof is done in considerable detail in [10]. It does, however, rely on the fact that the quotient $M_0 \to \Sigma_0$ is geometric, a fact that is derived in [10] from the explicit relations defining Σ_0 in $\mathbf{Gr(n, (n + 1)m}$. Regardless of the status of these relations, in light of Theorem 4.1, their theorem and proof holds. Since we shall not use the following fact, we only remark that one may also prove this in the same spirit as Kleiman's proof of the universality of the (general) Grassmann (see [19, especially Lemma, p. 283]).

Note that, in the definition of a family, the matrix $G(p)$ is given by global sections, rather than as an endomorphism $G \in \mathscr{H}om(E', E)$, for some bundle E' of rank m. The more naive definition is unworkable, as we shall explain shortly, and the more "rigidfied" version is used to obtain representability of $\mathscr{F}$. This definition is at least suggested by the discussion of the similarity problem in Mumford and Suominen (see especially pp. 177–178). In fact, as observed in [2], in the scalar input case Theorem 4.3 actually follows from the proof, given by Mumford and Suominen (Proposition 3, p. 177) that the moduli functor for nonderogatory matrices is representable. As is suggested, things can go bad pretty easily with the other definition. For example, one classifies the family $F_0 \to \Sigma_0$ by the identity $i \in \mathscr{H}om_k(\Sigma_0, \Sigma_0)$. However, according to the naive definition one has a family $F_0 \otimes \sigma_1 = (\Sigma_1 \otimes \sigma_1; F \otimes id, G \otimes id)$ which, for each of the trivializing neighborhoods W_α, is also classified by the identity morphism; yet $F_0 \otimes \sigma_1 \not\approx F_0$, unless $m = 1$. This pathology is circumvented by insisting that G be given by global sections of the underlying bundle.

The restriction on k is unnecessary and was made only so that $\mathscr{G}$ acts reductively. In fact, Theorem 2.2 remains valid for any reductive subgroup of $\mathscr{G}$, e.g., compact subgroups or the classical groups. The proof of Theorem 4.1, however, requires some alteration. Along these lines of generalization, it is, of course, also of interest to obtain canonical forms for machines that are not necessarily reduced or accessible. This is closely related to the problem of deciding whether arbitrary machines are equivalent mod $\mathscr{G}$, or modulo some subgroup. This generalizes the conjugacy problem, say, for the classical groups. We refer the reader to [2] for a discussion of these problems in the scalar input case.

5. Global Analysis of Linear Machines. Applications

In the first part of this section, we shall use the structure of the quotient $M_p \to \Sigma_p$ to tackle Problem C, viz., which system-theoretic properties are generic? Of course, by "generic" is meant that the set of isomorphism classes of machines with this property is Zariski open in Σ_p, for the appropriate $p \geqslant 0$. By Theorem 4.2, a criterion for the genericity of a property is easy to obtain; i.e., one may describe the open sets in Σ_p as precisely those sets V for which $\varphi_p^{-1}(V)$ is an invariant Zariski open subspace of M_p, and conversely. Thus if $\tilde{X}_p = M_p \cap M_m{}^*$ is the set of minimal machines and if X_p denotes $\varphi_p(\tilde{X}_p)$, then the following Proposition holds.

Proposition 5.1. *X_p is a Zariski open subvariety of Σ_p. Furthermore, by duality, accessibility is a generic property of reduced machines.*

One may find a statement of this result, for the scalar input case in [2]. Such a theorem was also proved by Hermann in the scalar input–output case, but by a different method, relying heavily on the Minimal realization theorem.

COROLLARY 5.1. (Hermann). *X_1 is a Zariski open subspace of k^{2n}.*

Proof (See [11, especially Chapters 5, 7, and 8). Now in the scalar input–output case, the frequency response function is rational, in the usual sense; i.e.,

$$\Phi(s) = f(s)/\chi(f, s), \tag{5.1}$$

where $f(s) = H(\beta(s) \cdot G)$, with $\beta(s)$ denoting the classical adjoint of $(F - s)$. Moreover, $\deg f(s) \leqslant n - 1$, while $f(s)$ and $\chi(F, s)$ have no common zeros (see [11]). In this case, the discrete inverse scattering problem may be described as follows: given a rational function $f(s)/g(s)$ such that $g(s)$ is monic of degree n, $\deg f(s) \leqslant n - 1$, and $f(s), g(s)$ have no roots in common, find a minimal (n-dimensional) scalar input–output machine (F, G, H) having $f(s)/g(s)$ as its frequency response function. Of course, one may represent the set of such rational functions geometrically in the obvious way, i.e., taking the coefficients of g first and then those of f as a subset of k^{2n}. Since the resultant $\rho(f, g)$ vanishes just in case f and g have a common root, this subspace Z_1 is Zariski open. Assuming the Minimal realization theorem, one has a bijection of Z_1 with X_1. This is essentially Hermann's proof. Q.E.D.

By Proposition 4.2, we can construct a moduli space for $\tilde{X}_p$ under the relation of system-theoretic equivalence. This construction assumes, at least provisionally, the "fundamental theorem of linear realization theory," since it is in virtue of this theorem that the moduli problem may be put in canonical form. We shall give an independent proof of this presently, but for now we just state

COROLLARY 5.2. *$\tilde{X}_p \to X_p$ is a geometric quotient. X_p is a reduced, irreducible, nonsingular quasi-projective prevariety of dimension $n(m + p)$. In particular, a (coarse) moduli space for the set of minimal machines, under the relation of system-theoretic equivalence, exists.*

We should mention here some of the unpublished work of R. Brockett. When $k = \mathbf{R}$, Brockett has shown that the connected components of X_1 are products of circles with Euclidean spaces. A computation of the topological picard group would, clearly, refine Brockett's results.

Now Kalman [17] asked, in particular, whether cyclicity is a generic property. A machine is cyclic just in case it can be controlled through a single input terminal; i.e., (F, G), with $G = (g_1, \ldots, g_m)$, is cyclic provided g_i

is a cyclic vector for F for some i. Again, by Theorem 4.2 and the definition of geometric quotient, we easily have

PROPOSITION 5.2. *Cyclicity is a generic property for machines in M_p and for machines in M_m*. In particular, the geometric assertions made in Corollary 5.2 hold.*

That cyclicity is generic for reduced, scalar input systems was proved in [2], by the same technique. Of course, there is a dual notion concerning the observation of systems. The corresponding property, which we call *total observability*, is generic—by duality.

In the earlier constructions of canonical forms, less restrictive conditions than cyclicity were made on G. For example, Popov's conditions on systems was accessibility and rank $G = m$ [24]. By the same reasoning we have

PROPOSITION 5.3. *Popov's condition is a generic property, thus a moduli space $\mathscr{P}$ for systems $(F, G) \in \tilde{\mathscr{P}}$ exists and is an irreducible quasi-projective variety, nonsingular of dimension nm.*

In fact, the local coordinates on $\mathscr{P}$ may be taken to be the restriction of those on Σ_0. These canonical forms, together with the total Kronecker invariants defining those nice α such that $\mathscr{P}$ is covered by $\mathscr{P} \cap W_\alpha$ are exactly the canonical forms derived by Popov.

COROLLARY 5.3. (Popov [24, Theorem 1]). *The total Kronecker invariants and the coordinate functions $R(F, G)_\alpha^{-1} R(F, G)$ are a complete set of invariants for $(F, G) \in \tilde{\mathscr{P}}$.*

We give some nonexamples. If $m > 1$, then one easily constructs (by avoiding cyclic machines) a κ so that the Zariski closure of M_κ, the set of (F, G) having total Kronecher invariant κ, is a proper subset of X. If M_κ were open, then, by irreducibility of X, M_κ would be dense as well, contradicting $\mathscr{C}\ell(M_\kappa) \neq X$. This has an interesting interpretation concerning feedback control laws. That is, it was shown in [15] that the total Kronecker invariants are complete invariants for the action of the group $\mathfrak{G}$ generated by (a) $\mathscr{G}$, (b) change of basis in A, and (c) the substitution $F - GL$, L arbitrary.

PROPOSITION 5.4. *Linear feedback equivalence is not a generic property.*

Obviously, an important notion to consider is that of stability, which pertains to the qualitative study of the solutions to the differential equations (1.4). For this reason, we shall restrict the discussion to the case $k = \mathbf{C}$. Stability is, thus, intimately tied with an analysis of the frequency–response function and, for this reason, may also be defined for discrete-time systems.

We shall show here that several types of stability are not generic properties. However, one should remember that the pertubations we allow are rather general, since we deal with Zariski open sets. For continuous-time systems, asymptotic stability is characterized by the location of the spectrum of F. Explicitly, this condition is that all eigenvalues have negative real parts [23, p. 576, Theorem 1]. For accessible prime machines defined over $\mathbf{R}$, the so-called "domain of stability" is thus $\mathbf{R}^- \times \mathbf{P}_\mathbf{m}(\mathbf{R})$, while over $\mathbf{C}$ this is just $\mathbf{H}^- \times \mathbf{P}_\mathbf{m}(\mathbf{C})$, where $\mathbf{H}^-$ is the left half-plane. In general, the Routh–Hurwitz conditions (see [7]) defined the domain of stability by polynomial inequalities in $\chi(F, s)$ and, using these, one easily shows that asymptotic stability is not a generic property for continuous-time accessible machines. Asymptotic stability for discrete-time machines is characterized by the condition that all the eigenvalues of F lie within the unit circle [23] and thus, for prime machines fails to be generic. The same statement holds for arbitrary n, clearly. By duality, this holds also for reduced machines and, as usual, for the intersection—the minimal machines.

Now, on $\tilde{X}_p$ the eigenvalues of F are precisely the poles of the frequency–response function. The location of these poles serves to characterize yet another type of stability, referred to as BIBO stability by Padulo and Arbib [23]; i.e., bounded inputs imply bounded outputs. Thus we have already shown that this (analytic) property is not generic for minimal machines. Since, in general, BIBO stability is characterized by the condition that the poles of Φ lie in $\mathbf{H}^-$, one may likewise show that BIBO stability is not generic for either accessible or reduced machines. Summarizing these results, we have

PROPOSITION 5.5. *Asymptotic stability and BIBO stability are* not *generic properties of linear machines, in either M_p or $M_m{}^*$.*

We now turn to Problem D. A solution will appear, in all generality, in [3]. Here we give a solution for scalar input–output minimal machines. Since cyclicity and total observability are generic, this already represents a wide class of machines. For the remainder of this section, k will denote a complete topological field with a valuation. Now in Hermann's proof of Corollary 5.1, it was noted that Z_1—the set of abstract frequency–response functions—carries a natural algebraic structure, viz., $Z_1 \subset k^{2n}$ is Zariski open, and is thus an analytic manifold of dim $2n$. By our proof of Corollary 5.1, viz., Proposition 5.1, X_1 is also an analytic, open submanifold of $\Sigma_1 = k^{2n}$. To solve the realization problem is to construct a map $\Psi: Z_1 \to X_1$ inverting the map $\Phi: X_1 \to Z_1$. Since Φ is polynomial map, Φ is analytic.

THEOREM 5.1. *Φ is étale. In fact, Φ has a global inverse Ψ on Z_1.*

Proof. We claim that a somewhat lengthy and unpleasant computation (see [3]) shows that the Jacobian of Φ never vanishes on X_1. By the Inverse function theorem (see Serre [25, LG 2.10]), Φ is étale; i.e., Φ has a local analytic inverse. Surprisingly, an inspection of the Jacobian yields an inverse for Φ. This is computed for $z = (g_0, \ldots, g_{n-1}, f_1, \ldots, f_n)$, where

$$g(s) = (-s)^n + \sum_{i=1}^{n} g_i s^{n+1} \qquad \text{and} \qquad f(s) = \sum_{i=1}^{n} f_i s^{n-i}$$

as follows. First, $F = C(g)$, the companion matrix for g, and $G = e_1$, so that (F, G) is already in "control canonical form." The entries of $H = (h_1, \ldots, h_n)$ are obtained as the solutions of the triangular set of linear equations

$$(\alpha_{n-j}) \cdot {}^{t}H = f_j, \tag{5.2}$$

where the components of the vector (α_{n-j}) are given by

$$(\alpha_{n-j})_i = \begin{cases} (-1)^i g_{(n-j)+(i+1)}, & i < j, \\ (-1)^{2n+1-j}, & i = j, \\ 0 & 1 > j. \end{cases} \tag{5.3}$$

In fact, this defines a global algebraic map Ψ. Assuming the claim, the proof is ended. Q.E.D.

REMARKS. This gives a new proof of the Minimal realization theorem and the Fundamental theorem, at least for scalar input–output machines. Actually, the proof extends, since here again $X_p \subset \Sigma_p \simeq k^{(p+1)n}$ is an open submanifold. In the general case, this corresponds to proving the theorem on a trivializing neighborhood, say W_α, for the bundle $\Sigma_p \to \Sigma_0$—which is exactly the technique used in [3]. Of course, the machine constructed above must be congruent mod $\mathscr{G}$ to the machine one constructs via Ho's algorithm. However, our algorithm is certainly new and appears, at least to us, to be more intuitive. In this connection, one should also note that, in contrast, our construction does not assume the a priori existence of a machine realizing $z = (g, f)$ but, rather, proves this fact as well. Indeed, assuming z is the response function of a minimal machine, one may easily construct (in the scalar input–output case) the "controllable canonical form" (F_c, G_c, H_c) realizing z (see [23, Chapter 7.1]). This is simpler than our form, which fits in better geometrically. In fact, one cannot use the geometry at hand to show that $\Phi(F_c, G_c, H_c) = z$ (which is fairly easy to show in any case) and, moreover, this technique does not generalize. Finally, we remark that applying Corollary 2.1 gives the sharper form of the Fundamental theorem (see Section 1).

The last problem we shall discuss is the development of a description of X_1, suitable for topological computations. This also gives criteria for the reducibility of an accessible machine, in the scalar input–output case (these results extend, as well). Of course, one such criterion is given by duality and the Cayley–Hamilton theorem, viz., the nonvanishing of

$$D(F,H) = \det({}^t(HF), \ldots, {}^t(HF^{n-1})). \tag{5.4}$$

The variety Z_1 was also defined as the co-locus of the resultant $\rho(f,g)$, where $\Phi(F,G,H) = f(s)/g(s)$. Putting (F,G,H) in canonical form, it is not hard to show

$$D_\Phi(F,H) = \Phi(D(F,H)) = \rho(f,g), \tag{5.5}$$

using some of the basic properties of $\rho(f,g)$ (see Lang [20, especially pp. 135–138]) and the fact that D_Φ and ρ cut out the same locus. Since we only need that the nonvanishing of D_Φ is equivalent to the nonvanishing of ρ—a fact we have already proved—we shall skip the computation. In particular, we can make use of the various formulas for $\rho(f,g)$ to compute $D(F,H)$. Now, if (F,G,H) is in canonical form, $D(F,H)$ depends only on the invariants $\chi_i(F)$ and H, and thus, in a symmetric way, on the spectrum of F and on H. This is also useful, computationally, if the eigenvalues of F are known. Let s_i be the ith symmetric polynomial (notation as in [20]) in the variables $t_1, \ldots, t_n$ and define σ_{ij} to be $s_{(n-j)+(i+1)}$. Furthermore, for $j = 1, \ldots, n$, define the polynomials

$$P_j(t,H) = (-1)^{j-1}h_j + (-1)^n \sum_{i=1}^{j-1} \sigma_{ij}(t)h_i \tag{5.6}$$

and

$$f(t,H) = \sum_{j=1}^{n} P_j(t,H)t^j. \tag{5.7}$$

Then we have

COROLLARY 5.4. *An accessible machine α is reduced if and only if $f(\lambda, H) \neq 0$ for each $\lambda \in \operatorname{spec}(F)$.*

Proof. σ is reduced just in case $\rho(f,g) \neq 0$, where $\Phi(\sigma) = f(s)/g(s)$. Since g is monic, this may be computed as

$$\rho(f,g) = \prod_j f(\lambda_j), \qquad \lambda_j \in \operatorname{spec}(F).$$

However, using (5.2), $f(\lambda_j)$ may be computed by (5.7). Q.E.D.

We note that $\rho(f, g)$ is computable as a (probably) more simple expression using the "controllable canonical form." In either situation, one obvious instance for which the spectrum of F is known is when the machine is already in its parallel sum decomposition. Here, however, the degree of $f(t, H)$ goes down. Finally, the spectrum of F topologically parameterizes the coefficients of $\chi(F, s)$ and, thus, one way to compute topological properties of X_1 is by Corollary 5.4 or any other expressions for the nonvanishing of $\rho(f, g)$.

ACKNOWLEDGENTS

We would like to thank Professors Michael Arbib and Ernie Manes for showing us this problem and for their helpful conversations, which sparked our interest. In this connection, we thank Professor R. E. Kalman for his correspondence and for making available his manuscripts [16 and 17]. The first author would like to state his appreciation to Professor Alan J. Silberger for his kindness and his generosity in providing the opportunity to carry out much of his research in the last sections in the stimulating atmosphere afforded by the Institute for Advanced Study. Finally, we thank Professors Hans Fischer, John Fogarty, Mike Gauger, and Steve Gersten for their help and interest at various points and we take this opportunity to congratulate Hans Fischer on his twentieth mathematical anniversary.

REFERENCES

1. M. ARBIB AND E. G. MANES, Foundations of Systems Theory: Decomposable Systems, COINS Tech. Rep. 73B-3, Univ. of Massachusetts.
2. C. I. BYRNES AND M. GAUGER, Decideability criteria for the similarity problem, with applications to the moduli of linear dynamical Systems, *Advan. in Math.* July (1977).
3. C. I. BYRNES AND N. E. HURT, On the moduli of linear dynamical systems, II (to appear).
4. J. DIEUDONNÉ AND J. CARRELL, Invariant theory, old and new, *Advan. in Math.* **4** (1970), 1–80.
5. S. EILENBERG, "Automata, Languages, and Machines," Vol. A, Academic Press, New York, 1974.
6. J. FOGARTY, "Invariant Theory," Benjamin, New York, 1969.
7. F. R. GANTMACHER, "The Theory of Matrices," Vol. 2, Chelsea, New York, 1959.
8. P. GRIFFITHS AND J. ADAMS, "Topics in Algebraic and Analytic Geometry," Princeton Univ. Press, Princeton, New Jersey, 1974.
9. A. GROTHENDIECK, Techniques de Construction et Théoremes d'Existence, *in* "Géométrie Albégrique," Vol. III: Préschemas Quotients, Sém. Bourbaki, Exp. 212, Benjamin, New York, 1966.
10. M. HAZEWINKEL AND R. E. KALMAN, Moduli and Canonical Forms for Linear Dynamical Systems, Rep. 7504/M, Erasmus Univ. Rotterdam Econometrisch Inst., 1976.
11. R. HERMANN, "Linear Systems Theory and Introductory Algebraic Geometry," Vol. 8, Interdisciplinary Math., Brookline, Massachusetts, 1974.
12. F. HIRZEBRUCH, "Topological Methods in Algebraic Geometry," 3rd ed., Die Grund. der Math. Wissen., Band 131, Springer-Verlag, Berlin and New York, 1966.

13. J. HUMPHREYS, "Linear Algebraic Groups," GTM 21, Springer-Verlag, Berlin and New York, 1975.
14. *IEEE Trans. Automatic Control* **AC-19**, No. C (1974), Special issue on system identification and time series analysis.
15. R. E. KALMAN, Kronecker invariants and feedback, *Conf. Ordinary Differential Equations, June, 1971.*
16. R. E. KALMAN, Global Structure of Classes of Linear Dynamical Systems, Lectures, NATO Advan. Study Inst. on Geometric and Algebraic Methods for Non-linear Systems, London, August–September, 1971.
17. R. E. KALMAN, Algebraic geometric description of the class of linear systems of constant dimension, *Annu. Princeton Conf. Informat. Sci. Syst., Princeton, New Jersey, March 1974* (to appear).
18. R. E. KALMAN, P. L. FALB, AND M. A. ARBIB, "Topics in Mathematics Systems Theory," McGraw-Hill, New York, 1969.
19. S. KLEIMAN, "Geometry on Grassmanians and Application to Splitting Bundles and Smooth Cycles," IHES Publ. Math. No. 36, 1969. [Vol. dédié au Prof. O. Zariski.]
20. S. LANG, "Algebra," Addison-Wesley, Reading, Massachusetts, 1971.
21. D. MUMFORD, "Geometric Invariant Theory," Ergeb. Math. Bd 34, Springer-Verlag, Berlin and New York, 1965.
22. D. MUMFORD AND K. SUOMINEN, Introduction to the theory of moduli, *Proc. Nordic Summer School in Math., 5th, Oslo* pp. 171–222 (1970).
23. L. PADULO AND M. A. ARBIB, "System Theory," Saunders, Philadelphia, Pennsylvania, 1974.
24. V. M. POPOV, Invariant description of linear time-invariant controllable systems, *SIAM J. Control* **10** (1972), 252–264.
25. J-P. SERRE, "Lie Algebras and Lie Groups," Benjamin, New York, 1969.
26. R. O. WELLS, Jr., "Differential Analysis on Complex Manifolds," Prentice–Hall, Englewood Cliffs, New Jersey, 1973.

AMS (MOS) subject classifications: primary 93B05, 93B10, 93B25, 14D20; secondary, 93BXX, 57E99.

STUDIES IN ANALYSIS
ADVANCES IN MATHEMATICS SUPPLEMENTARY STUDIES, VOL. 4

Hypoelliptic Second-Order Differential Operators with Complex Coefficients

KAZUHIRO YAMAMOTO

Department of Mathematics
Hokkaido University
Sapporo, Japan

1. INTRODUCTION

In this paper we shall prove hypoellipticity for some class of second-order differential operators with complex coefficients, which contains the operators considered in [2] and [5]. In [2], L. Hörmander gave a sufficient condition for hypoellipticity of second-order differential operators whose principal parts are written as the summation of square operators of a vector field. Furthermore, Olenik and Radkevič [5] treated more general real coefficient, second-order differential operators having the positive semi-definite principal part. In their papers it seemed that the assumption that operators are real played a crucial role. However, in the present paper we shall study operators with complex coefficients by an analogous method of the real case.

Let Ω be an open set of R^n and $L(x, D)$ a second-order differential operator

$$L(x, D) = X_\alpha(x, D)a^{\alpha\beta}(x)X_\beta(x, D) + X_0(x, D) + b(x), \tag{1.1}$$

where $a^{\alpha\beta}(x)$, $b(x) \in C^\infty(\Omega)$, $a^{\alpha\beta}(x) = a^{\beta\alpha}(x)$, $\alpha, \beta = 1, \ldots, N$, $D = \partial/i\,\partial x$, and $X_\alpha(x, iD)$, $\alpha = 1, \ldots, N$, $(\operatorname{Re} X_0)(x, iD)$, and $(\operatorname{Im} X_0)(x, iD)$ are vector fields over Ω. In (1.1), as well as the rest of this paper, we use the following convention of summation:

$$\sum_{\alpha,\beta=1}^{N} A^{\alpha\beta}a_\alpha b_\beta = A^{\alpha\beta}a_\alpha b_\beta \qquad \text{and} \qquad \sum_{\alpha=1}^{N} A^\alpha a_\alpha = A^\alpha a_\alpha,$$

where $A^{\alpha\beta}, A^\alpha, a_\alpha$, and b_β are functions or operators over $C^\infty(\Omega)$.

We shall define properly supported pseudodifferential operators $Y_k(x, D)$, $k = 0, \ldots, N + n$, having the following symbols:

$$Y_0(x, \xi) = (\operatorname{Im} X_0)(x, \xi), \qquad Y_\alpha(x, \xi) = (\operatorname{Re} a^{\alpha\beta})(x)X_\beta(x, \xi),$$
$$Y_{N+j}(x, \xi) = |\xi|^{-1}\,\partial(\operatorname{Re} a^{\alpha\beta})/\partial x_j(x)X_\alpha X_\beta(x, \xi),$$

ISBN 0-12-599150-9

where $\alpha = 1, \ldots, N$ and $j = 1, \ldots, n$. For a sequence $I = (i_1, \ldots, i_s)$, $0 \leqslant i_k \leqslant N + n$, with the length $|I| = s$, we denote the repeated commutator $[Y_{i_1}, [Y_{i_2}, \ldots, [Y_{i_{s-1}}, Y_{i_s}] \cdots]](x, D)$ by $Y_I(x, D)$.

DEFINITION. A second-order differential operator $L(x, D)$ with form (1.1) is of *rank n* over Ω if for any compact subset K of Ω there exists a positive integer $I(K)$ such that

$$\sum_{|I| \leqslant I(k)} |Y_I(x, \xi)|^2 \geqslant C_K |\xi|^2$$

for all $(x, \xi) \in K \times \{|\xi| \geqslant 1\}$. Here $Y_I(x, \xi)$ is the principal symbol of $Y_I(x, D)$.

We impose the following two assumptions on $L(x, D)$:

(A.1) For any compact set K of Ω there exists a proper, closed, convex cone $\Gamma_K \subset \{z \in C; \operatorname{Re} z > 0\} \cup \{0\}$ such that

$$\{(A(x)\eta, \eta) = a^{\alpha\beta}(x)\eta_\alpha \eta_\beta;\ (x, \eta) \in K \times R^N\} \subset \Gamma_K,$$

(A.2) For any compact set K of Ω there exists a constant C_K such that

$$|(\operatorname{Re} X_0)(x, \xi)|^2 \leqslant C_K (\operatorname{Re} a^{\alpha\beta})(x) X_\alpha X_\beta(x, \xi), \tag{1.2}$$

where $(x, \xi) \in K \times R^N$.

Under these two conditions, we get the following.

THEOREM S. *Let $L(x, D)$ be a second-order differential operator with form* (1, 1), *which satisfies assumptions* (A.1) *and* (A.2). *Then if $L(x, D)$ is of rank n over Ω, $L(x, D)$ is hypoelliptic, i.e., for $f, u \in \mathscr{D}'(\Omega)$ if $L(x, D)u = f$, then*

$$\operatorname{sing\,supp} u = \operatorname{sing\,supp} f.$$

As a consequence of our theorem, we obtain the following generalized result of that in [2].

COROLLARY 1. *Let $L(x, D)$ be a differential operator such that*

$$L(x, D) = \sum_{\alpha=1}^{N} X_\alpha{}^2(x, D) + ib^{\alpha\beta}(x) X_\alpha X_\beta(x, D)$$
$$+ iX_0(x, D) + b^\alpha(x) X_\alpha(x, D) + c(x),$$

where $X_\alpha(x, iD)$, $\alpha = 0, 1, \ldots, N$, are vector fields over Ω and $b^{\alpha\beta}$ and b^α, $\alpha, \beta = 1, \ldots, N$, are arbitrary, real-valued $C^\infty(\Omega)$ functions. Assume that *among the repeated commutators* iX_{α_1}, $[iX_{\alpha_1}, iX_{\alpha_2}]$,

$$[iX_{\alpha_1},[iX_{\alpha_2},iX_{\alpha_3}]],\ldots,[iX_{\alpha_1},[iX_{\alpha_2},[iX_{\alpha_3},\ldots,iX_{\alpha_k}]\cdots]],\cdots,$$

where $\alpha_l = 0, \ldots, N$, *there exist* n *which are linearly independent at any point in* Ω. *Then it follows that* $L(x, D)$ *is hypoelliptic.*

Clearly if $N = n$ and $X_\alpha(x, D) = \partial/i\,\partial x_\alpha$, then Theorem S is a general result of that in [5].

When $X_\alpha(x, D)$, $\alpha = 0, \ldots, N$, and properly supported, first-order, pseudo-differential operators with the real homogeneous symbols $X_\alpha(x, \xi)$ when $\alpha \geqslant 1$, then we also get the same theorem. However, in this case the following imposed condition on $L(x, D)$ instead of (1.2) is more tight. The imaginary part of the symbol of $X_0(x, D)$ is written by the following form:

$$\begin{aligned}(\operatorname{Im} X_0)(x,\xi) = {}& T_\alpha{}^0(x,\xi)(\operatorname{Re} a^{\alpha\beta})(x)X_\beta(x,\xi) + S_\alpha{}^0(x,\xi)(\operatorname{Im} a^{\alpha\beta})(x)X_\beta(x,\xi)\\ &+ \sum_{j=1}^N T_j^{-1}(x,\xi)(\partial(\operatorname{Re} a^{\alpha\beta})/\partial x_j)(x)X_\alpha X_\beta(x,\xi)\\ &+ \sum_{j=1}^N S_j^{-1}(x,\xi)(\partial(\operatorname{Im} a^{\alpha\beta})/\partial x_j)(x)X_\alpha X_\beta(x,\xi), \end{aligned} \tag{1.2$'$}$$

where $T_\alpha{}^0(x, \xi)$, $S_\alpha{}^0(x, \xi)$, $\alpha = 1, \ldots, N$, *and* $T_j^{-1}(x, \xi)$, $S_j^{-1}(x, \xi)$, $j = 1, \ldots, n$, are arbitrary, real, and positively homogeneous functions of degree 0 and -1, respectively.

By a convexity property for derivatives of a nonnegative function, we see that condition (1, 2)′ implies (1.2) and (1.2) is equivalent to a condition such that (1.2) holds for $(\operatorname{Re} L_1{}^s)(x, \xi)$ instead of $(\operatorname{Re} X_0)(x, \xi)$, where $L_1{}^s(x, \xi)$ is the subprincipal symbol of $L(x, D)$. Moreover if we define $Y_0(x, \xi) = (\operatorname{Im} L_1{}^s)(x, \xi)$, there is no influence on Theorem S.

The method of the proof of Theorem S employed here is that of [4], in which J. J. Kohn gave a simple proof of theorem in [2] by localized energy estimates of $L(x, D)$. In the final part of this paper we shall give a different generalized theorem of that in [2].

2. Preliminaries of the Proof of Theorem S

In this section under assumptions (A.1) and (A.2), we shall establish some inequalities.

First we have the following

LEMMA 2.1 *Let K be a compact subset of Ω and $u_\alpha(x)$, $v_\beta(x)$, $\alpha, \beta = 1, \ldots, N$, elements of $C_0^\infty(K)$. We assume that the symmetric matrix $A(x) = (a^{\alpha\beta}(x))_{\alpha, \beta=1,\ldots,N}$ satisfies assumption* (A.1). *Then we have*

$$|(a^{\alpha\beta}u_\alpha, v_\beta)| \leqslant C\{((\operatorname{Re} a^{\alpha\beta})u_\alpha, u_\beta) + ((\operatorname{Re} a^{\alpha\beta})v_\alpha, v_\beta)\},$$

where (,) is the standard inner product of $L^2(\Omega)$, and $C = C(K)$ does not depend on $u_\alpha(x)$ and $v_\beta(x)$.

Proof. Let η, ζ be elements of C^N. Then it is sufficient to show that for $x \in K$

$$|(A(x)\eta, \zeta)| \leqslant C\{((\operatorname{Re} A)(x)\eta, \eta) + ((\operatorname{Re} A)(x)\zeta, \zeta)\}, \tag{2.1}$$

where (,) is the standard inner product of C^N. Since

$$|(A(x)\eta, \zeta)| \leqslant |((\operatorname{Re} A)(x)\eta, \zeta)| \\ + |((C_K \operatorname{Re} A + \operatorname{Im} A)(x)\eta, \zeta)| + |((C_K \operatorname{Re} A - \operatorname{Im} A)(x)\eta, \zeta)|,$$

where, by (A.1), $C_K \operatorname{Re} A \pm \operatorname{Im} A \geqslant 0$ for $x \in K$, we have inequality (2.1) by Schwarz's inequality and the inequality $2xy \leqslant x^2 + y^2$, $x, y \geqslant 0$. This completes the proof of Lemma 2.1.

The following convex property for the derivatives of a nonnegative function is well known (cf. Lemma 1.7.1 in [5]).

LEMMA 2.2. *Let g be a nonnegative function belonging to $C^\infty(R^n)$ and K a compact set of R^n. Then there exists a constant C such that*

$$|\operatorname{grad}_x g(x)|^2 \leqslant Cg(x), \qquad x \in K,$$

where C depends on K and the second derivatives of $g(x)$.

Let $\tilde{Y}_\alpha(x, D)$, $\alpha = 1, \ldots, N$, and $\tilde{Y}_{N+j}(x, D)$, $j = 1, \ldots, n$, be properly supported pseudodifferential operators defined by the symbols $a^{\alpha\beta}(x)X_\beta(x, \xi)$ and $|\xi|^{-1}(\partial a^{\alpha\beta}/\partial x_j)(x)X_\alpha X_\beta(x, \xi)$, respectively. Then we have the following.

LEMMA 2.3. *The following statements hold.*

(i) *For any properly supported pseudodifferential operator Λ^s of order s, there exist properly supported pseudodifferential operators $\Lambda_k{}^s(x, D)$, $k = 0$,*

$1, \ldots, N+n$, *of order* s *such that*

$$[L, \Lambda^s](x, D) = \sum_{k=1}^{N+n} (\Lambda_k{}^s \tilde{Y}_k)(x, D) + \Lambda_0{}^s(x, D). \tag{2.2}$$

(ii) *There exist properly supported pseudodifferential operators* $\Lambda_\gamma{}^0(x, D)$, $\gamma = 0, \ldots, N$, *of order* 0 *such that for any compact set* K *of* Ω *and* $u \in C_0{}^\infty(K)$ *we have*

$$\sum_{k=1}^{N+n} (\|Y_k u\|^2 + \|\tilde{Y}_k u\|^2) + ((\operatorname{Re} a^{\alpha\beta}) X_\beta u, X_\alpha u)$$
$$\leqslant C\left\{\sum_{\gamma=0}^{N} |\operatorname{Re}(L\Lambda_\gamma{}^0 u, \Lambda_\gamma{}^0 u)| + \|u\|^2\right\}, \tag{2.3}$$

$$\sum_{k=1}^{N+n} (\|Y_k u\|^2 + \|\tilde{Y}_k u\|^2) \leqslant C\{\|Lu\|^2 + \|u\|^2\}, \tag{2.4}$$

where C *does not depend on* $u(x)$.

(iii) *For any* $s \in R$ *and for any properly supported pseudodifferential operator* $\Lambda^s(x, D)$ *of order* s *we have*

$$\|[L, \Lambda^s]u\|_{-s} \leqslant C(\|Lu\| + \|u\|), \tag{2.5}$$

where $u \in C_0{}^\infty(K)$.

Proof. Since

$$[X_\alpha a^{\alpha\beta} X_\beta, \Lambda^s] = [X_\alpha, \Lambda^s] a^{\alpha\beta} X_\beta + X_\alpha [a^{\alpha\beta}, \Lambda^s] X_\beta + X_\alpha a^{\alpha\beta} [X_\beta, \Lambda^s],$$

(2.2) is clear. To prove (2.3) first we shall compute $\sum_{\alpha=1}^N (\|Y_\alpha u\|^2 + \|\tilde{Y}_\alpha u\|^2)$. Applying Lemma 2.2 to $(\operatorname{Re} a^{\alpha\beta})(x)\eta_\alpha\eta_\beta$, $(C_K \operatorname{Re} a^{\alpha\beta} + \operatorname{Im} a^{\alpha\beta})(x)\eta_\alpha\eta_\beta$ and $(C_K \operatorname{Re} a^{\alpha\beta} - \operatorname{Im} a^{\alpha\beta})(x)\eta_\alpha\eta_\beta$, where $(C_K \operatorname{Re} A \pm \operatorname{Im} A)(x) \geqslant 0$ for $x \in K$, we have the following:

$$\sum_{\alpha=1}^{N} (\|Y_\alpha(x, D)u\|^2 + \|\tilde{Y}_\alpha(x, D)u\|^2) \leqslant C((\operatorname{Re} a^{\alpha\beta}) X_\alpha u, X_\beta u), \tag{2.6}$$

where $u \in C_0{}^\infty(K)$. By (1.1) we see that

$$((\operatorname{Re} a^{\alpha\beta}) X_\beta u, X_\alpha u) = \operatorname{Re}(Lu, u) - \operatorname{Re}(X_0 u, u)$$
$$+ \sum_{\alpha=1}^{N} \operatorname{Re}(\tilde{Y}_\alpha u, d_\alpha u) + \operatorname{Re}(d_0 u, u), \tag{2.7}$$

where $d_\alpha(x)$, $\alpha = 0, 1, \ldots, N$, are elements of $C^\infty(\Omega)$. Therefore by $\mathrm{Re}(i(\mathrm{Im}\, X_0)u, u) = O(\|u\|^2)$, (1.2), and (2.6) we obtain

$$((\mathrm{Re}\, a^{\alpha\beta})X_\alpha u, X_\beta u) \leqslant C\{|\mathrm{Re}(Lu, u)| + \|u\|^2\}. \tag{2.8}$$

Second we shall compute $\sum_{j=1}^n (\|Y_{N+j}u\|^2 + \|\tilde{Y}_{N+j}u\|^2)$. Under assumption (A.1) we have the following inequality:

$$\sum_{j=1}^{\eta} \{|(\partial a^{\alpha\beta}/\partial x_j)X_\alpha X_\beta v|^2 + |(\mathrm{Re}\,\partial a^{\alpha\beta}/\partial x_j)X_\alpha X_\beta v|^2\}$$
$$\leqslant C\left\{\sum_{\gamma=1}^{N} ((\mathrm{Re}\, a^{\alpha\beta})X_\alpha X_\gamma v, X_\beta X_\gamma v) + \sum_{\alpha,\beta=1}^{N} |[X_\alpha, X_\beta]v|^2\right\}, \tag{2.9}$$

where $v \in C_0^\infty(K)$ and $(\,,)$ is the standard inner product of C. When $n = N$, $X_\alpha = \partial/\partial x_\alpha$, and $A(x) \geqslant 0$, inequality (2.9) was proved in Lemma 1.7.1 of [5]. For our operators $A(x)$ and $X_\alpha(x, D)$ the proof of (2.9) is similar to that of Lemma 1.7.1 of [5]. Thus using (2.8) and (2.9) we have

$$\sum_{j=1}^{n} (\|\partial a^{\alpha\beta}/\partial x^j)X_\alpha X_\beta v\|^2 + \|(\mathrm{Re}\,\partial a^{\alpha\beta}/\partial x_j)X_\alpha X_\beta v\|^2)$$
$$\leqslant C\left\{\sum_{\gamma=1}^{N} |\mathrm{Re}(LX_\gamma v, X_\gamma v)| + \|v\|_1{}^2\right\}. \tag{2.10}$$

In (2.10) if we put $v = \Lambda^{-1}u$, where $\Lambda^{-1}(D)$ is a properly supported pseudodifferential operator having the symbol $|\xi|^{-1}$, then we have

$$\sum_{j=1}^{n} \{\|Y_{N+j}u\|^2 + \|\tilde{Y}_{N+j}u\|^2\} \leqslant C\left\{\sum_{\gamma=1}^{N} |\mathrm{Re}(LX_\gamma\Lambda^{-1}u, X_\gamma\Lambda^{-1}u)| + \|u\|^2\right\}.$$

This completes the proof of (2.3).

To prove (2.4) it is sufficient to estimate $(L\Lambda_\gamma{}^0u, \Lambda_\gamma{}^0u)$, $\gamma = 0, \ldots, N$. By (2.2) we have

$$(L\Lambda_\gamma{}^0u, \Lambda_\gamma{}^0u) = (\Lambda_\gamma{}^0Lu, \Lambda_\gamma{}^0u) + \sum_{k=1}^{N+n} (\Lambda_k{}^0\tilde{Y}_ku, \Lambda_\gamma{}^0u) + (\Lambda_0{}^0u, \Lambda_\gamma{}^0u),$$

where $\Lambda_k{}^0(x, D)$, $k = 0, \ldots, N + n$, are properly supported pseudodifferential operators of order 0. Therefore we see that

$$|\mathrm{Re}(L\Lambda_\gamma{}^0u, \Lambda_\gamma{}^0u)| \leqslant C_\delta(\|Lu\|^2 + \|u\|^2) + \delta \sum_{k=1}^{N+n} \|\tilde{Y}_ku\|^2,$$

which implies inequality (2.4) if δ is sufficiently small.

Inequality (2.5) is easily proved by (2.2) and (2.3). This completes the proof of Lemma 2.3.

3. Energy Estimate for $L(x, D)$ and the Proof of Theorem S

In this section we shall show an energy estimate for $L(x,D)$ under the condition that $L(x,D)$ is of rank n over Ω.

First we shall prove the following main inequality.

Proposition 3.1. *Let $L(x,D)$ be a second-order differential operator. We assume that $L(x,D)$ satisfies the assumptions of Theorem S. Then for any compact set K of Ω there exist a real number $\epsilon = \epsilon(K)$, $0 < \epsilon \leqslant \frac{1}{2}$ and a constant $C = C(K)$ such that*

$$\|u\|_\epsilon \leqslant C\{\|L(x,D)u\| + \|u\|\}, \qquad u \in C_0^\infty(K). \tag{3.1}$$

Proof. Since $L(x,D)$ is of rank n over Ω, if we consider the restriction operator of Y_I to a neighborhood of K, then by a strong Gårding inequality (cf., Theorem 1.3.3 of [3]) we have

$$\|u\|_\epsilon \leqslant C\left\{\sum_{|I| \leqslant I(K)} \|Y_I u\|_{\epsilon-1} + \|u\|\right\}, \qquad u \in C_0^\infty(K). \tag{3.2}$$

We have

$$\begin{aligned}\|Y_I(x,D)u\|_{\epsilon-1}^2 &= (Y_I u, \Lambda^{2\epsilon-1}u)\\ &= (Y_{i_1} Y_{I'} u, \Lambda^{2\epsilon-1}u) - (Y_{I'} Y_{i_1} u, \Lambda^{2\epsilon-1}u),\end{aligned}$$

where $I' = (i_2, \ldots, i_s)$ and $\Lambda^{2\epsilon-1}(x,D)$ is some properly supported pseudodifferential operator of order $2\epsilon - 1$. For simplicity we use the following notation: If the absolute value of a complex number A is dominated by $N_\epsilon(u) = \|Lu\|^2 + \|u\|^2 + \|Y_{I'}u\|_{4\epsilon-1}^2$, then we denote A by $O(N_\epsilon(u))$. Moreover, throughout this section, we mean that $\Lambda^s(x,D)$ is some properly supported pseudodifferential operator of order s.

Case I: $Y_{i_1} = Y_k$ with $1 \leqslant k \leqslant N + n$. Since the principal symbol of Y_{i_1} is real and that of $Y_{I'}$ is real or pure imaginary, we have

$$\begin{aligned}(Y_{i_1} Y_{I'} u, \Lambda^{2\epsilon-1}u) &= (\Lambda^{2\epsilon-1} Y_{I'} u, Y_{i_1} u) + (\Lambda^{2\epsilon-1} Y_{I'} u, u),\\ (Y_{I'} Y_{i_1} u, \Lambda^{2\epsilon-1}u) &= (Y_{i_1} u, \Lambda^{2\epsilon-1} Y_{I'} u) + (Y_{i_1} u, \Lambda^{2\epsilon-1}u).\end{aligned}$$

So we obtain, in case $Y_{i_1} = Y_k$ with $1 \leqslant k \leqslant N + n$, that $\|Y_I u\|_{\epsilon-1}^2$ is $O(N_\epsilon(u))$.

Case II: $Y_{i_1} = Y_0$. First we write

$$Y_0(x,D) = iL^* - iX_\alpha \overline{a}^{\alpha\beta} X_\beta + \Lambda_\alpha{}^0 \overline{a}^{\alpha\beta} X_\beta - i(\operatorname{Re} X_0) + \Lambda^0.$$

Thus by (A.2) and (2.4), we have

$$(Y_0 Y_{I'} u, \Lambda^{2\epsilon-1}u) = i(L^* Y_{I'} u, \Lambda^{2\epsilon-1}u) - i(X_\alpha \overline{a}^{\alpha\beta} X_\beta Y_{I'} u, \Lambda^{2\epsilon-1}u) + O(N_\epsilon(u)). \tag{3.3}$$

The first term of (3.3) is equal to

$$(Y_{I'}u, \Lambda^{2\epsilon-1}Lu) + (Y_{I'}u, [L, \Lambda^{2\epsilon-1}]u),$$

which is $O(N_\epsilon(u))$ by (2.5). Since $(X_\alpha \overline{a}^{\alpha\beta} X_\beta)^* = X_\alpha a^{\alpha\beta} X_\beta + \sum_{\alpha=1}^{N} \Lambda_\alpha{}^0 \tilde{Y}_\alpha + \Lambda^0$, the second term of (3.3) is equal to

$$\begin{aligned}
&(Y_{I'}u, X_\alpha a^{\alpha\beta} X_\beta \Lambda^{2\epsilon-1}u) + \sum_{\alpha=1}^{N} (Y_{I'}u, \Lambda_\alpha{}^0 \tilde{Y}_\alpha \Lambda^{2\epsilon-1}u) + O(N_\epsilon(u)) \\
&\quad = (\Lambda^{2\epsilon-1} Y_{I'}u, X_\alpha a^{\alpha\beta} X_\beta u) + (Y_{I'}u, [X_\alpha a^{\alpha\beta} X_\beta, \Lambda^{2\epsilon-1}]u) \\
&\qquad + \sum_{\alpha=1}^{N} (\Lambda_\alpha^{2\epsilon-1} Y_{I'}u, \tilde{Y}_\alpha u) + O(N_\epsilon(u)) \\
&\quad = (X_\alpha \Lambda^{2\epsilon-1} Y_{I'}u, a^{\alpha\beta} X_\beta u) + O(N_\epsilon(u)).
\end{aligned} \tag{3.4}$$

Here we use (2.2), (2.4), and (2.5). By Lemma 2.1 and (2.3) we have

$$\begin{aligned}
|(a^{\alpha\beta} X_\beta u, X_\alpha \Lambda^{2\epsilon-1} Y_{I'}u)| &\leqslant C\{((\operatorname{Re} a^{\alpha\beta}) X_\beta u, X_\alpha u) \\
&\qquad + ((\operatorname{Re} a^{\alpha\beta}) X_\beta \Lambda^{2\epsilon-1} Y_{I'}u, X_\alpha \Lambda^{2\epsilon-1} Y_{I'}u)\} \\
&\leqslant C\left\{\sum_{\gamma=0}^{N} |\operatorname{Re}(L\Lambda_\gamma^{2\epsilon}u, \Lambda_\gamma^{2\epsilon-1} Y_{I'}u)|\right\} + O(N_\epsilon(u)) \\
&\leqslant C \sum_{\gamma=0}^{N} \{|(Lu, \Lambda_\gamma^{4\epsilon-1} Y_{I'}u) \\
&\qquad + |([L, \Lambda_\gamma^{2\epsilon}]u, \Lambda_\gamma^{2\epsilon-1} Y_{I'}u)|\} + O(N_\epsilon(u)).
\end{aligned} \tag{3.5}$$

So by (2.5), (3.4), and (3.5) we obtain that $|(Y_0 Y_{I'}u, \Lambda^{2\epsilon-1}u)|$ is $O(N_\epsilon(u))$.

To deal with $(Y_{I'} Y_0 u, \Lambda^{2\epsilon-1}u)$ we write

$$Y_0(x, D) = -iL + iX_\alpha a^{\alpha\beta} X_\beta + i(\operatorname{Re} X_0) + \Lambda^0, \tag{3.6}$$

Thus we have

$$\begin{aligned}
(Y_{I'} Y_0 u, \Lambda^{2\epsilon-1}u) &= -i(Y_{I'} Lu, \Lambda^{2\epsilon-1}u) + i(Y_{I'} X_\alpha a^{\alpha\beta} X_\beta u, \Lambda^{2\epsilon-1}u) + O(N_\epsilon(u)) \\
&= (X_\alpha a^{\alpha\beta} X_\beta u, \Lambda^{2\epsilon-1} Y_{I'}u) + (X_\alpha a^{\alpha\beta} X_\beta u, \Lambda^{2\epsilon-1}u) + O(N_\epsilon(u)).
\end{aligned}$$

By the same computation in (3.4) and (3.5), we see that $(Y_{I'} Y_0 u, \Lambda^{2\epsilon-1}u) = O(N_\epsilon(u))$.

Thus for any $Y_{i_1}(x, D)$ we have

$$\|Y_I u\|_{\epsilon-1} \leqslant C\{\|Y_{I'}u\|_{4\epsilon-1} + \|Lu\| + \|u\|\}.$$

Inductively we then obtain

$$\|Y_I u\|_{\epsilon-1} \leqslant C\left\{\sum_{k=0}^{N+n} \|Y_k u\|_{4^{I(K)}\epsilon-1} + \|Lu\| + \|u\|\right\}.$$

Now we claim that

$$\|Y_k u\|_{-1/2} \leqslant C\{\|Lu\| + \|u\|\}. \tag{3.7}$$

If $1 \leqslant k \leqslant N + n$, then (3.7) is a consequence of (2.4). When $k = 0$, by using (3.6), Lemma 1.1, and (A.2), we have

$$\|Y_0 u\|^2_{-1/2} \leqslant |(Y_0 u, \Lambda^0 u)| \leqslant C\{\|Lu\|^2 + \|u\|^2\}.$$

Thus the desired estimate (3.1) is obtained for any $\epsilon \leqslant 2/4^{I(K)}$. This completes the proof of Proposition 3.1.

The following proposition is easily verified by Proposition 3.1, (2.3), and (2.4).

PROPOSITION 3.2. *For any $s \in R$ and for any compact subset K of Ω, there exists a constant $C = C(s, K)$ such that*

$$\|u\|_{s+\epsilon} \leqslant C\{\|Lu\|_s + \|u\|_s\}, \qquad u \in C_0^{\infty}(K). \tag{3.8}$$

Now we shall prove the localized estimate of (3.8). Let $\chi(x)$ and $\chi_1(x)$ be elements of $C_0^{\infty}(\Omega)$ such that $\chi_1 \equiv 1$ on $\operatorname{supp}\chi$. Then we denote this relation between χ and χ_1 by $\chi \Subset \chi_1$.

PROPOSITION 3.3. *For any $\chi(x)$, $\chi_1(x)$ with $\chi \Subset \chi_1$ and for any $s \in R$ and $N > 0$ there exist $\chi_2(x) \in C_0^{\infty}(\Omega)$ and a constant $C = C(\chi, \chi_1, s, N)$ such that*

$$\|\chi u\|_{s+\epsilon} \leqslant C\{\|\chi_1 Lu\|_s + \|\chi_1 u\|_s + \|\chi_2 u\|_{-N}\}$$

for all $u \in C^{\infty}(\Omega)$, where χ_2 is independent of N.

Proof. Using Proposition 3.2, we have

$$\|\chi u\|_{s+\epsilon} \leqslant C\{\|\chi Lu\|_s + \|[L, \chi]u\|_s + \|\chi u\|_s\}.$$

Since $[L, \chi] = \sum_{\alpha=1}^{N} a_\alpha(x)\tilde{Y}_\alpha(x, D) + a_0(x)$, where $\operatorname{supp} a_\alpha \subset \operatorname{supp}\chi$, $\alpha = 0, \ldots, N$, it is sufficient to show the following:

$$\sum_{k=1}^{N+n} \|\chi' \tilde{Y}_k u\|_s \leqslant C\{\|\chi_1 Lu\|_s + \|\chi_1 u\|_s + \|\chi_2 u\|_{-N}\}, \tag{3.9}$$

where $\chi \Subset \chi' \Subset \chi_1$ and $\chi_2 \in C_0^{\infty}(\Omega)$. We denote $\tilde{Y}_k$ by $Y_k + i\tilde{\tilde{Y}}_k$, where $\tilde{\tilde{Y}}_k = -i(\tilde{Y}_k - Y_k)$. Since

$$\begin{aligned}\sum_{k=1}^{N+n} \|\chi' \tilde{Y}_k u\|_s &\leqslant \sum_{k=1}^{N+n} \{\|\chi' Y_k u\|_s + \|\chi' \tilde{\tilde{Y}}_k u\|_s\} \\ &= \sum_{k=1}^{N+n} \{\|\Lambda^s \chi' Y_k u\| + \|\Lambda^s \chi' \tilde{\tilde{Y}}_k u\|\}\end{aligned} \tag{3.10}$$

and $\Lambda^s\chi' Y_k = Y_k\Lambda^s\chi' + [\Lambda^s\chi', Y_k](\chi_1 + (1-\chi_1))$, by a pseudolocal property of a pseudodifferential operator and (2.4), the right-hand side of (3.10) is dominated by

$$\sum_{\gamma=0}^{N} |(L\Lambda_\gamma{}^s\chi' u, \Lambda_\gamma{}^s\chi' u)| + \|\chi_1 u\|_s{}^2 + \|\chi_2 u\|_{-N}. \tag{3.11}$$

Here $\chi_2(x) \in C_0{}^\infty(\Omega)$ and we use a fact that all considered operators are properly supported. We also denote $L\Lambda_\gamma{}^s\chi'$ by

$$\Lambda_\gamma{}^s\chi' L + \sum_{k=1}^{N+n} (\tilde{Y}_k\Lambda_{\gamma,k}^s + \Lambda_{\gamma,0}^s)(\chi_1 + (1-\chi_1)),$$

where $\Lambda_{\gamma,k}^s(1-\chi_1) \in L^{-\infty}(\Omega)$, $k = 0, \ldots, N+n$. Thus (3.11) is dominated by

$$\begin{aligned} \sum_{\gamma,k} |(\Lambda_{\gamma,k}^s\chi_1 u, \tilde{Y}_k{}^*\Lambda_\gamma{}^s\chi' u)| + N_s(u) &\leqslant \sum_{k=1}^{N+n} \delta\|\chi'\tilde{Y}_k{}^* u\|_s{}^2 + C_\delta N_s(u) \\ &\leqslant \sum_{k=1}^{N+n} \delta\{\|\chi' Y_k u\|_s{}^2 + \|\chi'\tilde{\tilde{Y}}_k u\|_s{}^2\} + C_\delta N_s(u), \end{aligned} \tag{3.12}$$

where $N_s(u) = \|\chi_1 Lu\|_s{}^2 + \|\chi_1 u\|_s{}^2 + \|\chi_2 u\|_{-N}^2$. If δ is sufficiently small, then by (3.10), (3.11), and (3.12) we have the desired estimate (3.9). This completes the proof of proposition.

Let φ be an element of $C_0{}^\infty(R^n)$ with $\int\varphi\,dx = 1$. Then for $\delta < 0$ we define the convolution operator

$$S_\delta u(x) = \delta^{-n}\int u(y)\varphi((y-x)/\delta)\,dy.$$

Since S_δ is the properly supported pseudodifferential operators with the symbols $\hat{\varphi}(\delta\xi)$, the set $\{S_\delta(D); 0 < \delta \leqslant 1\}$ is a bounded set of $L^0(\Omega)$. Thus we have the following:

LEMMA 3.1. *Let χ and $\chi_1 \in C_0{}^\infty(\Omega)$ with $\chi \Subset \chi_1$ and $s \in R$, $N > 0$. Then there exists $\chi_2 \in C_0{}^\infty(\Omega)$ and a constant C such that for $u \in \mathscr{D}'(\Omega)$ with $\chi_1 Lu$, $\chi_1 u \in H_s(R^n)$*

$$\|S_\delta\chi u\|_{s+\epsilon} \leqslant C\{\|\chi_1 Lu\|_s + \|\chi_1 u\|_s + \|\chi_2 u\|_{-N}\},$$

where C and χ_2 does not depend on δ.

Proof of Theorem S. We assume that $Lu = f$ for $f, u \in \mathscr{D}'(\Omega)$. Then since $\operatorname{singsupp} u \supset \operatorname{singsupp} f$, we may prove that if $f_{|V} \in C^\infty(V)$ for an open set of Ω, we wish to prove that $u_{|V} \in C^\infty(\Omega)$. Let W be an open subset of V such that for some $\delta > 0$ $W_1 = \{x; \operatorname{dis}(x, W) < \delta\} \subset V$. Then we define $W_n =$

$\{x; \operatorname{dis}(x, W) < \delta/n\}$ and $\varphi_n \in C_0^\infty(W_{2n-1})$ with $\varphi_n \equiv 1$ on W_{2n}. Since $\varphi_1 \supset \varphi_2 \supset \cdots$, inductively applying Lemma 3.1 to $\varphi_{n-1}, \varphi_n, u$ as $\chi = \varphi_n$, $\chi_1 = \varphi_{n-1}$, we have $\varphi_n u \in H_{s+(n-1)\epsilon}(R^n)$. This means that $u_{|W} \in C^\infty(W)$. This completes the proof of Theorem S.

Finally we consider some hypoelliptic, second-order, differential operator that does not satisfy condition (A.1). Let $L(x, D)$ be a differential operator such that

$$L(x, D) = \sum_{j=1}^{N} X_j^2 + i \sum_{k=1}^{N_1} Y_k^2 + iX_0 + Y_0 + c, \tag{3.13}$$

where $X_j(x, iD)$, $j = 0, \ldots, N$, and $Y_k(x, iD)$, $k = 0, \ldots, N_1$, are vector fields over Ω and $c(x)$ belongs to $C^\infty(\Omega)$. We assume the following condition: For any compact set K of Ω,

$$\sum_{k=0}^{N_1} |Y_k(x, \xi)|^2 \leqslant C_K \sum_{j=1}^{N} |X_j(x, \xi)|^2, \qquad (x, \xi) \in K \times R^n. \tag{3.14}$$

Then we have the following:

THEOREM H. *Let $L(x, D)$ be a second-order differential operator with form* (3.16), *which satisfies condition* (3.17). *Assume that among the repeated commutators of $X_0(x, iD), \ldots, X_N(x, iD)$, there exist n that are linearly independent at every point of Ω. Then $L(x, D)$ is hypoelliptic.*

This theorem is verified by the same method of the proof of Theorem S.

REFERENCES

1. R. BEALS AND C. FEFFERMANN, On hypoellipticity of second order operators, *Comm. Partial Diff. Eq.* **1** (1976), 73–85.
2. L. HÖRMANDER, Hypoelliptic second order differential equations, *Acta Math.* **119** (1967), 147–171.
3. L. HÖRMANDER, Pseudo-differential operators and non-elliptic boundary problems, *Ann. of Math.* **83** (1966), 139–209.
4. J. J. KOHN, "Pseudo-Differential Operators and Hypoellipticity," C.I.M.E., Stresa, 1968.
5. O. A. OLENIK AND E. V. RADKEVIČ, "Second Order Equations with Non-negative Characteristic Form," Amer. Math. Soc., 1973.
6. C. ZUILY, Sur l'hypoellipticié des opérateurs différentiels du second order a coefficients réels, *J. Math. Pures Appl.* **55** (1976), 99–125.

AMS (MOS) subject classification: 35H05.

STUDIES IN ANALYSIS
ADVANCES IN MATHEMATICS SUPPLEMENTARY STUDIES, VOL. 4

Integral-Geometric Measures on Homogeneous Spaces in Hilbert Space

M. P. HEBLE

Department of Mathematics
University of Toronto
Toronto, Canada

INTRODUCTION

In this chapter our objective is to obtain some analogs of well-known traditional results concerning integral-geometric measures in homogeneous spaces (cf. [3, 21, 22]). The situations with which we shall deal are sufficiently nontrivial and the linear spaces in question are of sufficient practical interest (see, e.g., Dunford and Schwartz [7, Volume III, Chapter XV] for applications of such spaces).

Let $\mathscr{B}$ be an infinite-dimensional, real, Banach space, and $\mathscr{B}^n$ the direct sum of $\mathscr{B}$ with itself n times, where $n \geqslant 1$ is a positive integer:

$$\mathscr{B}^n = \mathscr{B} \oplus \cdots \oplus \mathscr{B}.$$

We shall generally use the notation of Dunford and Schwartz (referred to as D–S) [7, Vol. III, pp. 1960–1963]. The elements of $\mathscr{B}^n$ are ordered n-tuplets $x = [x_1, \ldots, x_n]$ of n elements in $\mathscr{B}$, and addition, scalar multiplication, and the norm are defined in terms of the corresponding concepts in $\mathscr{B}$ by

$$[x_1, \ldots, x_n] + [y_1, \ldots, y_n] = [x_1 + y_1, \ldots, x_n + y_n],$$

$$\alpha[x_1, \ldots, x_n] = [\alpha x_1, \ldots, \alpha x_n],$$

$$\|[x_1, \ldots, x_n]\| = \left(\sum_{i=1}^{n} \|x_i\|^2 \right)^{1/2}, \qquad \|x_i\| = \text{norm of } x_i \text{ in } \mathscr{B}.$$

With these definitions, $\mathscr{B}^n$ is indeed a Banach space. If $\mathscr{B}$ is also a Hilbert space, then the inner product in $\mathscr{B}^n$ is defined by

$$([x_1, \ldots, x_n], [y_1, \ldots, y_n]) = \sum_{i=1}^{n} (x_i, y_i),$$

ISBN 0-12-599150-9

where (x_i, y_i) is the inner product between x_i and y_i in $\mathscr{B}$; and in that case, clearly,

$$([x_1, \ldots, x_n], [x_1, \ldots, x_n]) = \|[x_1, \ldots, x_n]\|^2.$$

We shall consider in this chapter an *r-parameter Lie subgroup G* (cf. Eells [8], Maissen [17]) of the group of all transformations of $\tilde{h}^n$ onto itself, where $\tilde{h}$ is an infinite-dimensional, real, Hilbert space. We shall take as the parameter r-tuplet $a = [a_1, \ldots, a_r]$ a point in $\mathscr{B}^r$ where $\mathscr{B}$ is an infinite-dimensional real Banach space. When points of $\mathscr{B}^r$ are thus used as parameter r-tuplets, we shall call $\mathscr{B}^r$ the *parameter space.* Later in this chapter (see Section I.3) we shall widen our concept of *parameter r-tuple.*

The integral-geometric measures that we are going to be concerned with are *operator-valued measures* (cf. [7, Vol. II, pp. 891–892, Vol. III, pp. 1960–1962]). In Section III of this chapter we point out one such measure and its relation to a suitable Lie group of transformations of the type just mentioned. Both the measure itself in this particular case and its relation to a suitable Lie group which we point out do not seem to be at all trivial.

In Section I of this chapter we clarify our assumptions about the Lie group of transformations that we consider. Our attempt is to investigate integrability conditions for a set of differential equations involving first-degree differentials (which we call *Pfaffian differentials* and *Pfaffian systems of equations*, in analogy with classical Pfaffian systems). We obtain a necessary condition for such a Pfaffian system to be *completely integrable*—where we define a system to be completely integrable if through each point of the space passes one and only one integral "variety" depending on an appropriate number of parameters. Formulating a suitable definition of such a "variety" becomes a matter of importance; the definition which we use is adequate for our present purpose. The necessary condition for complete integrability is analogous to a classical condition (cf. Santalo [21, p. 86]), related to a condition of Frobenius.

In Section II we turn to the consideration of "measures" and "densities" on suitable sets of submanifolds of our space $\tilde{h}^n$, invariant under the "parameter group." The "measures" that we are looking for are operator-valued measures defined on suitable fields. We find that, as in the classical case dealt with by Chern, it is convenient for this purpose to turn our attention to the *space of parameters*, viz., $\mathscr{B}^r$ and suitable sets of submanifolds in $\mathscr{B}^r$. Here we use the result of Section I, and deduce that a necessary and sufficient condition for a certain differential form Ω to yield a density on our sets of submanifolds in $\mathscr{B}^r$, invariant under the parameter group, is that its exterior derivative $d\Omega = 0$. This is analogous to a result of Chern.

In Section III we exhibit one specific Lie group of transformations of the type discussed in Sections I and II, depending on a finite number of "parameters." The example which we elaborate is chosen with a view toward later use, and is patterned after the classical group of "Euclidean Motions" in n-dimensional Euclidean space. We call the group in our particular example the group of *Hilbertian motions* and denote it by $\mathscr{H}(n)$. We exhibit a finite set of suitable parameters in this group. We then use the condition obtained in Section II to prove that there exists a density on certain sets of varieties in $\tilde{h}^n$, these sets of varieties being the analog of Grassmanians, i.e., sets of h-dimensional subspaces in E^n. In this example we exhibit (Section III) an *operator jet-valued* function T_t $\{0 \leqslant t \leqslant 1\}$, with the property: $\mathsf{T}_t \cdot \mathsf{T}_{t'} = \mathsf{T}_{t+t'}$ $\{0 \leqslant t, t', t + t' \leqslant 1\}$. This property enables us to extend the domain of the real parameter t from $\{0 \leqslant t \leqslant 1\}$ to $\{0 \leqslant t < \infty\}$, and the operator jet-valued function thus defined on $\{0 \leqslant t < \infty\}$ has the semigroup property: $\mathsf{T}_t \cdot \mathsf{T}_{t'} = \mathsf{T}_{t+t'}$ $\{0 \leqslant t, t' < \infty\}$. Further we show that the operator jet-valued measure which we obtain in this case is countably additive.

I. Pfaffian Systems and Complete Integrability

This section consists of three parts. In Section I.1 we shall explain some general notation. In Section I.2 we shall explain our assumptions about the Lie group $\mathscr{G}$ and its parameter space and the *parameter group* $\mathbb{P}$. In Section I.3 we shall deal with the concept of $(r\text{–}h)$-variety and, in particular, with integral varieties of a Pfaffian system of differential equations.

1

Henceforth in this paper $\tilde{h}$ shall denote an infinite-dimensional, real, Hilbert space and $\mathscr{B}$ shall denote an infinite-dimensional, real, Banach space. In Section III we shall take for our space $\mathscr{B}$ the space $\mathscr{B}(\tilde{h})$, the algebra of all bounded linear operators of $\tilde{h}$ into itself.

Let $\mathscr{B}_i$ be the subspace in $\mathscr{B}^n$ consisting of vectors $x = [x_1, \ldots, x_n]$ in $\mathscr{B}^n$ with $x_j = 0$ for $j \neq i$, and P_i the projection of $\mathscr{B}^n$ onto $\mathscr{B}_i$. It will then be convenient to use the notation:

$$x = (x_1, \ldots, x_n) \qquad \text{with} \quad x_i \in \mathscr{B}_i, \quad 1 \leqslant i \leqslant n, \quad \text{where} \quad x \in \mathscr{B}^n$$

rather than the earlier

$$x = [x_1, \ldots, x_n] \in \mathscr{B}^n, \qquad x_i \in \mathscr{B}, \quad 1 \leqslant i \leqslant n.$$

If H_i is the linear isometric map of $\mathscr{B}_i$ onto $\mathscr{B}$ defined by

$$H_i x = x_i,$$

then $P_i[x_1, \ldots, x_n] = H_i^{-1}x_i$. For $1 \leqslant h < n$ we shall denote the subspace $\mathscr{B}_{h+1} \oplus \cdots \oplus \mathscr{B}_n$ of $\mathscr{B}^n$ by ${}_{n-h}\mathscr{B}$.

We shall have occasion to consider the matrix algebra $\mathscr{M}_n(\mathrm{B}(\mathscr{B}))$ of order n over $\mathrm{B}(\mathscr{B})$ (cf. [7, Vol. III, pp. 1961–1963]). This algebra is by definition the set of $(n \times n)$ operator matrices $A = ((a_{ij}))$, where each a_{ij} belongs to $\mathrm{B}(\mathscr{B})$, the algebra of bounded linear operators on $\mathscr{B}$ into itself, with matrix addition, matrix multiplication, and scalar multiplication defined, respectively, by

$$((a_{ij})) + ((b_{ij})) = ((a_{ij} + b_{ij}))$$

$$((a_{ij}))((b_{ij})) = \left(\left(\sum_{k=1}^{n} a_{ik}b_{kj}\right)\right)$$

$$\alpha((a_{ij})) = ((\alpha a_{ij})), \qquad \alpha \in \mathbb{R}.$$

The unit element in $\mathscr{M}_n(\mathrm{B}(\mathscr{B}))$ is $((e_{ij}))$ where

$$e_{ij} = \begin{cases} 0 & \text{if} \quad i \neq j, \\ e & \text{if} \quad i = j, \end{cases}$$

where e is the unit in $\mathrm{B}(\mathscr{B})$. We shall denote the unit (identity) element in $\mathrm{B}(\mathscr{B}^n)$ by I. Each matrix $((a_{ij})) \in \mathscr{M}_n(\mathrm{B}(\mathscr{B}))$ defines an element in $\mathrm{B}(\mathscr{B}^n)$ by the rule:

$$((a_{ij})) : x = [x_1, \ldots, x_n] \to ((a_{ij})) \cdot x = y = [y_1, \ldots, y_n],$$

$$y_i = \sum_{j=1}^{n} a_{ij}x_j, \qquad 1 \leqslant i \leqslant n.$$

Furthermore the correspondence $((a_{ij})) \to A$ thus set up is actually an algebraic isomorphism (cf. D–S [7, Vol. III, pp. 1961–1963]) between the algebras $\mathscr{M}_n(\mathrm{B}(\mathscr{B}))$ and $\mathrm{B}(\mathscr{B}^n)$; and defining the norm $\|((a_{ij}))\| \underset{\text{def}}{=} \|A\|$, this correspondence is seen to be an isometry.

A variant of this theme that will be used is as follows. For a given integer $h : 1 \leqslant h \leqslant r$, consider two direct-sum decompositions of $\mathscr{B}^h$:

$$\mathscr{B}^h = V_1 \oplus \cdots \oplus V_h = W_1 \oplus \cdots \oplus W_h,$$

where $V_i = \phi(\mathscr{B}_i)$, $1 \leqslant i \leqslant h$, $W_j = \psi(\mathscr{B}_j)$, $1 \leqslant j \leqslant h$, for some linear homeomorphisms $\phi, \psi : \mathscr{B}^h \xrightarrow{\text{onto}} \mathscr{B}^h$. We shall consider at times an operator matrix $A = ((a_{ij}))$, where $a_{ij} \in \mathscr{L}(V_j, W_i)$[1], as yielding a continuous linear operator $A : \mathscr{B}^h \to \mathscr{B}^h$, by the rule

$$y_i = \sum_{j=1}^{h} a_{ij}x_j, \qquad 1 \leqslant i \leqslant h,$$

[1] $\mathscr{L}(V_j, W_i)$ is the Banach space of continuous linear operators $T : V_j \to W_i$.

where $x = (x_1, \ldots, x_h) \in \mathscr{B}^h$, $x_i \in V_i$, $1 \leqslant i \leqslant h$, and $y = (y_1, \ldots, y_h) \in \mathscr{B}^h$, $y_j \in W_j$, $1 \leqslant j \leqslant h$. When we do use such an operator matrix we shall do so only with the additional hypothesis that $A = ((a_{ij}))$ yields a linear homeomorphism $\mathscr{B}^h \xrightarrow{\text{onto}} \mathscr{B}^h$.

We shall make use of exterior differential forms (cf. [2, 16]), for which purpose the following notation will be convenient (we shall mostly use the notation of Cartan [2, pp. 9–32]). For any Banach spaces $\mathscr{E}$, $\mathscr{F}$, and any integer $k \geqslant 1$, we shall denote by $\mathscr{L}_k(\mathscr{E};\mathscr{F})$ the $\mathscr{B}$-space of continuous k-linear mappings $\mathscr{E} \to \mathscr{F}$ ($\mathscr{L}_1(\mathscr{E};\mathscr{F}) = \mathscr{L}(\mathscr{E};\mathscr{F})$), and by $\mathscr{A}_k(\mathscr{E};\mathscr{F})$ the closed linear subspace in $\mathscr{L}_k(\mathscr{E};\mathscr{F})$ consisting of continuous k-linear alternating mappings $\mathscr{E} \to \mathscr{F}$ (for $k = 1, \mathscr{A}_1(\mathscr{E};\mathscr{F}) = \mathscr{L}(\mathscr{E};\mathscr{F})$). For a given open set $\mathscr{U} \subset \mathscr{E}$, and given integer $p: 0 \leqslant p \leqslant \infty$, we shall denote by $\Omega_k^{(p)}(\mathscr{U}, \mathscr{F})$ the vector space of differential forms of degree k, defined in $\mathscr{U}$ with values in $\mathscr{F}$ (or more briefly, differential k-forms defined in $\mathscr{U}$ with values in $\mathscr{F}$), of class p, i.e., mappings

$$\omega: \mathscr{U} \to \mathscr{A}_k(\mathscr{E};\mathscr{F})$$

which are of class C^p. If $\mathscr{U} = \mathscr{E}$, we shall simply talk of differential k-forms on $\mathscr{E}$ into $\mathscr{F}$ of class C^p.

We shall only consider the case where $\mathscr{E} = \mathscr{F} = \mathscr{B}^h$, for a given integer $h: 1 \leqslant h \leqslant r$. For $k = 1$, differential 1-forms of class C^p on $\mathscr{B}^h$ into $\mathscr{B}^h$ will be called *C^p-smooth Pfaffian forms on $\mathscr{B}^h$*; for an arbitrary integer $k > 0$, differential k-forms of class C^p on $\mathscr{B}^h$ into $\mathscr{B}^h$ will be called *C^p-smooth differential k-forms on $\mathscr{B}^h$*.

Next, let $\mathscr{E}, \mathscr{F}, \mathscr{G}$ be B-spaces and $\mathscr{K}$ a normed vector space. Let $f \in \mathscr{A}_p(\mathscr{E};\mathscr{F})$, $g \in \mathscr{A}_q(\mathscr{E};\mathscr{G})$. For defining exterior multiplication between f and g we introduce a bilinear continuous mapping $\Phi: \mathscr{F} \times \mathscr{G} \to \mathscr{K}$ with values in the normed vector space $\mathscr{K}$. This allows us to associate with f, g a mapping $h: \mathscr{E}^{p+q} \to \mathscr{K}$ by the definition

$$h(x_1, \ldots, x_{p+q}) = \Phi(f(x_1, \ldots, x_p), g(x_{p+1}, \ldots, x_{p+q})).$$

This mapping h is $(p + q)$-linear and continuous, but not (in general) alternating (it *is* alternating with respect to the first p variables *only*, and also alternating with respect to the last q variables *only*). The vector space of all such $(p + q)$-linear mappings h will be denoted by $\mathscr{A}_{p,q}(\mathscr{E};\mathscr{K})$. Next, a canonical procedure for associating with every $h \in \mathscr{A}_{p,q}(\mathscr{E};\mathscr{K})$ a mapping $\tilde{h} \in \mathscr{A}_{p+q}(\mathscr{E};\mathscr{K})$ is as follows. We define a bilinear continuous mapping

$$\phi_{p,q}: \mathscr{A}_{p,q}(\mathscr{E};\mathscr{K}) \to \mathscr{A}_{p+q}(\mathscr{E};\mathscr{K})$$

by

$$\phi_{p,q}(h) = \tilde{h} = \sum_{\sigma} \epsilon(\sigma)(\sigma h),$$

the summation being over all permutations σ of $[1, \ldots, (p+q)\}$ such that

$$\sigma(1) < \cdots < \sigma(p) \qquad \text{and} \qquad \sigma(p+1) < \cdots < \sigma(p+q)$$

and

$$(\sigma h)(x_1, \ldots, x_{p+q}) = h(x_{\sigma(1)}, x_{\sigma(2)}, \ldots, x_{\sigma(p+q)}).$$

(We shall omit the details concerning the use of this procedure for defining exterior multiplication $f \wedge g$ between $f \in \mathscr{A}_p(\mathscr{E}; \mathscr{F})$, $g \in \mathscr{A}_q(\mathscr{E}; \mathscr{G})$, and refer to Cartan [2] for all these details.)

Now let $a_{ij} \in \mathscr{L}(\mathscr{B}_i, \mathscr{B}_j)$, $1 \leqslant i, j \leqslant h$, and denote by $\bar{\omega}_i$ the continuous linear operator $\mathscr{B}_i \to \mathscr{B}_j$:

$$\bar{\omega}_i(x_i) = \sum_{j=1} a_{ij} x_i, \qquad 1 \leqslant i \leqslant h,$$

where $x_i \in \mathscr{B}_i$. The composite $\bar{\omega}_i \circ H_i^{-1}$ will be denoted by ω_i; then for each $x \in \mathscr{B}$, $\omega_i(x) \in \mathscr{L}(\mathscr{B}, \mathscr{B}^h)$. This then enables us to consider the h-fold exterior product of $\omega_1, \ldots, \omega_h$ to yield the differential h-form $\omega_1(\cdot) \wedge \cdots \wedge \omega_h(\cdot)$ defined on $\mathscr{B}^h$ with values in $\mathscr{B}^h$. For convenience, we shall call these, simply, *differential h-forms on* $\mathscr{B}^h$. A variant of this theme is as follows. Consider a direct sum decomposition of $\mathscr{B}^h$:

$$\mathscr{B}^h = V_1 \oplus \cdots \oplus V_h,$$

where the V_is are images of $\mathscr{B}_i$s under some linear homeomorphism ϕ: $\mathscr{B}^h \xrightarrow{\text{onto}} \mathscr{B}^h$. We shall also understand by the *Pfaffian form on* $\mathscr{B}^h$ (or differential 1-form on $\mathscr{B}^h$) a mapping of the form

$$\bar{\omega}_i(v_i) = \sum_{j=1}^{h} a_{ij} v_i, \qquad \text{where} \quad v_i \in V_i$$

and the $a_{ij} \in \mathscr{L}(V_i, V_j)$. Then let $\omega_i = \bar{\omega}_i \circ \phi^{-1} \circ H_i^{-1}$. If we consider h such Pfaffian forms $\omega_1, \ldots, \omega_h$, then we shall also understand by the differential h-form on $\mathscr{B}^h$ the h-fold exterior product of $\omega_1, \ldots, \omega_h$. The explicit context will be made clear on each occasion. In case we consider smooth differential k-forms $(1 \leqslant k \leqslant h)$, then we shall call these *C^p-differential k-forms on* $\mathscr{B}^h$, for some $p: 0 \leqslant p \leqslant \infty$.

2. *Assumptions about the Group* $\tilde{\mathscr{G}}$

In Sections I and II we shall assume $\tilde{\mathscr{G}}$ to be a topological group satisfying the following (A)–(D). We shall represent a point $x \in \tilde{\hbar}^n$ by its components (coordinates) in the respective summands $\tilde{\hbar}_i$, as: $x = (x_1, \ldots, x_n)$, $x_i \in \tilde{\hbar}_i$, $i = 1, \ldots, n$.

(A) *$\tilde{\mathscr{G}}$ is an r-parameter Lie group* (Maisen [17, 18]) *of C^1-diffeomorphisms of $\tilde{\hbar}^n \xleftrightarrow{\text{onto}} \tilde{\hbar}^n$, where the parameter r-tuple $a = (a_1, \ldots, a_r)$ is a point in the parameter space $\mathscr{B}^r$.*

To be more specific, we are assuming that: $\tilde{\mathscr{G}}$ is a topological group such that each element $g \in \tilde{\mathscr{G}}$ is a diffeomorphism of $\tilde{\hbar}^n \xleftrightarrow{\text{onto}} \tilde{\hbar}^n$:

(i) $f(g;x) = g(x) = x' \in \tilde{\hbar}^n, x \in \tilde{\hbar}^n, g \in \tilde{\mathscr{G}}$;

(ii) $g_1, g_2 \in \tilde{\mathscr{G}}, x \in \tilde{\hbar}^n \Rightarrow g_1(g_2(x)) = (g_1 g_2)x \in \tilde{\hbar}^n$;

(iii) $x' = g(x) = f(g;x)$ is simultaneously continuous with respect to both $x \in \tilde{\hbar}^n$ and $g \in \tilde{\mathscr{G}}$;

(iv) the coordinates $x' = (x_1', \ldots, x_n')$ of the transformed point are functions

$$x_i' = \phi_i(x_1, \ldots, x_n; a_1, \ldots, a_r), \qquad i = 1, \ldots, n,$$

which are C^1-smooth jointly with respect to the x_i and a_i and C^2-smooth with respect to the a_is for each fixed $x = (x_1, \ldots, x_n) \in \tilde{\hbar}^n$.

To each point a in the parameter space corresponds a transformation $T_a \in \tilde{\mathscr{G}}$, and we shall use interchangeably the two equivalent formulas:

$$x' = \phi(x; a) \qquad \text{or} \qquad x' = T_a x \tag{1}$$

to indicate that x' is the transform of x by T_a. With the first notation, property (ii) implies that given any two points a, b in the parameter space $\mathscr{B}^r$, there is a third point $c \in \mathscr{B}^r$, *independent of* x, such that

$$\phi(\phi(x; a); b) = \phi(x; c). \tag{2}$$

This establishes:

LEMMA 1.1. *To any two given points $a, b \in \mathscr{B}^r, \exists c \in \mathscr{B}^r$ such that c is independent of $x \in \tilde{\hbar}^n$ and*

$$T_b T_a = T_c, \tag{3}$$

$T_a, T_b, T_c \in \tilde{\mathscr{G}}$.

As is customary (cf. [21, p. 67]) we define a group $\mathscr{G}'$ of transformations of $\tilde{\hbar}^n \xleftrightarrow{\text{onto}} \tilde{\hbar}^n$ to be *transitive* with respect to the points of $\tilde{\hbar}^n$ if for each pair of points $x, x' \in \tilde{\hbar}^n$, $\exists T_a \in \tilde{\mathscr{G}}'$ such that $x' = T_a x$, and we define $\tilde{\mathscr{G}}$ to be *simply transitive* if this T_a is unique for every pair of points x, x'.

We shall now define the *parameter group*. We associate with each point a in the parameter space $\mathscr{B}^r$ a transformation $T_a \in \tilde{\mathscr{G}}$ as follows: To each $\alpha \in \mathscr{B}^r$, $\exists \alpha' \in \mathscr{B}^r$ such that

$$T_a : T_{\alpha'} = T_a T_\alpha. \tag{4}$$

LEMMA 1.2. *The set of all these transformations forms a group $\mathbb{P}$, which is simply transitive.*

Proof. From (4), we see that $T_a^{-1} \in \mathbb{P}$. Also

$$T_{\alpha'} = T_a T_\alpha, \qquad T_{\alpha''} = T_b T_{\alpha'} \Rightarrow T_{\alpha''} = T_b T_a T_\alpha,$$

i.e., with T_a, T_b, the product $T_b T_a$ also belongs to $\mathbb{P}$. The simple transivity of $\mathbb{P}$ follows from: $T_a = T_{\alpha'} T_\alpha^{-1}$.

DEFINITION 1.1. This group $\mathbb{P}$ is called the *parameter group*.

(B) *The transformation in $\mathscr{G}$ corresponding to $a_1 = a_2 = \cdots = a_r = 0$ is assumed to be the identity, i.e.,*

$$x_i = \phi_i(x_1, \ldots, x_n; 0, \ldots, 0), \qquad i = 1, \ldots, n. \tag{5}$$

Now suppose the parameters a_i take "infinitesimal values," which we shall assume to be, in the first approximation of the form (further explanation is given in the next two paragraphs):

$$\epsilon_i(a; da) = \sum_{k=1}^{r} A_k{}^i\, da_k, \tag{6}$$

where the $A_k{}^i(\cdot)$ are functions of $a \in \mathscr{B}^r$ and are continuous linear mappings: $A_k{}^i \in \mathscr{L}(\mathscr{B}_k, \mathscr{B}_i)$. Here the $A_k{}^i$ are understood to be operating on the da_k.

We shall mean by (6) that if the parameter point $a = (a_1, \ldots, a_r)$ is changed by the addition of a vector which we shall denote by $da = (da_1, \ldots, da_r)$ (for obvious conformity with the usual notation of differentials, in ordinary calculus), then the C^1-diffeomorphisms T_{a+da}, T_a are related by

$$T_a^{-1} T_{a+da} = T_\epsilon, \tag{7}$$

where $\epsilon = \epsilon(a; da) = [\epsilon_1, \ldots, \epsilon_r] = [\sum_{k=1}^r A_k{}^1(a)\, da_k, \ldots, \sum_{k=1}^r A_k{}^r(a)\, da_k]$ with the $A_k{}^i(a) \in \mathscr{L}(\mathscr{B}_k, \mathscr{B}_i)$ for $1 \leqslant i,\, k \leqslant r$ is an r-tuplet of elements in $\mathscr{B}^r$. Clearly $\epsilon \to 0$ as $da \to 0$. Here the matrix $((A_k{}^i))_{1 \leqslant i,\, k \leqslant r}$ is understood to represent an element in $\mathrm{B}(\mathscr{B}^r)$ as explained in Section I.1.

The preceding assumption (6) implies that in the first approximation:

$$x_i' = x_i + \sum_{j=1}^{r} D_{n+j}\phi_i \bigg|_{a=0} \cdot \epsilon_j, \tag{8}$$

where $D_{n+j}(\cdot)\phi_i|_{a=0}$ is the first partial derivative ([6] I, p. 172) of ϕ_i with respect to a_j at $a = 0$. In the first approximation shall mean that the difference between the left and the right sides is $o(\|da\|)$ for each $i = 1, \ldots, n$.

We may write this relation (8) as

$$\delta x_i = \sum_{j=1}^{r} \xi_{ij}(x) \cdot \epsilon_j(a; da), \tag{9}$$

where $\xi_{ij} = D_{n+j}\phi_i|_{a=0}$. We shall say that the ϵ_i define the infinitesimal transformation T_ϵ.

We shall next clarify our assumptions about "frames," for we intend to adapt this classical concept (cf. [3, 21]) of frames for our present purposes in the sequel.

(C) *By* frame *we shall mean either the original ordered n-tuple of Hilbert spaces* $\tilde{\hbar}_1, \tilde{\hbar}_2, \ldots, \tilde{\hbar}_n$ *with direct sum* $\tilde{\hbar}^n$ *or the image of this under a transformation:* $x \to \mathcal{O}x + \alpha$, *with* $\mathcal{O}$ *orthogonal:* $\tilde{\hbar}^n \xleftrightarrow{\text{onto}} \tilde{\hbar}^n$ *and* α *a fixed point in* $\tilde{\hbar}^n$ (α *independent of* x). *All the frames referred to in the sequel will be assumed to be of this type.*

We shall use the letter $\mathcal{R}$ to denote an arbitrary frame or the equivalent coordinate system.

(D) *We conceive of an* absolute *frame or coordinate system* $\mathcal{R}_0$ *which we shall refer to as an* initial *coordinate system. A transformation* $T_a \in \mathcal{G}$ *transforms* $\mathcal{R}_0$ *into a new frame* $\mathcal{R}_a$ *and we shall assume that*

$$\mathcal{R}_a = T_a \mathcal{R}_0. \tag{10}$$

By this statement (10) we shall mean the following. We consider a complete orthonormal system $\{\phi_\alpha^{(i)}\}_{\alpha \in I}$ in each $\tilde{\hbar}_i$ in the coordinate system $\{\tilde{\hbar}_1, \ldots, \tilde{\hbar}_n\}$. The "absolute" coordinate system $\mathcal{R}_0$ will then consist of the orthogonal subspaces $\tilde{\hbar}_1{}^0, \ldots, \tilde{\hbar}_n{}^0$ which are images of $\tilde{\hbar}_1, \ldots, \tilde{\hbar}_n$, respectively, under an orthogonal transformation $\mathcal{O}^0: \tilde{\hbar}^n \xleftrightarrow{\text{onto}} \tilde{\hbar}^n$ carrying the c.o.s. $\{\phi_\alpha^{(i)}\}_{\alpha \in I}$ into the c.o.s. $\{\phi_\alpha^{(i0)}\}_{\alpha \in I}$, $i = 1, \ldots, n$. Similarly the frame $\mathcal{R}_a$ will consist of the orthogonal subspaces $\tilde{\hbar}_1{}^a, \ldots, \tilde{\hbar}_n{}^a$ which are images, respectively, of $\tilde{\hbar}_1, \ldots, \tilde{\hbar}_n$ under an orthogonal transformation $\mathcal{O}^a: \tilde{\hbar}^n \xleftrightarrow{\text{onto}} \tilde{\hbar}^n$, carrying the c.o.s. $\{\phi_\alpha^{(i)}\}_{\alpha \in I}$ into the c.o.s. $\{\phi_\alpha^{(ia)}\}_{\alpha \in I}$ ("c.o.s." stands for "complete orthonormal system").

Let $x \in \tilde{\hbar}^n$ and let $x' = T_a x$. We represent the coordinates of x, x' with respect to $\mathcal{R}_0$ by $x_i{}^0$, $x_i'{}^0$, respectively, $i = 1, \ldots, n$, and the coordinates of the same points x, x' with respect to $\mathcal{R}_a$ by $x_i{}^a$, $x_i'{}^a$, respectively. Using $\{\phi_\alpha^{i0}\}_{\alpha \in I}$, $\{\phi_\alpha^{(ia)}\}_{\alpha \in I}$, the assumption stated in (10) means that

$$x_i'^a = x_i{}^0, \qquad i = 1, \ldots, n. \tag{11}$$

As a simple consequence of (10) we obtain:

LEMMA 1.3. $x_i^{(a)} = \phi_i^{-1}(x'^a, a) = \phi_i^{-1}(x^0, a).$ (12)

Here ϕ_i^{-1} is the ith component of the inverse of the function $\phi(x, a)$ which defines T_a [see (5)].

Next let $\mathcal{R}_b$ be an arbitrary frame, and $x' = T_a x$. With reference to $\mathcal{R}_0$ we have

$$x_i'^0 = \phi_i(x^0, a),$$

and according to the earlier assumption (1):

$$x_i^0 = \phi_i(x^b, b), \qquad x_i'^0 = \phi_i(x'^b, b).$$

Substituting this in the preceding equation we then get

$$\phi_i(x'^b, b) = \phi_i(\phi(x^b, b), a),$$

hence

$$x_i'^b = \phi_i^{-1}\{\phi[\phi(x^b, b), a], b\};$$

or, because the transformations are assumed to form a group,

$$x_i'^b = \phi_i(x^b, c).$$

Thus we conclude:

LEMMA 1.4. *The finite equations of the group are the same with reference to any coordinate system deduced from $\mathcal{R}_0$ by a transformation of the group $\tilde{\mathcal{G}}$.*

The preceding statements also tell us:

LEMMA 1.5. *The transformation T_a referred to the coordinate system $\mathcal{R}_b$ transforms into $T_b^{-1}T_aT_b$.*

We shall next define the *relative components* of a moving frame. The transformation carrying $\mathcal{R}_a$ into $\mathcal{R}_{a+da}$ is $T_{a+da}T_a^{-1}$, which when referred to $\mathcal{R}_a$ becomes

$$T_a^{-1}(T_{a+da}T_a^{-1})T_a = T_a^{-1}T_{a+da}.$$

This is an infinitesimal transformation as previously defined (see (6) and (7)). Let $\omega_i(a, da)$ be the corresponding values of the parameters as previously defined for this infinitesimal transformation. Then

$$x_i' = T_a^{-1}T_{a+da}x_i = x_i + \sum_{j=1}^{r} \xi_{ij}(x)\omega_j(a; da)$$

or

$$T_a^{-1}T_{a+da} = T_\omega. \tag{13}$$

We shall call the corresponding Pfaffian forms $\omega_j(a; da)$ of the r parameters a_i the *relative components* of $\mathscr{R}_a$ (or also the *relative components of the group* $\tilde{\mathscr{G}}$) or, the *forms of Maurer–Cartan.*

We next want to deduce a relation between the value of ϕ_i for a change δx in x and the value of ϕ_i for a small change da in a. For this purpose we note that we obtained

$$x_i' = x_i + \sum_{j=1}^{r} \xi_{ij}(x)\omega_j(a; da)$$

or

$$x + \delta x = T_a^{-1} T_{a+da} x,$$

so that

$$T_a(x + \delta x) = T_{a+da} x,$$

which can be written as (in the first approximation)

$$\phi_i(x + \delta x, a) = \phi_i(x, a + da), \qquad i = 1, \ldots, n. \tag{14}$$

This set of equations will be useful in Section III.

Three Properties of the Relative Components

The results of Chern ([21, pp. 89, 92]), which we have in mind here, heavily use three properties of the Pfaffian forms $\omega_i(a; da)$. In our case the counterparts of these properties are contained in Lemmas 1.6–1.8.

Note. In our notation each $\omega_i(a; da)$ is actually the value of a first-degree form at a specific argument. This slight difference between our notation and that of [2, pp. 18–24] should not cause confusion.

LEMMA 1.6. *The operator matrix* $((A_k{}^i(a)))_{1 \leqslant i, k \leqslant r}$ *in the preceding representation of the* ω_i *(see (13), (6)) yields a bijective linear mapping* $((A_k{}^i(a)))$: $\mathscr{B}^r \xleftrightarrow{\text{onto}} \mathscr{B}^r$.

Proof. We recall the equation (for a fixed $a \in \mathscr{B}^r$):

$$T_{a+da} = T_a T_\omega. \tag{7'}$$

This equation shows that given any values of $\omega_1(\cdot), \ldots, \omega_r(\cdot)$, there exist $da_i \in \mathscr{B}_i$, $i = 1, \ldots, r$ such that da and ω satisfy (7) with the $da_i \doteqdot 0$[2] if $\omega_i(\cdot) \doteqdot 0$. This proves the lemma.

[2] $\doteqdot$ means "approximately equal to."

These assumptions about $\widetilde{\mathscr{G}}$ and the parameter group $\mathbb{P}$ and the infinitesimal transformation T_ω and the consequent results now imply the following facts. To each element $a \in \mathscr{B}^r$ corresponds one and only one element $T_a \in \widetilde{\mathscr{G}}$. Also we have seen that each $T_a \in \widetilde{\mathscr{G}}$ determines a diffeomorphism $\mathscr{B}^r \xleftrightarrow{\text{onto}} \mathscr{B}^r$ by virtue of the mapping $\alpha \to \alpha'$ determined by T_a, i.e.,

$$\mathsf{T}_{\alpha'} = \mathsf{T}_a \mathsf{T}_\alpha .$$

We shall denote this diffeomorphism by S_a; then $S_a \in \mathbb{P}$. Furthermore, by our assumption about T_ω:

$$S_a(\alpha + d\alpha) - S_a(\alpha) = ((A_k{}^i(\alpha))) \cdot d\alpha + o(\|d\alpha\|), \tag{15}$$

and by (7) and Lemma 1.6, the operator matrix $((A_k{}^i(\alpha)))$ yields a linear homeomorphism $\mathscr{B}^r \xleftrightarrow{\text{onto}} \mathscr{B}^r$, and $((A_k{}^i(\alpha)))$ is C^1-smooth with respect to $\alpha \in \mathscr{B}^r$.

Let S_a be such a diffeomorphism. Then the differential $dS_a(0)$ at the point zero, i.e., at the origin, will map the subspaces $\mathscr{B}_i$ onto subspaces which will be denoted by $H_i(a)$, respectively, whose direct sum is $\mathscr{B}^r : H_1(a) \oplus \cdots \oplus H_r(a) = \mathscr{B}^r$ (the last part follows, e.g., by Palais [20, Theorem and Corollary on p. 22]). The next lemma supplements the results of Lemma 1.6.

Lemma 1.7. *The operator matrix* $((A_k{}^i(a)))_{1 \leqslant i, k \leqslant r}$ *in the previous representation of the* $\omega_i(a; da)$ *has the following property: the operator equation*

$$[A_1(a), \ldots, A_r(a)] = [B_1(a), \ldots, B_r(a)] \cdot ((A_k{}^i(a))),$$

for a given r-tuplet $[A_1(a), \ldots, A_r(a)]$ *with* $A_j(a) \in \mathscr{L}(\mathscr{B}_j, \mathscr{B}^r)$, *has a unique solution* $[B_1(a), \ldots, B_r(a)]$ *where* $B_j(a) \in \mathscr{L}(\mathscr{B}_j, \mathscr{B}^r)$.

Note. The matrix operations here are as explained earlier. For instance, the product of $[B_1, \ldots, B_r]$ by $((A_k{}^i))$ is the row-by-column product, i.e., $[B_1, \ldots, B_r] \cdot ((A_k{}^i)) = [\sum_{i=1}^r B_i A_1{}^i, \sum_{i=1}^r B_i A_2{}^i, \ldots, \sum_{i=1}^r B_i A_r{}^i]$.

Proof. This is an immediate consequence of the bijective property of the operator $((A_k{}^i(a)))$ established in Lemma 1.6.

Definition 1.2. A system of r differential forms $\Omega_1, \ldots, \Omega_r$ of the first degree will be called *independent* if these r forms have the properties of Lemma 1.7.

Lemma 1.8. *The relative components* $\omega_i(a, da)$ *of a moving frame are invariant with respect to the parameter group* $\mathbb{P}$.

Proof. Let c be a fixed point in the parameter space $\mathscr{B}^r$. Then given a point $a \in \mathscr{B}^r$, let a' be the transform of a by c, i.e. [see (4)],

$$T_{a'} = T_c T_a. \tag{4'}$$

Changing a to $a + da$, the right side then changes to

$$T_c T_{a+da} = T_c T_a T_\omega$$

by (7′).

On the other hand, let $a' + da'$ be the point such that

$$T_{a'+da'} = T_c T_{a+da}. \tag{16}$$

Then

$$T_{a'+da'} = \begin{cases} T_c T_a T_\omega, \\ T_{a'} T_\omega, \end{cases} \tag{17}$$

which implies

$$T_{a'}^{-1} T_{a'+da'} = T_\omega = T_a^{-1} T_{a+da}. \tag{18}$$

This proves the lemma.

LEMMA 1.9. *The only Pfaffian forms invariant with respect to the parameter group* $\mathbb{P}$ *are the linear combinations with constant coefficients of the* $\omega_i(a; da)$.

Note. These constant coefficients are elements of $\mathscr{L}(\mathscr{B}_i, \mathscr{B}^r)$, $1 \leqslant i \leqslant r$.

Proof. Any Pfaffian form $\Omega(a; da)$ may be written as (Lemma 1.6)

$$\Omega(a; da) = \sum_{i=1}^{r} A_i(a) \cdot \omega_i(a; da).$$

Then using Lemma 1.8:

$$\Omega(a; da) - \Omega(a'; da') = \sum_{i=1}^{r} \{A_i(a) - A_i(a')\}\omega_i(a; da).$$

The invariance of Ω implies

$$A_i(a) = A_i(a'), \qquad 1 \leqslant i \leqslant r.$$

Since the group $\mathbb{P}$ is transitive, the A_i are constants (elements of $\mathscr{L}(\mathscr{B}_i, \mathscr{B}^r)$, $1 \leqslant i \leqslant r$).

Notation 1.1. In the sequel we shall use the phrase, *linear combinations with constant coefficients of the* ω_i, meaning thereby linear combinations as in the preceding lemma. Furthermore, it will be convenient at times to consider linear combinations of ω_is such that $\Omega_i = \sum_{k=1}^{r} B_k{}^i \omega_k$, $i = 1, \ldots, r$, such that $((B_k{}^i))$ yields a bijective linear mapping $\mathscr{B}^r \xrightarrow{\text{onto}} \mathscr{B}^r$, and consider such Ω_is to be the forms of Maurer–Cartan.

3

At this point we need an appropriate definition of the concept of $(r-h)$-variety in $\mathscr{B}^r$, and also a wider concept of parameter r-tuplet than before (see Introduction).

Let $\mathbb{Q}$ be the set of all C^2-smooth diffeomorphisms $S: \mathscr{B}^r \xrightarrow{\text{onto}} \mathscr{B}^r$ with the properties: For $\alpha \in \mathscr{B}^r$

$$S(\alpha + d\alpha) - S(\alpha) = ((A_k{}^i(\alpha))\, d\alpha + o(\|d\alpha\|),$$

where

(i) for each $\alpha \in \mathscr{B}^r$, $((A_k{}^i(\alpha)))_{1 \leqslant i, k \leqslant r}$ yields a linear homeomorphism $\mathscr{B}^r \xrightarrow{\text{onto}} \mathscr{B}^r$,

(ii) $((A_k{}^i(\alpha)))$ is C^1-smooth with respect to $\alpha \in \mathscr{B}^r$, and

(iii) for each $\alpha \in \mathscr{B}^r$, $((A_k{}^i(\alpha)))$ satisfies the contention of Lemma 1.7.

DEFINITION 1.3. By a *smooth* $(r-h)$*-variety* in $\mathscr{B}^r$ we shall mean the image of ${}_{r-h}\mathscr{B} + w$ where $w \in \mathscr{B}^r$ under a diffeomorphism $S: \mathscr{B}^r \xrightarrow{\text{onto}} \mathscr{B}^r$, where $S \in \mathbb{Q}$.

Our definition of "variety" agrees with traditional and current usage of the word "variety" (cf. Santalo [21], De Rham [5], Kuiper [14, 15], Moulis [19], Whitney [23, pp. 304–307]).

We now want to define appropriate "coordinates" or "parameter r-tuples" for points in $\mathscr{B}^r$ with reference to the varieties V_i^h, W_j^{r-h} (see below), and also to express a given $(r-h)$-variety be means of convenient vector equations. To this end, we proceed as follows.

Let $S \in \mathbb{Q}$, $S(\mathscr{B}^h) = V^h$, $S({}_{r-h}\mathscr{B}) = W^{r-h}$. Then both V^h, W^{r-h} are submanifolds in $\mathscr{B}^r$. Since S is a diffeomorphism of $\mathscr{B}^r$ onto $\mathscr{B}^r$, therefore the tangent space $T(\mathscr{B}^r)_0$, i.e., $\mathscr{B}^r$, is linearly homeomorphic under $dS(0)$ to the tangent space of $S(\mathscr{B}^r)$ ($= \mathscr{B}^r$!), i.e., to $\mathscr{B}^r$ at $a = S(0)$, and $dS(0)(\mathscr{B}^h)$ is linearly homeomorphic to $T(V^h)_a$, as also $dS(0)({}_{r-h}\mathscr{B})$ is linearly homeomorphic to $T(W^{r-h})_a$. Hence, $T(V^h)_a$, $T(W^{r-h})_a$ the tangent spaces of V^h, W^{r-h} at $a = S(0)$, are direct summands in $\mathscr{B}^r$ (cf. Palais [20], p. 22, Theorem and Corollary).

Further, the diffeomorphism S is transversal to both the submanifolds V^h and W^{r-h} at 0. It follows that $\mathscr{B}^r$ is isomorphic to the product manifold $V^h \times W^{r-h}$. By induction, $\mathscr{B}^r$ is diffeomorphic to the product manifold $V_1{}^h \times \cdots \times V_h{}^h \times W_1^{r-h} \times \cdots \times W_{r-h}^{r-h}$, where $V_i^h = S(\mathscr{B}_i)$, $1 \leqslant i \leqslant h$, $W_j^{r-h} =$

$S(\mathscr{B}_{h+j})$, $1 \leqslant j \leqslant r-h$. We shall denote a point in this product manifold by $(y_1, \ldots, y_h; y_{h+1}, \ldots, y_r)$, with $y_i \in V_i^h$, $1 \leqslant i \leqslant h$, $y_{h+j} \in W_j^{r-h}$, $1 \leqslant j \leqslant r-h$.

DEFINITION 1.4. The r-tuples $(y_1, \ldots, y_r)$ corresponding to points $y \in \mathscr{B}^r$, where $y_i \in V_i^h$, $1 \leqslant i \leqslant h$, $y_{h+j} \in W_j^{r-h}$, $1 \leqslant j \leqslant r-h$, will be called *coordinates* or *parameter r-tuples* determined by the varieties V_i^h and W_j^{r-h}.

Remark 1.1. We note here for future reference that we can also define *local coordinates* or *local parameter r-tuples determined* by the varieties V_i^h and W_j^{r-h}. This can be done by using the exponential map $\exp: N_e \to U_e$, where e is the identity element of $\mathbb{P}$, U_e a neighborhood of e, and N_e a neighborhood of 0 in the tangent space T_e at e (see, e.g., Maissen [17, p. 245], for properties of this exponential map). Since we shall not use these "local" coordinates in this chapter, we shall omit the details here.

We shall denote by d_j the differential of the mapping $S^{-1}|_{V_j^h}$, $1 \leqslant j \leqslant h$, and by d_{h+j} the differential of the mapping $S^{-1}|_{W_j^{r-h}}$, $1 \leqslant j \leqslant r-h$. Then we note that by the Theorem and Corollary on p. 22 of Palais [20] and by the remarks in Section I.1 that the operator matrix $(\{P_i \circ d_j(S^{-1}|_{V_j^h})\})_{1 \leqslant i, j \leqslant h}$ yields a linear homeomorphism of $T(V^h)_a$ (respectively, $T(W^{r-h})_a$) with $\mathscr{B}^h$ (respectively, ${}_{r-h}\mathscr{B}$), where P_i is the projection of $\mathscr{B}^r$ onto $\mathscr{B}_i$, $1 \leqslant i \leqslant r$.

Let $S \in \mathbb{Q}$ and let P_i be the projection of $\mathscr{B}^r$ onto $\mathscr{B}_i$, $1 \leqslant i \leqslant h$. Then the $(r-h)$-variety $V = S({}_{r-h}\mathscr{B} + w)$ will be given by the vector equations:

$$P_i \circ S^{-1}(y) - \xi_i \equiv P_i \circ S^{-1}((y_1, \ldots, y_h; y_{h+1}, \ldots, y_r)) = 0, \quad 1 \leqslant i \leqslant h, \tag{19}$$

where $\xi_i = P_i(w)$, $1 \leqslant i \leqslant h$, and y is a variable point on V. We can write these vector equations as

$$F_i(y_1, \ldots, y_r; \xi_1, \ldots, \xi_h) = 0, \qquad 1 \leqslant i \leqslant h. \tag{20}$$

These equations have Property $\mathbb{B}$.

PROPERTY $\mathbb{B}$. *The functions F_i are C^2-smooth jointly with respect to $y = (y_1, \ldots, y_r)$ and $\xi = (\xi_1, \ldots, \xi_h)$ varying over $\mathscr{B}^r = V_1{}^h \times \cdots \times V_h{}^h \times W_1^{r-h} \times \cdots \times W_{r-h}^{r-h}$ and over $\mathscr{B}^h = \mathscr{B}_1 \oplus \cdots \oplus \mathscr{B}_h$, respectively; the operator matrix $((d_j F_i))_{1 \leqslant i, j \leqslant h}$ yields a linear homemorphism: $H^h \xrightarrow{\text{onto}} \mathscr{B}^h$, where $H^h = T(V^h)_a$ and the operator matrix $((d_{r+j} F_i))_{1 \leqslant i, j \leqslant h}$ yields a linear homeomorphism $\mathscr{B}^h \xrightarrow{\text{onto}} \mathscr{B}^h$.*

Next, let $\mathscr{F}$ be a family of smooth $(r-h)$-varieties with Property $\mathbb{A}$.

PROPERTY $\mathbb{A}$. *Through each point of the space $\mathscr{B}^r$ there passes one and only one variety of the family $\mathscr{F}$.*

Then it is clear that this family $\mathscr{F}$ can be parametrized by the points $\xi = (\xi_1, \ldots, \xi_h) \in \mathscr{B}^h$, and furthermore the functions F_i in the vector equations $F_i = 0$ $(1 \leqslant i \leqslant h)$ which represent the varieties of $\mathscr{F}$, are C^2-smooth jointly with respect to $y \in \mathscr{B}^r$, $\xi \in \mathscr{B}^h$, and have the further properties stated in Property $\mathbb{B}$.

Notation 1.2. Henceforth, a family $\mathscr{F}$ of smooth $(r - h)$-varieties V satisfying the hypothesis $\mathbb{A}$ will be denoted by h vector equations

$$F_i(y;\xi) = 0, \qquad i = 1, \ldots, h, \tag{21}$$

where now ξ varying over $\mathscr{B}^h$ is taken to parametrize the family $\mathscr{F}$, for each fixed $\xi \in \mathscr{B}^h$, $y \in \mathscr{B}^r$ varies over the particular variety V corresponding to the parameter value ξ, and the F_i have Property $\mathbb{B}$.

We shall next formulate a satisfactory definition of complete integrability of a system of differential equations obtained by equating to zero h differential forms of the first degree: $\Omega_1, \ldots, \Omega_h$. Our definition agrees with the traditional definition (cf. Santalo [21], Dieudonné [6, I, IV]).

DEFINITION 1.5. The system of differential equations

$$\Omega_1 = 0, \ldots, \Omega_h = 0, \tag{22}$$

where $\{\Omega_1, \ldots, \Omega_h\}$ form a system of Pfaffian forms, is called *completely integrable* iff there exists a family $\mathscr{F}$ of smooth $(r - h)$-varieties $V = V_{r-h}$ satisfying Property $\mathbb{A}$ (see 1.16) and such that the differential equations (22) are identically satisfied when the point moves on a variety $V \in \mathscr{F}$, and under these circumstances we call these varieties $V \in \mathscr{F}$ the *integral varieties* of the system (22).

Conversely suppose we are given a family $\mathscr{F}$ of smooth $(r - h)$-varieties $V = V_{r-h}$ such that $\mathscr{F}$ has Property $\mathbb{A}$. We shall adhere to the preceding notation (see 1.16). Then the varieties of $\mathscr{F}$ can be considered as the integral varieties of the system of differential equations

$$\sum_{j=1}^{r} d_j F_i \, dy_j = 0, \qquad i = 1, \ldots, h, \tag{23}$$

where the ξ_i (see Notation 1.2) are supposed to be replaced by functions of y defined by (21).

DEFINITION 1.6. Consider two direct sum decompositions of $\mathscr{B}^h$ $(1 \leqslant h \leqslant r)$:

$$\mathscr{B}^h = V_1 \oplus \cdots \oplus V_h = W_1 \oplus \cdots \oplus W_h,$$

where the V_is and the W_js are, respectively, images of $\mathscr{B}_i$s under some linear homeomorphisms $\phi, \psi: \mathscr{B}^h \xrightarrow{\text{onto}} \mathscr{B}^h$. Let

$$(\omega): \omega_1 = 0, \ldots, \omega_h = 0, \tag{24}$$

$$(\Omega): \Omega_1 = 0, \ldots, \Omega_h = 0 \tag{25}$$

be two completely integrable Pfaffian systems defined on $\mathscr{B}^h$

$$\omega_i = \sum_{k=1}^{h} A_k{}^i\, da_k, \qquad da_k \in V_k, \quad 1 \leqslant k \leqslant h,$$

$$\Omega_i = \sum_{l=1}^{h} B_l{}^i\, db_l, \qquad db_l \in W_l, \quad 1 \leqslant l \leqslant h,$$

with $A_k{}^i \in \mathscr{L}(V_k, \mathscr{B}^h)$, $B_l{}^i \in \mathscr{L}(W_l, \mathscr{B}^h)$, and such that the operator matrices $((A_k{}^i))$, $((B_l{}^j))$ both yield linear homeomorphisms $\mathscr{B}^h \xrightarrow{\text{onto}} \mathscr{B}^h$. The two Pfaffian systems (ω) and (Ω) will be said to be *equivalent* if

$$\Omega_j(\cdot) = \sum_{m=1}^{h} \Omega_m{}^j \omega_m, \qquad j = 1, \ldots, m,$$

where the $\Omega_m{}^j(\cdot)$ are C^1-smooth linear operators: $\Omega_m{}^j(\cdot) \in \mathscr{L}(\mathscr{R}(\omega_m), \mathscr{B}^h)^3$ such that the operator matrix $((\Omega_m{}^j))_{1 \leqslant j, m \leqslant h}$ yields a linear homeomorphism $\mathscr{B}^h \xrightarrow{\text{onto}} \mathscr{B}^h$.

We saw previously that any differential form $\Omega(a; da)$ of the first-degree invariant under the parameter group $\mathbb{P}$ is a linear combination with constant coefficients (see Notation 1.1) of the $\omega_i(a; da)$ (Lemma 1.9). The next result is a more general one.

THEOREM 1.1. *If the system*

$$\Omega_1 = 0, \ldots, \Omega_h = 0 \tag{26}$$

formed by h independent (Definition 1.2) Pfaffian forms (i.e., differential forms of the first degree) is completely integrable and is invariant under $\mathbb{P}$, *then it is equivalent to the system formed by equating to zero h linear combinations with constant coefficients (Notation 1.1) of* $\omega_i(a; da)$.

Proof. Recall Definition 1.5 and note that the integral varieties of (26) are given by (Notation 1.2)

$$F_i(y; \xi) \equiv F_i(y_1, \ldots, y_r; \xi_1, \ldots, \xi_h) = 0, \qquad i = 1, \ldots, h \tag{27}$$

[3] $\mathscr{R}(\omega_m)$ is the range of $\mathscr{B}^h$ under ω_m.

with the F_i satisfying Properties $\mathbb{A}$ and $\mathbb{B}$. By our very choice, the equations $F_i = 0$ are already solved for the "parameters" ξ_i, $1 \leqslant i \leqslant h$. We therefore write equations (27) as

$$\phi_i(y) = \xi_i, \qquad i = 1, \ldots, h, \tag{28}$$

where we have put $\phi_i(y) \equiv P_i \circ S^{-1}(y)$ (see Definition 1.5), with $y = (y_1, \ldots, y_r)$ varying over $\mathscr{B}^r = V_1{}^h \times \cdots \times V_h{}^h \times W_1^{r-h} \times \cdots \times W_{r-h}^{r-h}$ corresponding to $S \in \mathbb{Q}$ and $\xi = (\xi_1, \ldots, \xi_h) \in \mathscr{B}^h$. Also by our very choice, the given system (26) is equivalent to the system

$$d\xi_1 = 0, \quad \ldots, \quad d\xi_h = 0. \tag{29}$$

For convenience we shall write equations (28) as $\phi_i(y', y'') = \xi_i$, $1 \leqslant i \leqslant h$, where we put $y = (y', y'')$ with

$$y' = (y_1, \ldots, y_h) \in V_1{}^h \times \cdots \times V_h{}^h, \tag{30}$$

$$y'' = (y_{h+1}, \ldots, y_r) \in W_1^{r-h} \times \cdots \times W_{r-h}^{r-h}. \tag{31}$$

We can now apply the Implicit function theorem [6, Vol. I, p. 273] since

$$\phi: V^h \times W^{r-h} \to \mathscr{B}^h$$

is a mapping (onto), $D_y\phi$ exists and is continuous with respect to $y \in \mathscr{B}^r$, and for each $y' \in V^h$, $D_{y'}\phi: H^h(a) \to \mathscr{B}^h$ is a linear homeomorphism $H^h(a) \xrightarrow{\text{onto}} \mathscr{B}^h$. Using the Implicit function theorem, we then solve the equations

$$\phi_i(y', y'') = \xi_i, \qquad 1 \leqslant i \leqslant h$$

for y' in terms of ξ, y'' with y', y'' as in (30), (31), viz.,

$$\psi_i(\xi, y'') = y_i', \qquad 1 \leqslant i \leqslant h, \tag{32}$$

where the ψ_i are also C^2-smooth in all their variables ξ_i, y_j''. This suggests that we should now regard the $(r - h)$-integral varieties of (26) as "parametrized" by the h-tuplets $(y_1', \ldots, y_h') \in V^h$ instead of by the h-tuplets $(\xi_1, \ldots, \xi_h) \in \mathscr{B}^h$ (the latter being the parametrization earlier), and that we should consider the r-tuplets

$$(\xi_1, \ldots, \xi_h; y_{h+1}, \ldots, y_r)$$

as coordinates (Definition 1.4) for points in $\mathscr{B}^r$.

Therefore we next note that the r-tuplet $(\xi_1, \ldots, \xi_h, \lambda_1, \ldots, \lambda_{r-h})$, where the ξ_is are as previously determined [i.e., $(\xi_1, \ldots, \xi_h) \in \mathscr{B}^h$] and we put

$$y_{h+1} = \lambda_1, \ldots, y_r = \lambda_{r-h},$$

is clearly a parameter r-tuplet in $\mathbb{P}$ because the product manifold $\mathscr{B}^h \times \mathscr{W}^{r-h}$ is C^2-diffeomorphic to the product manifold $\mathscr{B}^h \times {}_{r-h}\mathscr{B}$, which is C^2-

diffeomorphic to $\mathscr{B}^r$. We then change the parameters in $\mathbb{P}$ to the parameter r-tuplets $(\xi_1, \ldots, \xi_h, \lambda_1, \ldots, \lambda_{r-h})$. The relative components ω_i will then be expressible as

$$\omega_i = \omega_i(\xi, d\xi; \lambda, d\lambda).$$

Let λ^0, ξ^0 be a fixed point of the parameter space. Then we form h linear combinations of $\omega_i(\xi^0, d\xi^0; \lambda^0, d\lambda^0)$ (with coefficients which are C^1, with respect to ξ^0, λ^0) such that in these linear combinations the differentials $d\lambda_1{}^0, \ldots, d\lambda_{r-h}^0$ do not appear and further such that if we express these linear combinations as $\alpha\omega$, then we have

$$\alpha\omega = \beta\, d\xi,$$

where β is a linear homeomorphism $\mathscr{B}^h \xrightarrow{\text{onto}} \mathscr{B}^h$, which is C^1-smooth with respect to (ξ, λ). We shall denote these by

$$\omega_k{}^*(\lambda^0, \xi^0, d\xi^0) = \sum_i \alpha_i{}^k(\xi^0, \lambda^0)\omega_i(\xi^0, d\xi^0; \lambda^0, d\lambda^0), \qquad 1 \leqslant k \leqslant h.$$

Consider the Pfaffian forms

$$\omega_i(\lambda, d\lambda; \xi, d\xi) = \sum_i \alpha_i{}^k(\xi^0, \lambda^0)\omega_i(\xi, d\xi; \lambda, d\lambda), \qquad 1 \leqslant k \leqslant h.$$

These are linear combinations with constant coefficients of the original relative components ω_i.

Going back, the system

$$\Omega_i(y^0, dy^0) = 0, \qquad i = 1, \ldots, h,$$

or its equivalent

$$d\xi_i{}^0 = 0, \qquad i = 1, \ldots, h,$$

which by assumption is invariant under the group $\mathbb{P}$ will be equivalent to the system

$$\Omega_i(y, dy) = 0, \qquad i = 1, \ldots\ h.$$

On the other hand, the system $d\xi_i{}^0 = 0$ $(i = 1, \ldots, h)$ is equivalent to (by our definition of the ω_i):

$$\omega_i(\xi^0, d\xi^0, \lambda^0, d\lambda^0) = 0, \qquad i = 1\ \ldots, h.$$

Next, this system, owing to the invariance of the relative components under the parameter group $\mathbb{P}$, is equivalent to the system

$$\omega_i(\xi, d\xi; \lambda, d\lambda) = 0, \qquad i = 1, \ldots, h.$$

As corollaries we obtain:

COROLLARY 1.1. *If the system* $\Omega_1 = 0, \ldots, \Omega_h = 0$ *is completely integrable, then (with the preceding notation) each* Ω_i *is expressible as*

$$\Omega_i = \sum_{k=1}^{h} \gamma_{ik}\, d\xi_k, \qquad i = 1, \ldots, h,$$

the coefficients γ_{ik} *being, in general, functions of* ξ_i.

COROLLARY 1.2. *A necessary condition for the system* $\Omega_1 = 0, \ldots, \Omega_h = 0$ *to be completely integrable is that the exterior derivatives* $\Omega_i' = 0$ *when the equations of the system are taken into account.*

II. DENSITY AND MEASURES IN HOMOGENEOUS SPACES

Next we turn to the question of introducing convenient measures on sets of submanifolds Π of $\mathscr{k}^n$, these measures being invariant with respect to a given group $\tilde{\mathscr{G}}$ of the kind previously described.

Let $\tilde{\mathscr{G}}$ be a r-parameter Lie group, the parameters being as described [viz., $a = (a_1, \ldots, a_r) \in \mathscr{B}^r$]. The group $\tilde{\mathscr{G}}$ is also assumed to be such that it transforms the elements Π *transitively*. The transformations of $\tilde{\mathscr{G}}$ that leave invariant a certain fixed element Π form a subgroup g. We shall suppose that the subgroup g is a closed Lie subgroup depending on a certain set of elements belonging to a $(r - h)$-variety in $\mathscr{B}^r$.

For convenience we shall denote both the subgroup as well as the $(r - h)$-variety related to it, by the same symbol, viz., g_{r-h}.

Now consider transformations $T_s \in \mathscr{G}$ such that T_s does not belong to g_{r-h}. Then the set $T_s g_{r-h}$ or sg_{r-h}, for short, will denote the image of g_{r-h} under the transformation s of $\mathbb{P}$. These varieties sg_{r-h} fill all of the parameter space, and two of them cannot have a common point without being identical. For

$$\begin{aligned} a \in sg_{r-h} \cap tg_{r-h} &\Rightarrow a = sa_1 = ta_2 \qquad \text{for} \quad \text{some } a_1, a_2 \in g_{r-h} \\ &\Rightarrow s = t(a_2 a_1^{-1}) \Rightarrow sg_{r-h} = t(a_2 a_1^{-1}) g_{r-h} = tg_{r-h} \\ &\Rightarrow sg_{r-h}, tg_{r-h} \text{ are identical.} \end{aligned}$$

From the earlier discussion it follows that the varieties sg_{r-h} are integral varieties of a completely integrable Pfaffian system, and the totality of these varieties sg_{r-h} is invariant with respect to the parameter group $\mathbb{P}$.

Hence by Theorem 1.1 the left-hand sides of the Pfaffian system corresponding to the varieties sg_{r-h} are linear combinations with constant coefficients of the relative components $\omega_1, \ldots, \omega_h$ of $\tilde{\mathscr{G}}$. These relative components are determined up to a linear transformation with constant coefficients (see earlier notation). Hence we may suppose the integrable Pfaffian system whose integral varieties are sg_{r-h} to be formed by the first h relative components, i.e., $\omega_1 = 0, \ldots, \omega_h = 0$.

Now let

$$\phi_i(a_1, \ldots, a_r) = \xi_i, \qquad i = 1, \ldots, h$$

be h independent integrals of the above Pfaffian system, that is, to each point $(\xi_1, \ldots, \xi_h) \in \mathscr{B}^h$ corresponds an integral variety. We may also consider the ξ_i as the coordinates of a point ξ in a certain homogeneous space, which we may also denote by $E_h(\xi)$, viz., $\tilde{\mathscr{G}}(r)/g_{r-h}$. To each point ξ in this homogeneous space, there corresponds an integral variety of the above system and also a geometrical element, viz., a submanifold that will be a transform $T_s\Pi$ of the original fixed element Π. Hence we may consider the h-tuple $(\xi_1, \ldots, \xi_h)$ as the coordinates of the elements Π of $\mathscr{B}^h$.

At this point we need to consider certain operator-valued measures and integrals. We first consider a suitable field Σ of subsets such that each $E \in \Sigma$ is a union of integral varieties of completely integrable systems $\{\omega_1^{(\alpha)} = 0, \ldots, \omega_h^{(\alpha)} = 0\}$ $\alpha \in I$, I being some index set and ωs being previously defined. Write $\Omega_h^{(\alpha)} = \omega_1^{(\alpha)} \wedge \cdots \wedge \omega_h^{(\alpha)}$ and also $\mathbf{\Omega}_h(E) = (\boldsymbol{\omega})_1 \wedge \cdots \wedge (\boldsymbol{\omega})_h(E) = \bigcup_{\alpha \in I} \Omega_h^{(\alpha)}$. We shall further suppose that $\mathbf{\Omega}_h(\cdot)$ is an additive set function on the sets of Σ. The *existence* of such a situation is exhibited in Section III, *independently* of the proof of Theorem 2.1. Furthermore let f be a bounded function from X to $\mathbb{R}$ (or X to $\mathbb{C}$), where X is a set in Σ, and we suppose f is measurable with reference to Σ. Then the integral of f with reference to μ defined as in Dunford and Schwartz [7, Vol. II, pp. 891–892], is what we shall denote by

$$\int_X f(x)\, d\boldsymbol{\omega}_1 \wedge \cdots \wedge d\boldsymbol{\omega}_h(x) \qquad \text{or} \qquad \int_X f\, d\boldsymbol{\omega}_1\, d\boldsymbol{\omega}_2 \cdots d\boldsymbol{\omega}_h.$$

We return to our present problem of finding a measure for the elements Π, invariant under $\mathscr{G}$, which is seen to be equivalent to that of finding a measure for the points ξ of $E_h(\xi)$ invariant with respect to the group $\mathbb{P}$. Given a set X of points ξ, we consider a "measure" that we want to be able to write as an integral

$$\int_X f\, d\boldsymbol{\omega}_1 \cdots d\boldsymbol{\omega}_h,$$

where the integral is to be understood in the sense of the preceding paragraph; that is to say, we first decide on a suitable field Σ of sets of points ξ in $E_h(\xi)$, X is a set in Σ, and f a bounded real-valued function on X measurable with reference to Σ.

Now we have seen that the Pfaffian system $\omega_1 = 0, \ldots, \omega_h = 0$ is equivalent to the system $d\xi_1 = 0, \ldots, d\xi_h = 0$. Hence each $d\xi_i$ must be a linear combination, with coefficients in general depending on $a_1, \ldots, a_r$ of the $\omega_i(y, dy)$. Thus $d\xi_i = \sum_{k=1}^{h} \gamma_{ik}\omega_k$, $i = 1, \ldots, h$.

We next note that the function

$$\mathbf{\Omega}_h(\cdot): \Sigma \to 2^{\Lambda^h(B)},$$

the value of which at an element in Σ is the collection of differential forms:

$$\mathbf{\Omega}_h(\cdot) = \boldsymbol{\omega}_1 \wedge \boldsymbol{\omega}_2 \wedge \cdots \wedge \boldsymbol{\omega}_h(\cdot)$$

which is invariant with reference to $\mathbb{P}$, but may not always be a measure on Σ, because it depends on the r parameters y_i. Each point $y = (y_1, \ldots, y_r)$ determines an element Π; but each element Π has associated with it *all the points* a of the corresponding variety sg_{r-h}. In order that $\mathbf{\Omega}_h$ may be a measure for the elements Π, invariant under $\tilde{\mathscr{G}}$, its value must not change when the points a are displaced on the manifolds sg_{r-h}.

We shall now prove the following theorem, the object of which is to obtain a suitable criterion for $\mathbf{\Omega}_h$ to be a measure for the elements Π, invariant under $\tilde{\mathscr{G}}$.

THEOREM 2.1. *A necessary and sufficient condition for $\mathbf{\Omega}_h$ to be a density (i.e., element of measure) for the elements Π, is that its exterior derivative vanish*: $d\mathbf{\Omega}_h = 0$ (*i.e.*, $d\Omega_h^{(\alpha)} = 0$ *for each* $\Omega_h^{(\alpha)} \in \mathbf{\Omega}_h$).

Proof. We recall that the submanifold g_{r-h} and its left cosets fill up the Lie group manifold $\tilde{\mathscr{G}}$ in such a way that through each point of $\tilde{\mathscr{G}}$ passes one and only one of these submanifolds (i.e., left cosets). Thus the system

$$\omega_1 = 0, \ldots, \omega_h = 0$$

is completely integrable and equivalent to a system of the form

$$d\xi_1 = 0, \ldots, d\xi_h = 0,$$

where $\xi_i = \xi_i(y_1, \ldots, y_r)$ are functions of y such that the manifolds sg_{r-h} are represented by ξ_i a constant ($i = 1, \ldots, h$). Now, as before, we make in $\mathscr{G}$ the change of local coordinates

$$(y_1, \ldots, y_r) \to (\xi_1, \ldots, \xi_h, \lambda_1, \ldots, \lambda_{r-h}),$$

where $\lambda_1 = y_{h+1}, \ldots, \lambda_{r-h} = y_r$. As noted above, the two systems

$$\omega_1 = 0, \ldots, \omega_h = 0 \quad \text{and} \quad d\xi_1 = 0, \ldots, d\xi_h = 0$$

are equivalent.

More specifically proceeding exactly as in the proof of our earlier Theorem 1.1 we form h linear combinations of $\omega_i(\xi_0, d\xi^0; \lambda^0, d\lambda^0)$ (with coefficients α^k which are C^1-smooth with respect to ξ^0, λ^0) such that in these linear combinations, the differentials $d\lambda_1{}^0, \ldots, d\lambda_{r-h}^0$ *do not appear* and, further, such that if we express these linear combinations as $\alpha\omega$, then we have

$$\alpha\omega = \beta\, d\xi,$$

where β is a linear homeomorphism $\mathscr{B}^h \xrightarrow{\text{onto}} \mathscr{B}^h$. Since by virtue of our earlier assumptions we already have

$$\omega = A \begin{bmatrix} d\xi \\ d\lambda \end{bmatrix},$$

therefore it follows that

$$\alpha_1 \begin{bmatrix} \omega_1 \\ \vdots \\ \omega_h \end{bmatrix} = \beta \begin{bmatrix} d\xi_1 \\ \vdots \\ d\xi_h \end{bmatrix},$$

where

$$\alpha_1 = \left[\begin{array}{c|c} \multicolumn{2}{c}{\alpha} \\ \hline 0_{r-h,\,h} & 1_{r-h} \end{array}\right].$$

Without loss of generality we assume β to be the identity $1_h\colon \mathscr{B}^h \to \mathscr{B}^h$. The matrix α_1 yields a linear homeomorphism: $\mathscr{B}^r \xrightarrow{\text{onto}} \mathscr{B}^r$ such that Φ is C^1 in ξ, λ. Then Φ^{-1} is again C^1 in ξ, λ, and

$$\Phi^{-1} \begin{bmatrix} d\xi_1 \\ \vdots \\ d\xi_h \end{bmatrix} = \begin{bmatrix} \omega_1 \\ \vdots \\ \omega_h \end{bmatrix}.$$

Now a slight adaptation of known arguments, viz., the proof of Theorem 2.9.1, p. 29 in Cartan [2] (or of [1]; this classical argument assumes the "change of variable" mapping Φ to be C^{n+1} and the mapping Φ^* of differential forms, obtained by this "change of variable" mapping Φ, maps C^n differential forms into C^n differential forms and their wedge products into wedge products), yields that

$$\Omega_h = \omega_1 \wedge \cdots \wedge \omega_h = \psi(\xi; \lambda)(d\xi_1 \wedge \cdots \wedge d\xi_h),$$

where we have written $\psi(\xi; \lambda)$ for Φ^{-1}. As noted, $\psi(\xi, \lambda)$ is C^1-smooth in ξ and λ. Next we note that in order for Ω_h to be a density of a measure on $\tilde{\mathscr{G}}/g_{r-h}$, its value should not change when the point y is displaced along the manifold sg_{r-h}, i.e., along the variety $\phi_i(y) = \xi_i$ $(1 \leqslant i \leqslant h)$. That is to say, in order for Ω_h to be a density of a measure on the manifolds sg_{r-h}, it is necessary and sufficient that $\delta\Omega_h = 0$ under a change in the variable *along* sg_{r-h}, where we denote by $\delta\Omega_h$ a change in Ω_h corresponding to a change in the variable y. We now change from the parameters y to the parameters ξ, λ as explained in the proof of Theorem 1.1. Then a change in the ξ_i and λ_j, denoted by $\delta\xi_i$ and $\delta\lambda_j$, will cause a change $\delta\Psi$ in $\Psi(\xi;\lambda)$ and we then have

$$\delta(\omega_1 \wedge \cdots \wedge \omega_h) = \delta\Psi(\xi;\lambda)(d\xi_1 \wedge \cdots \wedge d\xi_h).$$

Since Ψ is C^1-smooth with respect to ξ_i and λ_j, we have

$$\begin{aligned}\delta\Psi(\xi;\lambda) &= \Bigg(\sum_{i=1}^{h} (d/d\xi_i)\Psi(\xi,\lambda)\,\delta\xi_i + \sum_{j} (d/d\lambda_j)\Psi(\xi,\lambda)\,\delta\xi_j \\ &\qquad + o\Bigg(\sum_{j=1}^{r-h} \|d\lambda_j\|\Bigg)\Bigg)(d\xi_1 \wedge \cdots \wedge d\xi_h) \\ &= \Bigg(\sum_{j=1}^{r-h} (d/d\lambda_j)\Psi_{\xi,\lambda}(\xi;\lambda)\,\delta\lambda_j + o\Bigg(\sum_{j=1}^{r-h} \|\delta\lambda_j\|\Bigg)\Bigg)(d\xi_1 \wedge \cdots \wedge d\xi_h)\end{aligned}$$

for as a point $y = y(\xi_1, \ldots, \xi_h; \lambda_1, \ldots, \lambda_{r-h})$ varies on sg_{r-h} (i.e., on the variety $\phi_i(y) = \xi_i$, $1 \leqslant i \leqslant h$), the coordinates ξ_i are constant and hence $\delta\xi_i = 0$, $1 \leqslant i \leqslant h$. Hence we find

$$(\omega_1 \wedge \cdots \wedge \omega_h)' = \sum_{j=1}^{r-h} (d/d\lambda_j)\Psi(\xi,\lambda)(\delta\lambda;(d\xi_1 \wedge \cdots \wedge d\xi_h)),$$

where on the left side, the prime denotes the derivative. Next we take the composite of this with the map $\phi_{1,p}: \mathscr{A}_{1,p}(\mathscr{B}^h;\mathscr{B}^h) \to \mathscr{A}_{p+1}(\mathscr{B}^h;\mathscr{B}^h)$ (see Section I.1; also Cartan [2], pp. 20–21). We immediately conclude that $d\mathbf{\Omega}_h = 0$ is necessary and sufficient for $\mathbf{\Omega}_h$ to be a density of a measure on the collection of manifolds $\{sg_{r-h}\}$, i.e., the collection of varieties $\{\phi_i(y) = \xi_i,$ $1 \leqslant i \leqslant h\}$, $\xi = (\xi_1, \ldots, \xi_h) \in \mathscr{B}^h$. This completes the proof of the theorem.

III. The Group of Hilbertian Motions $\mathscr{H}(n)$

Our objective in this section will be to consider one nontrivial example of a Lie group of transformations, and also a measure on certain subsets of h-varieties in $\tilde{h}^n$ of the type discussed in the preceding sections. Our choice is motivated with a view toward further use, and this example is suggested by the classical group of Euclidean motions in real Euclidean n-space, which depends on $\frac{1}{2}n(n+1)$ real parameters (cf. Santalo [21]).

In this section we shall assume $n > 1$. The group that we deal with here will be called the *group of Hilbertian motions* in $\tilde{h}^n$, and denoted by $\mathscr{H}(n)$. It consists of transformations of $\tilde{h}^n \xrightarrow{\text{onto}} \tilde{h}^n$, of the type

$$x \to \alpha + \mathcal{O}x, \qquad \alpha \in \tilde{h}^n,$$

where $\mathcal{O}$ is an orthogonal transformation: $\tilde{h}^n \xrightarrow{\text{onto}} \tilde{h}^n$, α being a fixed vector.

The sets on which we exhibit the measures of this section are collections of h-varieties in $\tilde{h}^n$. We shall retain the earlier notation explained in Section 1. We shall also use a representation of the elements $x \in \tilde{h}^n$ by means of the unit spheres in $\tilde{h}_i$.

For our present purpose we shall construct certain basic differential elements which have been called "relative components" before, and the measures in question will be constructed from exterior products of these relative components.

We first want to exhibit a convenient set of parameters in $\mathscr{H}(n)$. This is done in the next theorem. In the remaining part of this section we shall take for the parameter space $\mathscr{B}$ of our Lie group in question the algebra $\mathrm{B} = \mathrm{B}(\tilde{h})$. For $\mathscr{A} \in \mathrm{B}(\tilde{h}^n)$ we define the *exponential mapping* $A \to \exp(A)$, by

$$\exp(A) = I + A + \frac{1}{2!} A^2 + \cdots + \frac{1}{n!} A^n + \cdots .$$

Let M^s [4, *p*. 7; 18, p. 73] be the set of skew-symmetric matrices in $\mathscr{M}_n(\mathrm{B})$ and $\mathcal{O}(\tilde{h}^n)$ the group of all orthogonal transformations $\tilde{h}^n \xrightarrow{\text{onto}} \tilde{h}^n$.

THEOREM 3.1. *There exists a neighborhood U of* $\mathbf{0}$ *in the algebra $\mathscr{M}_n(\mathrm{B})$ of $(n \times n)$ operator matrices where $\mathrm{B} = \mathrm{B}(\tilde{h})$, such that U is mapped diffeomorphically onto a neighborhood of I in $\mathcal{O}(\tilde{h}^n)$.*

Proof. The algebraic isomorphism $((a_{ij})) \to A$ from $\mathscr{M}_n(\mathrm{B})$ to $\mathrm{B}(\tilde{h}^n)$ has already been explained earlier (Section I.1). We now represent a matrix $A = ((a_{ij}))$ in $\mathscr{M}_n(\mathrm{B})$ by the "point" of B^{n^2} (where $\mathrm{B} = \mathrm{B}(\tilde{h})$), whose coordinates are the components a_{ij} of the matrix A, these components being arranged in some order (we shall keep this order fixed throughout). The series $\sum_{p=0}^{\infty} \|A\|^p/p!$ is convergent, where $\|A\|$ is the norm of A in $\mathrm{B}(\tilde{h}^n)$. Hence the components $y_{ij}(A)$ of $\exp(A)$ are analytic functions (according to the definition of analyticity; cf. Maissen [17, p. 248]) $F_{ij}(\cdots a_{kl} \cdots)$ of the components of A.

The terms of index <2 in the series for $F_{ij}(\cdots a_{kl} \cdots)$ are clearly $\delta_{ij}I + a_{ij}$, where $\delta_{ij} = 0$ for $i \neq j$ and 1 for $i = j$. Hence the $(n^2 \times n^2)$ Jacobian $D = ((D_{kl}))$ of the derivatives of the F_{ij} with respect to their n^2 arguments, at $a_{kl} = 0$ $(1 \leqslant k, l \leqslant n)$, is equal to the identity I. By virtue of the Implicit function theorem (Dieudonné [6, Vol. I, Theorem 10.2.5, p. 273]) there is an open neighborhood of $((a_{ij})) = ((o_{ij}))$, $((o_{ij}))$ being the zero operator, which is mapped topologically onto an open neighborhood of $((y_{ij})) = ((\delta_{ij}I))$.

The first part of the proposition now follows if $\|A\|$ is sufficiently small. For the second part, denoting by $\mathcal{O}$ an element of $\mathcal{O}(\tilde{\hbar}^n)$, close to the identity I, and $A \in \mathcal{U}$ such that $\exp(A) = \mathcal{O}$, we find

$$\exp({}^tA) = (\exp(A))^* = \mathcal{O}^* = \mathcal{O}^{-1} = (\exp(A))^{-1} = \exp(-A),$$

where ${}^tA$ denotes the transpose of A. Hence ${}^tA = -A$, which implies that A is skew-symmetric, and also this last argument is reversible. This completes the proof.

As a simple corollary, we obtain the following.

COROLLARY 3.1. *The group $\mathcal{H}(n)$ of Hilbertian motions in $\tilde{\hbar}^n$ depends on $r = \frac{1}{2}n(n+1)$ parameters.*

Proof. The group $\mathcal{O}(\tilde{\hbar}^n)$ is shown to depend on $r' = \frac{1}{2}n(n-1)$ parameters varying over $\mathrm{B}^{r'}$, where $\mathrm{B} = \mathrm{B}(\tilde{\hbar})$. Any transformation in $\mathcal{H}(n)$ is of the type $x \to \alpha + \mathcal{O}(x)$, $\alpha \in \tilde{\hbar}^n$, $\mathcal{O} \in \mathcal{O}(\tilde{\hbar}^n)$. Hence, adding to the above r parameters of $\mathcal{O}(\tilde{\hbar}^n)$, the n parameters which represent the coordinates of a point $\alpha = (\alpha_1, \ldots, \alpha_n)$ with respect to the frame $\{\tilde{\hbar}_1, \ldots, \tilde{\hbar}_n\}$, we obtain $r = \frac{1}{2}n(n+1)$ parameters for $\mathcal{H}(n)$, varying over $\tilde{\hbar}^n \oplus \mathrm{B}^{r'}$. Global parameters are established by using known results (see [7, III, p. 1963]).

Although to start (Section I) a general description of the Lie group, etc., is given (Assumptions (A)–(D)), at times it is convenient to proceed a little differently. Thus we shall next explain a representation of points in $\tilde{\hbar}^n$ by means of unit spheres in $\tilde{\hbar}_i$, and any r independent (Definition 1.2) linear combinations of the ω_i and ω_{ij} defined below can be assumed to be the forms of Maurer–Cartan. Let C_i denote the unit sphere in $\hbar_i : \|x_i\| = 1$, $i = 1, \ldots, n$. Then for any point $x \in \hbar^n$, $x = \sum_{k=1}^n x_k$, where $x_i \in \hbar_i$, $i = 1, \ldots, n$, we write

$$x = \sum_{k=1}^{n} t_k S_k(x), \tag{33}$$

where $S_k(\cdot)$ is the nonlinear function which maps each point $x = (x_1, \ldots, x_n)$ into the unique point on the unit sphere C_k in which the line joining 0 to x_k in $\hbar_k$ cuts C_k and $t_k = \|x_k\|$. We shall identify $(-t)S_k(x)$ with $t_k S_k(-x)$, where $t_k \geqslant 0$, $x \in \hbar^n$. This representation can conveniently be considered as a "frame"; and in this representation we call the t_k the "coordinates" of the point x with reference to the frame in question.

We now introduce an inner product in the collection of such linear combinations of $S_j(x)$ with nonnegative coefficients t_j. Let $x = (x_1, \ldots, x_n)$.

Then define:

$$(t_jS_j(x), t_kS_k(x')) \underset{\text{def}}{=} \delta_{jk}t_jt_k \cos(\theta_{jk}(x, x')), \tag{34}$$

where

$$\cos\theta_{jk}(x, x') = (x_j, x_k')/\|x_j\|\,\|x_k'\|, \quad \delta_{jk} = 0 \text{ if } j \neq k, \ = 1 \text{ if } j = k. \tag{35}$$

We shall call $S_j(x)$ "the jth unit sphere directed by the vector x_j" or "the jth unit sphere with orientation x_j." For our final calculations we shall need only the inner products between "like orientated" jth and kth oriented unit spheres, i.e., the inner products: $(t_jS_j(x), t_kS_k(x))$ [as distinct from the inner products between "unlike oriented" jth and kth unit spheres: $(t_jS_j(x), t_kS_k(x'))$].

We express the required differentials in convenient form, as follows. The earlier equation I (14) yields

$$A + \sum_{i=1}^{n} (t_i + \delta t_i)S_i(x) = A + dA + \sum_{i=1}^{n} t_i(S_i(x) + dS_i(x)).$$

We denote the coordinates of $A + dA$ with respect to $\mathscr{R}_a$ by ω_{ik}, the components of dS_i:

$$dA(x) = \sum_{k=1}^{n} \omega_k t_k S_k(x), \tag{36}$$

$$dS_i(x) = \sum_{k=1}^{n} \omega_{ik} t_k S_k(x). \tag{37}$$

Now the point A and the $S_k(\cdot)$ depend only on the parameters a_i in question and not on the variable point $x \in \mathscr{h}^n$. Also, the ω_i, ω_{ik} are Pfaffian forms in da_i. So from the preceding we obtain:

$$\sum_{i=1}^{n} \delta t_i t_i S_i(x) = \sum_{k=1}^{n} \omega_k t_k S_k(x) + \sum_{i,k=1}^{n} t_i \omega_{ik} S_k(x) \tag{38}$$

which implies

$$t_k\,\delta t_k = t_k\omega_k + \sum_{i=1}^{n} t_i\omega_{ik}. \tag{39}$$

Since this is true for all nonnegative real numbers t_k, we have

$$\delta t_k = \omega_k + \sum_{i=1}^{n} t_i\omega_{ik}. \tag{40}$$

Next the conditions expressing the relations between ω_{ij} and ω_{ji} (cf. the classical case, Santalo [23, p. 74]) are of considerable importance. To obtain these, we note that since

$$(t_iS_i(x), t_jS_j(x)) = \delta_{ij}t_it_j \tag{41}$$

for all nonnegative t_i and t_j where δ_{ij} is the Kronecker delta ($\delta_{ij} = 0$ if $i \neq j$, $= 1$ if $i = j$), we therefore have

$$(S_i(x), dS_j(x)) = -(dS_i(x), S_j(x)),$$

i.e.

$$\omega_{ij} = -\omega_{ji}. \tag{42}$$

For convenience, henceforth, we shall write $S_k(\cdot)$ instead of $S_k(x)$, for the reason mentioned, viz., we do not have to consider terms like $(S_j(x), S_k(x'))$. Then

$$\omega_k = (dA(\cdot), S_k(\cdot)). \tag{43}$$

These $\frac{1}{2}n(n + 1)$ independent parameters determine this particular group.

We proceed to verify that in the case of this group the relative components do satisfy the condition obtained earlier, for a density on the h-varieties which are h-subspaces of $\mathscr{h}^n$, $1 \leqslant h < n$ (see Santalo [21, pp. 118–119]).

As noted,

$$dA(\cdot) = \sum_{k=1}^{n} \omega_k S_k(\cdot), \tag{44}$$

$$dS_i(\cdot) = \sum_{k=1}^{n} \omega_{ik} S_k(\cdot). \tag{45}$$

From these, using exterior differentiation [2, pp. 20–24; 16, pp. 111–114] we obtain, first,

$$(dA)' = \sum_{k=1}^{n} \{\omega_k' S_k(\cdot) - [\omega_k, dS_k]\} = 0, \tag{46}$$

i.e.,

$$\sum_{k=1}^{n} \omega_k' S_k(\cdot) - \sum_{k,l} [\omega_k, \omega_{kl}] S_l(\cdot) = 0,$$

which again yields, by taking inner products with $S_k(\cdot)$,

$$\omega_k' = \sum_{h=1}^{n} [\omega_h, \omega_{hk}]. \tag{47}$$

Second, using $(dS_k(\cdot))' = 0$, we obtain

$$\sum_k \{\omega_{ik}' S_k(\cdot) - [\omega_{ik}, dS_k(\cdot)]\} = 0,$$

that is,

$$\sum_k \omega'_{ik} S_k(\cdot) - \sum_{h,l} [\omega_{ih}, \omega_{hl}] S_l(\cdot) = 0.$$

and then taking inner products with $S_k(\cdot)$,

$$\omega'_{ik} = \sum_{h=1}^{n} [\omega_{ih}, \omega_{hk}]. \tag{48}$$

These equations of structure of the group $\mathscr{H}(n)$ are completely analogous to the equations of structure in the classical case (see Santalo [21, p. 83, equations (17.24), (17.25)]).

For this group $\mathscr{H}(n)$ we now consider the possibility of a density in our present sense, for the sets of h-varieties which are transforms under $\mathscr{H}(n)$, of the subspace $\mathscr{h}^h$ for $1 \leqslant h < n$; this density has to be invariant under this group. Denote by $\mathscr{V}'$ such an h-variety.

Choose as fixed h-variety $\mathscr{V}$, which is determined by the point A and the h "orthonormal units" $S_k(\cdot)$, $k = 1, \ldots, h$. Then from these equations of structure we get

$$\begin{aligned} \omega_i &= 0 \qquad \text{for} \quad i = h+1, \ldots, r, \\ \omega_{ik} &= 0 \qquad \text{for} \quad i \leqslant h, \quad k > h. \end{aligned} \tag{49}$$

Using these relations and $\omega_{ik} + \omega_{ki} = 0$, we obtain

$$\left[\prod_{i=h+1}^{n} \omega_i \cdot \prod_{\substack{i=1 \\ k=h+1}}^{\substack{k=n \\ i=h}} \omega_{ik} \right]' = 0 \tag{50}$$

which means that the sets of varieties $\mathscr{V}'$ have a density given by

$$d\Omega_h = \left[\prod_{i=h+1}^{n} \omega_i \cdot \prod_{\substack{i=1 \\ k=h+1}}^{\substack{k=n \\ i=h}} \right] \omega_{ik} \;, \tag{51}$$

where ω_i, ω_{ik} have the values previously expressed.

We shall exhibit one convenient sigma field Σ on which the above expression will give us a legitimate measure. Also such a measure on Σ is *not unique* and it will be convenient to exhibit on Σ another measure, the two measure spaces being isomorphic. We shall consider the special case where the point $A = 0$ for all the frames. In this case the density for these h-varieties is given by

$$d\mu = \prod \omega_{ij}$$

the product being taken over admissible i and j as explained.

We now use the functions δ, $\hat{\delta}$, $\hat{d}$ which are defined for pairs of closed linear subspaces in $\tilde{\mathscr{h}}^n$ with values in nonnegative real numbers and have been investigated thoroughly (cf. Kato [13, pp. 197–199]). In particular, we shall use the following:

(i) $0 \leqslant \hat{\delta} \leqslant 1; 0 \leqslant \hat{d} \leqslant 2$ (pp. 197–198 of [13]);

(ii) the completeness of the class of closed linear subspaces of $\mathscr{h}^n$ with the metric $\hat{d}$ (Theorem 2.1, p. 199 of Kato [13]);

(iii) the equivalence of $\hat{d}$ and $\hat{\delta}$ (p. 198, (2.12) of Kato [12]).

We shall order the h-varieties $\mathscr{V}'$ in $\mathscr{h}^n$ by the values of the function $\frac{1}{2}\hat{d}(\mathscr{V}_0, \mathscr{V}')$. Let Σ be the σ-field generated by the sets of the type

$$\mathrm{B} = [\mathscr{V}' \text{ an } h\text{-variety} \,|\, \tfrac{1}{2}\hat{d}(\mathscr{V}_0, \mathscr{V}') < t],$$

where $0 \leqslant t \leqslant 1$ or $0 \leqslant t < 1$ according to whether or not there exists an h-variety (subspace in this case) which is perpendicular to $\mathscr{h}^h$ in $\mathscr{h}^n$. For such a set

$$\mathrm{B} = [\mathscr{V}' \text{ an } h\text{-variety} \,|\, \tfrac{1}{2}\hat{d}(\mathscr{V}, \mathscr{V}') < t]$$

we determine the operator-valued measure $\Omega(B)$ as follows.

Since $\sup_{\mathscr{V}' \in B} \frac{1}{2}\hat{d}(\mathscr{V}, \mathscr{V}') = t$, we consider a finite sequence of h-varieties in B:

$$\{\mathscr{V} = \mathscr{V}_0, \mathscr{V}_1', \ldots, \mathscr{V}_p'\}$$

such that $\frac{1}{2}\hat{d}(\mathscr{V}, \mathscr{V}_p') \to t$ as $p \to \infty$. For such a finite sequence, consider one operator $\prod \omega_{ij}^k$ mapping $\mathscr{V}'_{k-1}$ onto $\mathscr{V}_k'$; we can denote this also by

$$d\mu'_{k-1} = \prod \omega_{ij}^k, \qquad i, j \text{ admissible},$$

the ω_{ij}^ks forming the density when $\mathscr{V}'_{k-1}$ is the initial (fixed) subspace and the "change" $d\mathscr{V}'_{k-1}$ is from $\mathscr{V}'_{k-1}$ to $\mathscr{V}_k'$.

At this point we need the following proposition.

PROPOSITION 3.1. *The mapping $d\mathscr{V}'_{k-1}$ is a linear homeomorphism*

$$\mathscr{V}'_{k-1} \xrightarrow{\text{onto}} \mathscr{V}_k'.$$

Proof. The proof is a repetition of the first part of Theorem 3.1. For completeness we shall give the proof here, with a slight change of notation.

In our case we are dealing with *real* spaces; hence the $\omega = ((\omega_{ij}))$ form a skew-symmetric matrix with coefficients which are operators. We represent a matrix $d\mu = ((\omega_{ij}))$ by the point of $\mathscr{L}^{n^2}$ where $\mathscr{L} = \mathscr{L}(\tilde{h}, \tilde{h})$, whose coordinates are the coefficients ω_{ij}, these coefficients being arranged in some fixed order. Now the series $\sum_0^\infty \|d\mu\|^p/p!$ converges, where $\|d\mu\|$ is the norm of the operator $d\mu$ (see Dunford and Schwartz [7, Vol. III, pp. 1960–1962]).

Hence if we define the exponential function $\exp(d\mu)$ by

$$\exp(d\mu) = \sum_{p=0}^{\infty} (d\mu)^p/p!,$$

the coefficients $y_{ij}(d\mu)$ of $\exp(d\mu)$ are analytic functions $F_{ij}(\ldots, \omega_{kl}, \ldots)$ of the coefficients of $d\mu$.

The terms of index <2 in the "expansion" of $F_{ij}(\ldots, \omega_{kl}, \ldots)$ are clearly $\delta_{ij} + \omega_{ij}$. Hence the $(n^2 \times n^2)$ derivative matrix $D = ((D_{kl}))$ of the derivatives of the F_{ij} with reference to their n^2 arguments, at $\omega_{kl} = 0$ $(1 \leqslant k, l \leqslant n)$, is equal to the identity I. By virtue of the Theorem on implicit functions (Theorem 10.2.5, p. 273 of Dieudonne [6, Vol. I]), there is an open neighborhood of $((\omega_{ij})) = ((0_{ij}))$ the 0_{ij} being the zero operator, which is mapped topologically onto an open neighborhood of $y_{ij} = \delta_{ij}$.

From this the proposition follows, if $\| d\mu_k \|$ is sufficiently small.

Next, let A_p be the mapping

$$A_p = d\mu_p' \, d\mu_{p-1}' \cdots d\mu_1'$$

which maps $\mathscr{V}_0$ one–one onto $\mathscr{V}_p'$. Now, since $\frac{1}{2}\hat{d}(\mathscr{V}_0, \mathscr{V}_p') \to t$ as $p \to \infty$ and the function $\hat{d}$ has the completeness property mentioned above (Kato [13, p. 199]), therefore there exists a subspace $\mathscr{V}'$ such that $\frac{1}{2}\hat{d}(\mathscr{V}_0, \mathscr{V}') = t$. Furthermore $\mathscr{V}' = \lim_{p\to\infty} A_p(\mathscr{V})$ *pointwise*. Hence we now define $T_t = \lim_{p\to\infty} A_p$ *pointwise*. Also T_t must be a homeomorphism: $\mathscr{V}_0 \xrightarrow{\text{onto}} \mathscr{V}'$. It follows that $\mathscr{V}'$ must be an h-variety.

Next we note that $\frac{1}{2}\hat{d}(\mathscr{V}_0, \mathscr{V}_p') \to t$ implies

$$\hat{d}(\mathscr{V}_p', \mathscr{V}_q') \to 0 \qquad \text{as} \quad p, q \to \infty. \tag{52}$$

As shown in [13], p. 198, this condition is equivalent to

$$\hat{\delta}(\mathscr{V}_p', \mathscr{V}_q') \to 0 \qquad \text{as} \quad p, q \to \infty. \tag{53}$$

We shall make use of both these conditions (20) and (21) in proving the next proposition.

PROPOSITION 3.2. *The product A_p converges in operator norm, i.e., $\lim_{p\to\infty} A_p$ exists in operator norm.*

Proof. For $1 < p < q$ integers, let $B_{pq} = \prod_{n=p+1}^{q} a_n$ (successive products being left by multiplication), where we write a_k for $d\mu_k$. To say that the product converges in operator norm is equivalent to asserting that if

$$P_N = \prod_{n=1}^{N} a_n,$$

then $\|p_N - p_{N'}\| \to 0$ as $N, N' \to \infty$. Now for $N' > N$,

$$\prod_1^{N'} a_n - \prod_1^{N} a_n = \prod_1^{N} a_n\{1 - B_{N,N'}\},$$

i.e.,

$$\left\{\prod_1^N a_n\right\}^{-1} \cdot (p_{N'} - p_N) = \{1 - B_{N,N'}\}.$$

Hence $\|p_{N'} - p_N\| \to 0$ is equivalent to $\|B_{N,N'} - 1\| \to 0$ as $N, N' \to \infty$.

Now consider the condition

$$\|B_{pq} - 1\| \to 0 \qquad \text{as} \quad p, q \to \infty. \tag{54}$$

Since $\|B_{pq} - 1\| = \sup_{\|u\|=1,\, u \in \mathscr{V}_p'} \|B_{p,q}u - u\|$, clearly we have to investigate the condition

$$\sup_{\substack{\|u\|=1 \\ u \in \mathscr{V}_p'}} \|B_{pq}u - u\| \to 0 \qquad \text{as} \quad p, q \to \infty. \tag{55}$$

Now

$$\|B_{p,q}u - u\| \leqslant \|B_{pq}u - v\| + \|u - v\|, \tag{56}$$

where we understand that $u \in \mathscr{V}_p'$ and $v \in \mathscr{V}_q'$ with $\|u\| = 1$, $\|v\| = 1$. This is true for each such $u \in \mathscr{V}_p'$, $v \in \mathscr{V}_q'$. Hence

$$\sup_{\substack{\|u\|=1 \\ u \in \mathscr{V}_p'}} \|B_{pq}u - u\| \leqslant \sup_{\substack{\|u\|=1 \\ v \in \mathscr{V}_p'}} \inf_{\substack{\|v\|=1 \\ v \in \mathscr{V}_q'}} \|B_{pq}u - v\| + \sup_{\substack{\|u\|=1 \\ u \in \mathscr{V}_p'}} \inf_{\substack{\|v\|=1 \\ v \in \mathscr{V}_q'}} \|u - v\|. \tag{57}$$

First $\hat{d}(\mathscr{V}_p', \mathscr{V}_q') \to 0$ implies that the second term on the right side in (25) tends to 0 as p, $q \to \infty$.

Next

$$\|B_{pq}u - v\| = \|B_{pq}u\| \cdot \left\| \frac{1}{\|B_{pq}u\|} \{B_{pq}u - v\} \right\|.$$

Therefore

$$\sup_{\substack{\|u\|=1 \\ u \in \mathscr{V}_p'}} \inf_{\substack{\|v\|=1 \\ v \in \mathscr{V}_q'}} \|B_{pq}u - v\| = \sup_{\substack{\|u\|=1 \\ u \in \mathscr{V}_p'}} \inf_{\substack{\|v\|=1 \\ v \in \mathscr{V}_q'}} \left[\|B_{pq}u\| \cdot \frac{1}{\|B_{pq}u\|} \|\{B_{pq}u - v\}\| \right]$$

$$= \|B_{pq}\| \sup_{\substack{\|u\|=1 \\ u \in \mathscr{V}_p'}} \inf_{\substack{\|v\|=1 \\ v \in \mathscr{V}_q'}} \left[\frac{1}{\|B_{pq}u\|} \|\{B_{pq}u - v\}\| \right]. \tag{58}$$

A simple application of the Uniform boundedness principle implies that $(\|B_{pq}\|)$ is bounded with reference to p and q. The equivalence of $\hat{d}(\mathscr{V}_p', \mathscr{V}_q') \to 0$ and $\hat{\delta}(\mathscr{V}_p', \mathscr{V}_q') \to 0$, together with (20) and (21), now show that the right side in (26) tends to 0 as $p, q \to \infty$, i.e.,

$$\sup_{\substack{\|u\|=1 \\ u \in \mathscr{V}_p'}} \inf_{\substack{\|v\|=1 \\ v \in \mathscr{V}_q'}} \|B_{pq}u - v\| \to 0 \qquad \text{as} \quad p, q \to \infty. \tag{59}$$

Hence, finally, by virtue of (20), (27), and (25), the proof of Proposition 3.4 is complete.

We have thus shown the existence of *one* system of such operators $\{T_t\}$, $0 \leqslant t < 1$, obtained as above, satisfying: $T_t \cdot T_{t'} = T_{t+t'}$ for $0 \leqslant t,\ t' < t + t' < 1$.[4] Each such system thus obtained is defined to be a *family* of diffeomorphisms.

Such a family is now extended into a semigroup of operators $\{T_t\}$, $0 \leqslant t < \infty$, by the definition

$$T_0 \underset{\text{def}}{=} I \quad \text{(identity)}, \qquad T_1 \underset{\text{def}}{=} T_{1-} \quad \text{(existence shown already)}.$$

$$\text{For } 1 \leqslant t \leqslant 2, \quad T_t = T_1 \cdot T_\delta, \qquad \delta = t - 1;$$

and by induction:

$$\text{For } n \leqslant t < n + 1, \quad T_t = T_n \cdot T_\delta, \qquad \delta = t - n.$$

Clearly this semigroup is uniformly continuous at the origin:

$$\|T_t - T_0\| \to 0 \qquad \text{as} \quad t \to 0_+.$$

We now consider all possible semigroups $\{T_t; 0 \leqslant t < \infty\}$ obtained in this manner, each satisfying:

$$\begin{aligned} T_t \cdot T_{t'} &= T_{t+t'} && \text{for} \quad 0 \leqslant t, t' < \infty, \\ T_0 &= I, \\ \|T_t - T_0\| &\to 0 && \text{as} \quad t \to 0_+. \end{aligned}$$

Next, for a fixed $t \in \mathbb{R}$, we shall denote by T_t the collection of all the operators T_t obtained from all the semigroups $\{T_t; 0 \leqslant t < \infty\}$ obtained above, and by $\mathscr{V}_t = \mathsf{T}_t \mathscr{V}_0$ the total collection of all the h-varieties $\mathscr{V}_t$ such that $T_t \mathscr{V}_0 = \mathscr{V}_t$. Further we note that for any $T_t \in \mathsf{T}_t$ and any $T_{t'} \in \mathsf{T}_{t'}$ we have $T_t \cdot T_{t'} \in \mathsf{T}_{t+t'}$; this property shall be expressed as

$$\mathsf{T}_t \cdot \mathsf{T}_{t'} = \mathsf{T}_{t+t'}.$$

[4] We shall assume that the metric $\hat{d}$ is multiplied by a suitable constant, so that, denoting the new modified metric again by $\hat{d}$, we have: $\sup[\hat{d}(V_0, V') \mid V' \text{ an } h\text{-subspace}] = 1$.

The total collection of h-varieties $\{\mathsf{T}_t\mathscr{V}_0; 0 \leqslant t < 1\}$ (or $\{\mathsf{T}_t\mathscr{V}_0; 0 \leqslant t \leqslant 1\}$ will be considered with the metric $\hat{d}$ of Kato mentioned above.

The total collection $\{\mathsf{T}_t; 0 \leqslant t < \infty\}$ is *uniformly equicontinuous*, i.e.,

$$\|\mathsf{T}_{t+h} - \mathsf{T}_t\| \to 0 \qquad \text{as} \quad h \to 0,$$

uniformly for all $T_t \in \mathsf{T}_t$, $T_{t+h} \in \mathsf{T}_{t+h}$, where by $\|\mathsf{T}_\alpha - \mathsf{T}_\beta\|$ we mean the collection of all the norms $\|T_\alpha - T_\beta\|$, $T_\alpha \in \mathsf{T}_\alpha$, $T_\beta \in \mathsf{T}_\beta$. Hence by known results [7, Vol. I, VIII 1.2, p. 615] about semigroups, we know that $\mathsf{T}_t = e^{t\mathbb{C}}$ where $\mathbb{C} = \lim_{h\to 0_+}(\mathsf{T}_h - I)/h$, this limit being understood to be in operator norm, uniformly with respect to $T_h \in \mathsf{T}_h$.

We shall call the collection $\{\mathsf{T}_t; 0 \leqslant t < \infty\}$ a *uniformly equicontinuous semigroup jet.*[5] Our σ-field of h-varieties and the measure on this σ-field, obtained by Chern's techniques in the preceding, are then defined as follows (where we use the semigroup jet $\{\mathsf{T}_t\}$).

We define the measure on the set

$$B_{t,t'} = [\mathscr{V}' \text{ an } h\text{-variety} \,|\, t \leqslant \tfrac{1}{2}\hat{d}(\mathscr{V}_0, \mathscr{V}') < t'], \qquad 0 \leqslant t, t' < 1,$$

to be equal to $\mathsf{T}_{t'} - \mathsf{T}_t = [e^{t'A} - e^{tA} \,|\, A \in \mathbb{C}]$; in particular the measure on

$$B_{0,t} = [\mathscr{V}' \text{ an } h\text{-variety} \,|\, \tfrac{1}{2}\hat{d}(\mathscr{V}_0, \mathscr{V}') < t]$$

is $\mathsf{T}_t - I$. For the σ-field Σ we take the σ-field generated by such sets $B_{t,t'}$, $0 \leqslant t$, $t' < 1$. Define T_1 by T_{1-}. Then the total measure is $\mathsf{T}_1 - \mathsf{T}_0 = \mathsf{T}_1 - I$. Furthermore, from the construction of the function T_t it is clear that $\mathsf{T}_t \cdot \mathsf{T}_{t'} = \mathsf{T}_{t+t'}$ where $0 \leqslant t, t', t + t' < 1$ (or $0 \leqslant t, t', t + t' \leqslant 1$).

To show the countable additivity of this measure, it is enough to show that if

$$(t_1, t_2), (t_3, t_4), \ldots, (t_{2n-1}, t_{2n}), \ldots$$

is a sequence of disjoint intervals in $[0, 1]$, then

$$\sum_{n=1}^{\infty} (e^{t_{2n}\mathbb{C}} - e^{t_{2n-1}\mathbb{C}})$$

is convergent in operator norm.

We note that

$$(e^{t_{2p}\mathbb{C}} - e^{t_{2p-1}\mathbb{C}}) = \int_{t_{2p-1}}^{t_{2p}} \{(d/ds)e^{s\mathbb{C}}\}\, ds.$$

[5] If we define $\mathsf{T}_{-t} = \mathsf{T}_t^{-1}$ $(t > 0)$, then clearly $\{\mathsf{T}_t\}_{-\infty < t < \infty}$ is a group. However for the measure-theoretical purpose which we have in view it is the semigroup property that is relevant. Hence this name appears reasonable.

Hence we see that $(q > p)$

$$\left\| \sum_{n=p}^{n=q} (e^{t_{2n}\mathbb{C}} - e^{t_{2n-1}\mathbb{C}}) \right\| \leqslant \sum_{n=p}^{n=q} \| e^{t_{2n}\mathbb{C}} - e^{t_{2n-1}\mathbb{C}} \|$$

$$\leqslant \sum_{n=p}^{n=q} \|\mathbb{C}\| e^{\|\mathbb{C}\|} (t_{2n} - t_{2n-1})$$

$$\to 0 \quad \text{as} \quad p, q \to \infty.$$

We shall summarize our results of this section as follows.

THEOREM 3.2. *Consider the σ-field of h-subspaces in $\tilde{\hbar}^n$ previously defined. The measure $\{d\mathsf{T}_t\}$ on Σ, which is obtained above by using Chern's techniques of Sections I and II, is a Borel measure with values in the uniformly equicontinuous semigroup jet $\{e^{t\mathbb{C}}; 0 \leqslant t < \infty\}$, where $\mathbb{C}$ is a jet of bounded linear operators: $\tilde{\hbar}^n \to \tilde{\hbar}^n$. This measure is invariant under $\mathcal{O}(\tilde{\hbar}^n)$.*

ACKNOWLEDGMENT

A summary of the results of this chapter was presented at the 31st Ontario Mathematical Meeting at Waterloo, Ontario. The author has profited from a careful study of the work of many other investigators concerning measures and integration in infinite-dimensional linear spaces (cf. [9–11]; the last two papers together contain a more complete bibliography; cf. also [12]). I would also like to express my thanks to W. H. Greub and to J. T. Schwartz for suggesting improvements in the chapter. It is our intention to continue further work along the present lines. As far as this author is aware, the results of this chapter are new.

REFERENCES

1. L. AUSLANDER AND R. E. MCKENZIE, "Introduction to Differentiable Manifolds," McGraw–Hill, New York, 1963.
2. H. CARTAN, "Differential Forms," Houghton–Mifflin, Boston, Massachusetts, 1970.
3. S. S. CHERN, On integral geometry in Klein spaces, *Ann. of Math.* **43** (1942), 178–189.
4. C. CHEVALLEY, "Lie Groups," Princeton Univ. Press, Princeton, New Jersey, 1946.
5. G. DE RHAM, "Variétés différentielles." Hermann, Paris, 1960.
6. J. DIEUDONNÉ, "Treatise on Analysis," Vols. I–IV, Academic Press, New York, 1972.
7. N. DUNFORD AND J. T. SCHWARTZ, "Linear Operators," Vols. I, II, III, Wiley (Interscience), New York, 1971.
8. J. EELLS, JR., A setting for global analysis, *Bull. Amer. Math. Soc.* **7** (1966), 751–807.
9. J. EELLS, JR., Integration in Banach manifolds, *Biennial Seminar of Canad. Math. Congr., 13th, September 1971.*
10. I. M. GELFAND AND G. E. SHILOV, "Generalised Functions," Vols. 1–5, Academic Press, New York, 1964–1968.

11. L. GROSS, "Classical Analysis on a Hilbert Space," Chapter 4, pp. 51–60 (*Proc. Conf. Theory and Appl. Anal. Function Space, Dedham, Massachusetts, 1963*) (W. T. Martin and I. E. Segal, eds.), M.I.T. Press, Cambridge, Massachusetts, 1964.
12. A. GUICHARDET, "Symmetric Hilbert Spaces and Related Problems," Lecture Notes in Math. Vol. 261, Springer-Verlag, Berlin and New York, 1972.
13. T. KATO, "Perturbation Theory for Linear Operators," Springer-Verlag, Berlin and New York, 1966.
14. N. H. KUIPER, Les variétés Hilbertiénnes, *Semin. Mathe. Superieure, Montreal, 1969.*
15. N. H. KUIPER, The differential topology of separable Banach manifolds, Lecture at Internat. Congr. of Math., Nice, 1970.
16. S. LANG, "Differentiable Manifolds" (rev. ed.), Addison–Wesley, Reading, Massachusetts, 1972.
17. B. MAISSEN, Lie Gruppen mit Banachraumen als parameterraume, *Acta Math.* (1963), 229–269.
18. D. MONTGOMERRY AND L. ZIPPIN, "Topological transformation Groups," Tracts in Pure and Appl. Math., No. 1 Wiley (Interscience), New York, 1955.
19. N. MOULIS, Approximation de fonctions différentiables sur certains espaces de Banach, *Grenoble Univ. Annales Inst. Fourier*, 1971.
20. R. S. PALAIS, Lectures on the Differential Topology of Infinite Dimensional Manifolds, Brandeis Univ., 1964–1965.
21. L. A. SANTALO, "Introduction to Integral Geometry," Actualités Sci. Ind. No. 1198, Paris.
22. L. A. SANTALO, Integral Geometry, *in* "Studies in Global Geometry and Analysis," (S. S. Chern, ed.), Vol. 4, pp. 147–193, MAA Studies in Math. 1967.
23. H. WHITNEY, "Complex Analytic Varieties," Addison–Wesley, Reading, Massachusetts, 1972.

AMS (MOS) subject classifications: 28A45, 28A70, 58A15.

STUDIES IN ANALYSIS
ADVANCES IN MATHEMATICS SUPPLEMENTARY STUDIES, VOL. 4

Finite Rickart C^* Algebras and Their Properties[†]

DAVID HANDELMAN

Mathematics Department
University of Ottawa
Ottawa, Ontario
Canada

INTRODUCTION

Kaplansky [15, p. 249] has asked whether a Rickart C^* algebra (there referred to as an $\aleph_1$–AW^* algebra) has the property that left projections are equivalent to right projections. We prove this conjecture in the finite case ($xx^* = 1$ implies $x^*x = 1$), and here, in fact, equivalent projections are unitarily equivalent. In this case, techniques exclusively from the theory of von Neumann regular rings are used. Corresponding to each finite Rickart C^* algebra is a regular overring; this overring is used to show, for example, that any simple homomorphic image is an AW^* factor of finite type, and all matrix rings are also finite. While a great many results on finite AW^* algebras apply in a modified form to finite Rickart C^* algebras, some examples are presented to illustrate the limitations.

Some purely algebraic results on $\aleph_0$-continuous regular rings are also proved en passant.

The organization of this article is as follows. Sections 1 and 2 deal with the construction of the regular ring (in the finite case), and in Section 3 is proved the results on $\aleph_0$-continuous rings. Section 4 contains the main results, as consequences of the existence of the regular ring. Further results about matrix rings over Rickart C^* being Rickart, and extending the LP $\sim$ RP results to the nonfinite case will be considered in a future article.

[†] Supported in part by a NATO Postdoctoral Fellowship from the National Research Council of Canada.

ISBN 0-12-599150-9

1. Preliminaries

In order to read Sections 1–3 it really is necessary that the reader have some familiarity with maximal rings of quotients [18, p. 94ff.] and regular rings (in the sense of von Neumann; see [18, p. 68ff.]).

All rings are associative and possess 1. If S is a subset of a ring R, we denote the right annihilator of S in R, by $S^{\mathrm{r}R}$, or S^{r} if no ambiguity will result. That is,

$$S^{\mathrm{r}} = \{r \in R \mid Sr = (0)\}.$$

Similar notation is used for the left annihilator: $S^{\mathrm{l}R}$ or S^{l}.

Let R be a ring with an involution $*$. An element p of R is called a *projection* if $p = p^2 = p^*$. The ring R is a *Rickart $*$ ring* [2], if for all r in R, there exists a projection p such that

$$r^{\mathrm{r}} = pR.$$

The existence of the involution allows us to find a projection q such that $r^{\mathrm{l}} = Rq$. (Find q such that $(r^*)^{\mathrm{r}} = qR$.) If $r^{\mathrm{r}} = pR$, then $r(1-p) = r$; and $1-p$ is the "smallest" projection with the property that it fixes r on the right. We shall call $1-p$ the *right projection* of r in R, that is $\mathrm{RP}(r) = 1-p$, where $r^{\mathrm{rl}} = R(1-p)$. Similarly, we define the left projection by $\mathrm{LP}(r) = 1-q$, where $r^{\mathrm{lr}} = (1-q)R$.

The collection of projections may be partially ordered:

$$p \leqslant q \qquad \text{if} \quad pq = p \qquad \text{(applying the involution } p = pq = qp).$$

Equivalently, $p \leqslant q$ if $pR \subseteq qR$. In fact, the projections form a lattice by [2, p. 14, Proposition 7]. However, the lattice operations described there are in terms of left annihilators; it will be convenient to express the lattice operations via right annihilators.

Lemma 1.1. *Let R be a Rickart $*$ ring.*

(a) *If $\{r_i\}$ is a finite subset of R, there exists a projection e such that $eR = \{r_i\}^{\mathrm{r}}$.*

(b) *The collection of right annihilators of finite subsets may be naturally identified with the collection of projections, and may be given the structure of a lattice by means of: If p, q are projections, there exist projections $p \wedge q$ and $p \vee q$ such that*

$$(p \wedge q)R = PR \cap qR, \qquad (p \vee q)R = (pR + qR)^{\mathrm{lr}}.$$

Proof. (a) We may assume that $\{r_i\}$ has just two elements. Say $r_1{}^{\mathrm{r}} = pR$, $r_2{}^{\mathrm{r}} = qR$. Then $\{r_1, r_2\}^{\mathrm{r}} = pR \cap qR$, so we must show that $pR \cap qR$ is generated by a projection. Set $e = RP((1 - q)p)$. Now $Rp \supset R(1 - q)p)$, so taking right annihilators, then left annihilators $Rp \supset Re$, whence $e \leqslant p$, so $e = ep = pe$, and $p - e$ is a projection in pR. If $x \in (p - e)R$, $px = x$, and $x = (p - e)x$, so $ex = 0$, whence $x \in e^r = ((1 - q)p)^r$; thus $(1 - q)px = 0$, so $(1 - q)x = 0$, that is, $x \in qR$, and so $(p - e)R \subset pR \cap qR$. If $x \in pR \cap qR$, $px = qx = x$, so $(1 - q)px = (1 - q)x = 0$, whence $x \in ((1 - q)p)^r = e^r$, so $ex = 0$, and thus $x = (p - e)x$, so $x \in (p - e)R$. Hence $pR \cap qR = (p - e)R$.

(b) By (a), there exists a projection f such that $fR = pR \cap qR$. If g is any projection such that $g \leqslant p$, $g \leqslant q$, then $gR \subseteq pR$, $gR \subseteq qR$, so $gR \subseteq pR \cap qR$, so $gR \subseteq fR$, i.e., $g \leqslant f$, so f is the greatest lower bound of p and q. Now $(pR + qR)^l = \{p, q\}^l = R(1 - p) \cap R(1 - q)$. By the left-hand version of (a), there exists a projection h such that $\{p, q\}^l = Rh$, so $(pR + qR)^{lr} = (1 - h)R$. If k is a projection with $p \leqslant k$, $q \leqslant k$, then $pR + qR \subseteq kR$. Taking left annihilators, $R(1 - h) \subseteq Rh$, so $(1 - h)R \subseteq kR$, whence $(1 - h) \leqslant k$. Thus $1 - h$ is the least upper bound of p, q, and so the projections form a lattice. The identification of this lattice with that of the right annihilators of finite sets is clear. ■

The lattice obtained in Lemma 1.1 is of course complemented, in fact orthocomplemented, since $p \vee (1 - p) = 1, p \wedge (1 - p) = 0$.

Let e, f be idempotents in *any* ring R. We say e is *equivalent* to $f (e \sim f)$ if $eR \simeq fR$ as right R-modules; equivalently, there exist $x \in eRf$, $y \in fRe$ such that $xy = e$, $yx = f$. If e, f are projections in a $*$ ring, then e is $*$-*equivalent* to $f (e \overset{*}{\sim} f)$, if there exists w such that $ww^* = e$, $w^*w = f$. An element w such that ww^* is idempotent is called a *partial isometry*.

A C^* algebra that is also a Rickart $*$ ring, is called a Rickart C^* algebra [2]. Thus Kaplansky's question [15, p. 249] expressed in this notation is: If T is a Rickart C^* algebra and $t \in T$, does it follow that $\mathrm{LP}(t) \overset{*}{\sim} \mathrm{RP}(t)$? This will be answered in a future article in the monotone σ complete case. However, in the finite case ($xx^* = 1$ implies $x^*x = 1$), a different proof with strong consequences is available as a subsidiary result to the construction of the regular ring.

In this and the next section we shall construct, for a finite Rickart C^* algebra T, a $*$-regular ring R containing T such that the involution extends to R, and all the projections of R lie in T. In Section 3, it will be shown that equivalence, perspectivity, and unitary equivalence of projections in R are identical concepts.

A lattice is $\aleph_0$-complete if every countable set has an infimum and a supremum. A lattice is *upper* $\aleph_0$-*continuous* if it is $\aleph_0$-complete, complemented, modular, and satisfies the following: For any ascending sequence $\{p^n\}_{n \in \mathbb{N}}$, q arbitrary,

$$q \wedge (Vp^n) = V(q \wedge p^n).$$

Lower $\aleph_0$-continuity is dually defined. The lattice is *$\aleph_0$-continuous* if it is both upper and lower $\aleph_0$-continuous.

LEMMA 1.2. *Any Rickart C^* algebra T has the following properties*:

(1) *The lattice of projections of T is $\aleph_0$-complete.*

(2) *T satisfies the parallelogram law* ((P): [2, Section 13]): *if e, f are projections, then $e \vee f - f \overset{*}{\sim} e - e \wedge f$.*

(3) *For any element t of T, there exists an upward directed sequence of projections $\{p^n\}_{n \in \mathbb{N}}$ $(p^n \leqslant p^{n+1})$ such that*

$$\{p^n\} \subset tT \qquad \text{and} \qquad \{p^n\}^1 = t^1.$$

Proof

(1) This is [2, p. 45, Lemma 2].

(2) Either [19, Theorem 2] or [2, p. 68, Theorem 1].

(3) By [2, p. 47, Proposition 4], there exist projections e_n $(n \in \mathbb{N})$ lying in the bicommutant of tt^* such that $e_n \in tt^*T$, and

$$\|tt^* - tt^*e_n\| \leqslant 1/n.$$

As e_n lies in the bicommutant of tt^*, $tt^*e_n = e_n tt^*$, so $\|tt^* - e_n tt^*\| \leqslant 1/n$. Suppose $st = 0$. Then $stt^* = 0$, so $se_n = 0$ for all n. Suppose $se_n = 0$ for all n. Then $\|stt^*\| = \|stt^* - se_n tt^*\| \leqslant \|s\|/n$, for all n, so $stt^* = 0$, and thus $(st)(st)^* = 0$, so $st = 0$. Hence

$$\{e_n\}^1 = t^1.$$

Now the e_n all lie in the bicommutant of a single element, so they lie in a commutative ring—hence they commute. It is an easy (Boolean algebra) exercise to find projections p^n such that $p^n T = \sum_{i=1}^n e_i T$. (The computations may also be done in the bicommutant). ■

Properties (1)–(3) of Lemma 1.2 are all that are necessary (with finiteness) to prove $LP \overset{*}{\sim} RP$ holds. From this point T will denote a finite Rickart $*$ ring satisfying (1)–(3) of Lemma 1.2.

If R is any ring, a right ideal is *essential* (or large) if it has nonzero intersection with every nonzero right ideal of R. If $A \subset B$, for A, B right (left) modules then A is *essential* in B, if every nonzero submodule of B has nonzero intersection with A. A ring is *nonsingular* if every essential right (left) ideal has zero left (right) annihilator. Any Rickart $*$ ring is nonsingular: $r^{\mathrm{r}} \cap r^*R = (0)$. Nonsingularity of the ring implies if A, B are right ideals and A is essential in B, then $A^1 = B^1$.

A Rickart $*$ ring R is *strongly modular* [12] if $r^{\mathrm{r}} = (0)$ implies rR is essential (as a right ideal). We shall say a Rickart $*$ ring is *symmetrically strongly modular* if $z = z^*$, $z^{\mathrm{r}} = 0$ imply jointly zR is essential. In the case of Baer $*$

rings, the two concepts coincide (in [12], it was desirable that the definition *not* depend on the involution).

COROLLARY 1.1. *If the finite Rickart* $*$ *ring* T *satisfies* (1)–(3) *of Lemma* 1.2, *then* T *has the following properties*:

(a) *Every nonzero right ideal contains a nonzero projection.*
(b) *The lattice of projections of* T *is modular.*
(c) *The lattice of projections of* T *is* $\aleph_0$*-continuous.*
(d) *For any countable set* $\{r_i\}_{i\in\mathbb{N}} \subset T$, *there exist projections* p, q *such that*

$$\{r_i\}^{\mathrm{r}} = pT \qquad \text{and} \qquad \{r_i\}^{\mathrm{l}} = Tq$$

if the r_i *are all projections, then* $\sup\{r_i\} = 1 - p$.
(e) T *is symmetrically strongly modular.*
(f) *If* $\{r_i\}$ *is a countable subset of* T, *then* $\sum r_i R$ *is essential in* $\{r_i\}^{\mathrm{lr}}$ *and* $\sum Rr_i$ *is (left) essential in* $\{r_i\}^{\mathrm{rl}}$.
(g) *If* p, q *are projections which are perspective in the projection lattice of* T, *then* $p \overset{*}{\sim} q$.

Proof (a) A trivial consequence of Lemma 1.2, part (3).

(b) The parallelogram law [Lemma 1.2, part (2)] and finiteness jointly imply modularity—see the proof of [17, Theorem 67]. (Baerness, or any of the other additional assumptions there is not required.)

(c) The lattice of projections of T is an orthocomplemented modular $\aleph_0$-complete [Lemma 1.2, part (1)] lattice, so by [1, Theorem 9.5], the lattice is $\aleph_0$-continuous.

(d) Let p_i be the projections so that $r_i^{\mathrm{r}} = p_i T$. Then $\{r_i\}^{\mathrm{r}} = \bigcap p_i T$; so it suffices to show $(\inf\{p_i\})T = \bigcap p_i T$. Set $p = \inf\{p_i\}$. Then $p \leqslant p_i$, so $pT \subseteq p_i T$, and thus $pT \subseteq \bigcap p_i T$. If there exists an element r in $\bigcap p_i T$ such that $rT \cap pT = (0)$, then, by (a), there exists a projection $h \in rT$, so $hT \cap pT = (0)$. Now $hT \subseteq p_i T$, so $h \leqslant p_i$, from which $h \leqslant p$, by the definition of greatest lower bound. Thus $hT \subseteq pT$, a contradiction. We have thus shown that pT is essential in $\bigcap p_i T = \{r_i\}^{\mathrm{r}}$. Since T is a Rickart $*$ ring, T is nonsingular, so $(pT)^{\mathrm{l}} = \{r_i\}^{\mathrm{rl}}$, and so

$$pT = (pT)^{\mathrm{lr}} = \{r_i\}^{\mathrm{rlr}} = \{r_i\}^{\mathrm{r}}.$$

To obtain q, consider $\{r_i^*\}^{\mathrm{r}}$. To see that $\sup\{r_i\} = 1 - p$ if the r_i are projections, observe that $\{r_i\}^{\mathrm{r}} = pT$ and $\{r_i\}^{\mathrm{l}} = Tp$ (since $r_i = r_i^*$). Then $\{r_i\} \subset \{r_i\}^{\mathrm{lr}} = (1 - p)T$, so $r_i \leqslant 1 - p$ for all i. If $r_i \leqslant h$ for some projection h, then $\{r_i\} \subset hT$, so taking left then right annihilators,

$$(1 - p)T = \{r_i\}^{\mathrm{lr}} \subset (hT)^{\mathrm{lr}} = hT$$

and so $1 - p \leqslant h$, so $1 - p$ is indeed the supremum.

(e) Choose z in T such that $z = z^*$ and $z^{\mathrm{r}} = (0)$. Then $z^{\mathrm{l}} = (0)$. By Lemma 2.3, part (3), there exists an ascending sequence of projections $\{p^n\}$ in zT with $\{p^n\}^{\mathrm{l}} = (0)$. It suffices to show $\bigcup p^n T$ is an essential right ideal. If $cT \cap (\bigcup p^n T) = 0$ for some nonzero c in T, by (a), there exists a nonzero projection h with $hT \cap (\bigcup p^n T) = (0)$. Set $p = \sup p^n$. By (d), $p = 1$, but by upper $\aleph_0$-continuity,

$$h = h \wedge p = h \wedge (V p^n) = V(h \wedge p^n) = 0,$$

as $(h \wedge p^n)T = hT \cap p^n T = (0)$. This contradicts the existence of such a c, so $\bigcup p^n T$, and hence zT must be essential.

(f) Observe that the proofs of [12, Lemma 2.2(a), Lemma 2.5] only require symmetric strong modularity in a Rickart $*$ ring. Thus (f) holds for finite subsets. Now, the proof of [12, Lemma 2.6] for countably generated ideals requires only $\aleph_0$-continuity, proving the result.

(g) This is a trivial application of the parallelogram law. If p, q are perspective, there exists a projection x such that $1 = p \vee x = q \vee x$, and $0 = p \wedge x = q \wedge x$. Thus we have $1 - x = p \vee x - x \overset{*}{\sim} p - p \wedge x = p$, and similarly $q \overset{*}{\sim} 1 - x$. Thus $p \overset{*}{\sim} q$. ∎

The properties required most are Corollary 1.1, parts (f) and (d).

2. Construction of the Regular Ring

Throughout this section, T will denote a finite Rickart $*$ ring satisfying (1)–(3) of Lemma 1.2, and thus having all seven properties from Corollary 1.1.

If R is a regular ring, the collection of principal right ideals of R is a complemented modular lattice, which we shall denote by $L(R)$. A ring R is $\aleph_0$-*continuous* if it is regular and $L(R)$ is an $\aleph_0$-continuous lattice. We will construct an $\aleph_0$-continuous regular ring containing T as a subring, so that the involution extends, and all the projections of R lie in T. This is a coordinatization theorem of a particularly strong form.

The standard construction for regular overrings is the maximal ring of quotients [18, p. 94 on]. In the case of Baer $*$ rings, the maximal ring of quotients is the "correct" choice (see [12, Section 2 and Proposition 5.3]). In the present situation, however, this is too large—we must choose a special subring, which cannot in general be constructed.

Recall that for nonsingular rings the maximal ring of (right) quotients may be viewed as the collection of pairs [18, p. 97, Proposition 6]: (f, E) such that E is an essential right ideal and $f: E \to T$ is a right T-module homomorphism subject to the equivalence relation $(f, E) \equiv (g, E')$ if and only if $f = g$ on $E \cap E'$. Addition is defined by

$$(f, E) + (g, E') = (f + g, E \cap E')$$

and multiplication by

$$(f, E) \cdot (g, E') = (fg, g^{-1}E),$$

where $g^{-1}E = \{r \in E' \mid g(r) \in E\}$.

It is routine to verify that $E \cap E'$ and $g^{-1}E$ are essential, so addition and multiplication make sense. Then T is embedded in the maximal ring of quotients by the mapping $t \mapsto (t, T)$ (t acts by left multiplication). If one is following the terminology of [18], it should be pointed out that in a nonsingular ring, *dense* is the same as *essential*.

A right ideal I of T is *essentially countably generated* (briefly, ecg), if there exists a countable collection $\{t_i\}_{i \in \mathbb{N}} \subset I$ such that $\sum t_i T$ is essential in I.

We would like to define a subring of the maximal ring of quotients

$$Q_H(T) = \{(f, E) \text{ in the maximal ring of quotients} \mid E \text{ is essential and ecg}\}.$$

For arbitrary T, this does not appear to be a ring; but in the case of our T, we show it is. The crucial step is in showing that if $(g, E) \in Q_H(T)$, and I is ecg, then $g^{-1}I = \{r \in E \mid g(r) \in I\}$ is ecg.

To prove this, and other results, we require the result that any ecg right (left) ideal contains as an essential submodule, a right (left) ideal generated by a commuting countable collection of projections (equivalently, a countable increasing chain of projections).

LEMMA 2.1 [12, Lemma 2.4]. *Let A, B, C be submodules of a nonsingular module. If A is essential in B, then $A + C$ is essential in $B + C$.*

LEMMA 2.2. *For the ring T, suppose A, B are right (left) ideals, A is* ecg, *and $A \subset B$. Then A is essential in B if and only if $A^{\mathrm{l}} = B^{\mathrm{l}}$ ($A^{\mathrm{r}} = B^{\mathrm{r}}$).*

Proof. If A is essential in B, then $A^{\mathrm{l}} = B^{\mathrm{l}}$ ($A^{\mathrm{r}} = B^{\mathrm{r}}$) follows from the nonsingularity of T. Conversely, if $A^{\mathrm{l}} = B^{\mathrm{l}}$ ($A^{\mathrm{r}} = B^{\mathrm{r}}$), by Corollary 1.1 (f), A is essential in A^{lr} (A^{rl}). But $A \subset B \subset B^{\mathrm{lr}} = A^{\mathrm{lr}}$ ($A \subset B \subset B^{\mathrm{rl}} = A^{\mathrm{rl}}$), so A is essential in the intermediate ideal B. ∎

PROPOSITION 2.1. *Let A be an essentially countably generated right (left) ideal of T. There exists a countable orthogonal collection of projections $\{p_i\}_{i \in \mathbb{N}}$ in A such that $\bigoplus p_i T$ is essential in A ($\bigoplus T p_i$ is essential in A).*

Proof. We first note that if r belongs to T, then rT contains a sequence of projections $\{g^n\}$ such that $\{g^n\}^1 = r^1$; by Lemma 2.2, $\bigcup g^n T$ is essential in rT. Thus if $\sum_{i \in \mathbb{N}} r_i T$ is essential in A, there exists a countable collection of projections $\{e_j\}$ in A with $\sum e_j T$ essential in A. The next step is to find projections $\{q_k\}_{k \in \mathbb{N}}$ such that $\sum q_k T = \bigoplus q_k T$, and $\bigoplus q_k T$ is essential in $\sum e_j T$.

Define $q_1 = e_1$. Suppose $\{q_k\}_{k=1}^n$ have been defined so that $\sum^n q_k T = \bigoplus^n q_k T$, and $\bigoplus^n q_k T$ is essential in $\sum^n e_i T$. We show a projection q_{n+1} may be found inside $e_{n+1}T$ so that $q_{n+1}T \cap (\sum^n q_k T) = (0)$, and $\sum^{n+1} q_k T$ is essential in $\sum^{n+1} e_i T$. Set $q = \sup\{q_k | k \leqslant n\}$. Then by Corollary 1.1 (f), $qR = (\sum^n q_k T)^{\mathrm{lr}}$, and so by Lemma 2.2, $\sum^n q_k T$ is essential in qT (naturally q need not belong to $\sum e_i T$). Now there exists a projection t with $tT = qT \cap e_{n+1}T$; in particular, $t \in e_{n+1}T$, so $q_{n+1} = e_{n+1} - t$ is a projection contained in $e_{n+1}T$, orthogonal to t. Now $(\sum^n qT) \cap (q_{n+1}T) \subset qT \cap q_{n+1}T = (0)$ (since $tT = qT \cap e_{n+1}T$). Further, $\sum^n q_k T + q_{n+1}T$ is essential in $qT + q_{n+1}T$ by Lemma 2.1; but $qT + q_{n+1}T = qT + e_{n+1}T$. As

$$\textstyle\sum^{n+1} q_k T \subset \sum^n q_k T + e_{n+1}T \subset qT + e_{n+1}T,$$

and the first term is essential in the third, so the first is essential in the second. By the induction assumption and Lemma 2.1, $\sum^{n+1} q_k T$ is essential in $\sum^{n+1} e_i T$. This completes the induction argument. Set $A^m = \bigoplus^m q_k T$ (for any m), $B^m = \sum^m e_i T$. Then A^m is essential in B^m, and as unions of chains preserve essentiality,

$$\bigoplus_{k \in \mathbb{N}} q_k T = \bigcup A^m \quad \text{is essential in} \quad \bigcup B^m = \textstyle\sum e_i T.$$

So we have reduced the problem to the case $\bigoplus_{k \in \mathbb{N}} q_k T$. We shall orthogonalize. We first observe that $(\sum^n q_k)T$ is essential in $\sum^n q_k T$: by Lemma 2.2, it suffices to show $(\sum^n q_k)^1 \subset (\sum^n q_k T)^1$. Suppose $t(\sum^n q_k) = 0$. Then $\sum^n q_k t^* = 0$. Since $\sum q_k T = \bigoplus q_k T$, $q_k t^* = 0$ for $k \leqslant n$, so $tq_k = 0$, and thus $t(\bigoplus^n q_k T) = 0$.

Set $t_1 = q_1$. Suppose there exists a countable collection of commuting projections $\{t_i\} \subset \bigoplus^n q_k T$ such that $\sum t_i T$ is essential in $\bigoplus^n q_k T$. Set $t = \sup t_i$. Then $\sum t_i T$ is essential in tT by Corollary 1.1 (d, f), and also by Lemma 2.2 and Corollary 1.1 (d), $t = \sup q_k$. As $(\sum^{n+1} q)T$ is essential in $\bigoplus^{n+1} q_k T$, $(\sum^{n+1} q_k)T$ is essential in $(t \vee q_{n+1})T$ $[=(\bigvee^{n+1} q_k)T]$. Now $(\sum^{n+1} q_k)T$ is principal, so by Lemmas 1.2(3) and 2.2, there exists a sequence of ascending projections $\{h^m\} \subset (\sum^{n+1} q)T$ such that $\bigcup h^m T$ is essential in $(t \vee q_{n+1})T$, so $\sup h^m = t \vee q_{n+1}$. By upper $\aleph_0$-continuity,

$$(t \vee q_{n+1}) \wedge (1 - t) = (Vh^m) \wedge (1 - t) = V(h^m \wedge (1 - t)). \tag{1}$$

Set $g^m = h^m \wedge (1 - t)$; then $g^m \in h^m T \subset \sum q_k T$, and as $g^m \leqslant 1 - t$, $g^m t = tg^m = 0$, so $g^m t_i = t_i g^m = 0$ for all i, m. By (1), $\sup g^m = (t \vee q_{n+1}) \wedge (1 - t)$.

From the modular law

$$t \vee ((t \vee q_{n+1}) \wedge (1 - t)) = t \vee q_{n+1},$$

so $t \vee (\bigvee g^m) = t \vee q_{n+1}$. Thus we have $\{g^m, t_i\}$ is a commuting collection of projections in $\sum^{n+1} q_k T$, with $\sup\{g^m, t_i\} = t \vee (\sup g^m) = t \vee q_{n+1}$, and by Corollary 1.1 (d, f), $\sum g^m T + \sum t_i T$ is essential in $(t \vee q_{n+1})T$; as $\sum^{n+1} q_k T$ is an intermediate right ideal, $\sum g^m T + \sum t_i T$ is essential in $\sum^{n+1} q_k T$. Thus the original set of commuting projections has been *augmented* to make a larger but still countable commuting collection. Repeating this process for all n and taking the union of the collections of projections, and as essentiality commutes with unions, we obtain the result that A contains an essential countable commuting set of projections, say, $\{f_i\}_{i \in \mathbb{N}}$. Define p^n by $p^1 = f_1, \ldots, p^n = p^{n-1} + f_n - p^{n-1} f_n$; as the f_is all commute, the p^ns are all projections and $p^n \leqslant p^{n+1}$; also $p^n = \sup_{i \leqslant n} f_i$, and $p^n \in \sum^n f_i T$ (in fact $p^n T = \sum^n f_i T$). It easily follows that $\{p_i = p^i - p^{i-1}\}$ is the desired orthogonal collection ($p^0 = 0$). ■

We remind the reader that our immediate objective is to prove that if E, E' are ecg, and $f: E \to T$ is a right T module homomorphism, then $f^{-1}E' = \{r \in E \mid f(r) \in E'\}$ is ecg.

LEMMA 2.3. *Let A be a nonsingular ring, I, J, L, K_i be right A-ideals, and $f: I \to A$ a right A-module homomorphism.*

(a) *If $K_i \subset K_{i+1}$ ($i \in \mathbb{N}$), then $f^{-1}(\bigcup K_i) = \bigcup f^{-1}(K_i)$.*
(b) *If J is essential in L, then $f^{-1}J$ is essential in $f^{-1}L$.*
(c) *If J is essential in L and $L \subset I$, then $f(J)$ is essential in $f(L)$.*

Proof. All are trivial. ■

LEMMA 2.4. *Suppose I, J are essentially countably generated right ideals of T with I essential, and let $f: I \to T$ be a right T-module homomorphism. Then $f^{-1}J = \{r \in I \mid f(r) \in J\}$ is essentially countably generated.*

Proof. By Proposition 2.1 there exists an ascending sequence $\{p^n T\}_{n \in \mathbb{N}}$ of projection-generated right ideals such that $\bigcup p^n T$ is essential in J. By Lemma 2.3 (b), it suffices to show $f^{-1}(\bigcup p^n T)$ is ecg. By Lemma 2.3 (a), the problem reduces to showing that if p is a projection, then $f^{-1}(pT)$ is ecg.

Now $r \in f^{-1}(pT)$ if $r \in I$ and $f(r) = pf(r)$; i.e., $(1 - p)f(r) = 0$. Essential inside I we may find $\bigcup q^n T$ ($q^n \leqslant q^{n+1}$ are projections); by Lemma 2.3 (c), we may assume $I = \bigcup q^n T$. Now, we claim

$$(\bigcup q^n R) \cap \sum_{n \in \mathbb{N}} ((1 - p)f(q^n))^r = f^{-1}(pT). \tag{2}$$

If $r \in \bigcup q^n R$ and $f(r) \in pT$, then $r = q^n r$ some n and $(1-p)f(r) = 0$, so $(1-p)f(q^n)r = 0$, proving $\supseteq$ in (2). Conversely, if $r = \sum_{j=1}^{n} r_j$, where each r_j satisfies $(1-p)f(q^j)r_j = 0$, there exists an integer $k \geqslant n$, such that $q^k r = r$ and, because $f(q^i) = f(q^i)q^i$, we have

$$0 = (1-p)f(q^j)r_j = (1-p)f(q^j)q^j r_j = (1-p)f(q^j)q^j q^k r_j = (1-p)f(q^j)q^k r_j;$$

we may thus assume each $r_j \in \bigcup q^n R$. Then $f(r_j)$ is defined and belongs to pT, whence $f(r) = \sum f(r_j) \in pT$.

Thus we have shown equality in (2), but $\sum((1-p)f(q^n))^{\mathrm{r}}$ is countably generated (by projections, since there exist projections t_n such that $(1-p)f(q^n)^{\mathrm{r}} = t_n T$), so we may find ascending projections b^n with $\bigcup b^n T$ essential in $\sum((1-p)f(q^n))^{\mathrm{r}}$.

We have thus reduced the problem to showing $(\bigcup q^n T) \cap (\bigcup b^n T)$ is countably generated. However, as we are here dealing with set unions and intersections, we have

$$(\bigcup g^n T) \cap (\bigcup b^n T) = \bigcup_n \left(\bigcup_j (q^j T \cap b^n T) \right).$$

Each $q^j T \cap b^n T$ is principal, so the whole union is countably generated. ∎

We may finally verify that $Q_H(T)$, viewed as a subset of the maximal ring of quotients of T, is actually a ring. Suppose (f, E), $(g, E') \in Q_H(T)$. Then the element of the maximal ring of quotients defined by $(fg, g^{-1}E)$ belongs to $Q_H(T)$ by Lemma 2.4, so multiplication is defined.

If E, E' are ecg, then if $i: E \to R$ is the identity, we have $i^{-1}E' = E \cap E'$, so by Lemma 2.5, $E \cap E'$ is ecg. Thus addition defined by

$$(f, E) + (g, E') = (f + g, E \cap E')$$

makes sense in $Q_H(T)$.

Finally, T as a right T-module is principal, so there is an embedding $T \to Q_H(T)$, $t \mapsto (t, T)$. Thus we have obtained $Q_H(T)$ as an intermediate ring between T and its maximal ring of quotients. We identify T with its image in $Q_H(T)$.

Next we would like to show that $Q_H(T)$ is a regular ring. To avoid notational difficulties, if $f \in Q_H(T)$, we will denote the right annihilator of f in $Q_H(T)$ by $f^{\mathrm{r}\varrho}$ (similarly $f^{\mathrm{l}\varrho}$ will denote the left annihilator).

LEMMA 2.5. *Suppose A, B are rings, with A nonsingular, $A \subset B$, and B contained in the maximal ring of right quotients of A.*

(a) *If E is any right A-submodule of B and E is essential in a right B-ideal F (viewed as an A-module), then EB is essential as a right B-submodule of F.*

(b) *For any set* $S \subset B$, $S^{r_B} \cap A$ *is essential in* S^{r_B}, *viewed as an A-module. If* $S \subseteq A$, $S^{r_B} \cap A = S^{r_A}$.

Proof. All are well known and easy to verify. ■

LEMMA 2.6. *For all elements* (f, E) *of* $Q_H(T)$, *there exists a projection p in T such that*

$$f^{r_Q} = pQ_H(T).$$

Proof. The right ideal (0) is ecg, so $f^{-1}(0)$ is ecg by Lemma 2.4. Thus there exists an ascending sequence $\bigcup_{n \in \mathbb{N}} p^n T \subset f^{-1}(0) = \{t \in E \mid f(t) = 0\}$ with p^n projections, and the inclusion is essential. Set $p = \sup p^n$. Now by Lemma 2.5, $f^{-1}(0) = f^{r_Q} \cap T$ is essential in f^{r_Q}, so applying Lemma 2.5 $\bigcup p^n Q_H(T)$ is essential in f^{r_Q}, as either a $Q_H(T)$- or a T-module. We also have $\bigcup p^n Q_H(T)$ essential in pQ_H (since $\bigcup p^n T$ is essential in pT). Now by Lemma 2.4, with multiplication taken in $Q_H(T)$, we have $(0) = f \cdot (\bigcup p^n T)$ essential in $f \cdot (pQ_H(T))$, so $f \cdot p = 0$. Thus $pQ_H \subset f^{r_Q}$, and hence pQ_H is Q_H-essential in f^{r_Q}. Taking left Q-annihilators, by Lemma 2.2, $Q_H(T) \cdot (1 - p) = f^{r_Q l_Q}$, so taking right annihilators, $pQ_H(T) = f^{r_Q}$. ■

LEMMA 2.7. *The ring* $Q_H(T)$ *is regular.*

Proof. Choose $(f, E) \in Q_H(T)$. There exists a projection p in T such that $\{t \in E \mid f(t) = 0\} = pT \cap E$ ($f^{r_Q} = pQ_H$ for this p). As E is essential, $E \cap pT$ is essential in pT, and so

$$E' = (E \cap pT) \oplus (E \cap (1 - p)T)$$

is essential. As all of E, pT, $(1 - p)T$ are ecg, so is E ecg. Since E' is essential in E, $f(E') = f(E \cap (1 - p)T)$ is essential in $f(E)$. Now $f(E')$ is ecg, so there exists a projection q in T such that $f(E')$ is essential in qT (find $\bigcup p^n T$ essential in $f(E')$, p^n projections, and set $q = \sup p^n$). Define $E'' = f(E') \oplus (1 - q)T$. We have $(E'')^{l_T} \subset (f(E'))^l \cap (1 - q)^l = (0)$, so E'' is essential by Lemma 2.2, and obviously E'' is ecg. Define a homomorphism g on E'' by

$$g(1 - q) = 0 \qquad g(f(e)) = e \quad \text{if} \quad e \in E \cap (1 - p)T.$$

This makes sense, since f restricted to $E \cap (1 - p)T$ is one-to-one. It is clear that g extends to a module homomorphism $E'' \to T$, so $(g, E'') \in Q_H(T)$. If $r \in E \cap pT$, then $f(r) = 0$, so $(fgf - f)(r) = 0$; if $r \in E \cap (1 - p)T$,

$$(fgf - f)(r) = f(r) - f(r) = 0,$$

and thus fgf agrees with f on E', so $fgf = f$ in $Q_H(T)$. Thus $Q_H(T)$ is regular. ■

$Q_H(T)$ is *not* the regular ring we want—for one thing, it is not clear that the involution extends to $Q_H(T)$. Instead we find a subring

$$R = \{f \in Q_H(T) \mid T:f \text{ is essential and ecg as a left ideal}\},$$

where $T:f = \{t \in T \mid tf \in T\}$. Because Lemma 2.5 applies as well to left ideals, the same arguments involved in showing $Q_H(T)$ is a ring, also show R is a ring. By Lemma 2.5 again, R may be viewed as the collection of triples $\{(E', f, E)\}$, where E is essential and ecg as a right ideal and $f: E \to T$ is a right T-module homomorphism; E' is essential and ecg as a left ideal and $T \leftarrow E':f$ is a left module homomorphism, and $[(e')f] \cdot e = e' \cdot f(e)$ for all $e' \in E', e \in E$.

Now if E is essential (ecg) as a right ideal, then E^* is essential (ecg) as a left ideal. So we may define an involution $*$ on R by

$$(E', f, E)^* = (E^*, f^*, E'^*),$$

where $f^*(e'^*) = ((e')f)^*$ and $(e^*)f^* = (f(e))^*$. Then T is obviously embedded in R by $t \mapsto (T, t, T)$, and it is clear that the involution on R extends that on T. Also, this extended involution is unique, for $f^\#(e'^*)$ must equal $((e')f)^*$, and so $(f^\# - f^*)E^* = 0$, whence by nonsingularity $f^\# = f^*$.

A ring R with involution $*$ is *$*$-regular* if for all r in R there exists a projection p such that $rR = pR$. From [17, Ex. 5, p. 38]), R is $*$-regular if and only if R is regular and $rr^* = 0$ implies $r = 0$.

THEOREM 2.1. *Let T be a finite Rickart $*$ ring satisfying*:

(a) *the lattice of projections of T is $\aleph_0$-complete*;
(b) *the parallelogram law*;
(c) *for all r in T, there exists a countable upward ascending chain of projections $\{p^n\}$ in rT such that $r^1 = \{p^n\}^1$.*

Then there exists an embedding from T into a $*$-regular ring R with the following properties:

(1) *The involution on R extends the involution on T.*
(2) *Every projection of R lies in T.*
(3) *The inclusion $T \to R$ induces a lattice isomorphism between the projection lattice of T and the lattice of principal right ideals of R (R "coordinatizes" the projection lattice).*
(4) *R is an $\aleph_0$-continuous ring.*
(5) *The left projection of an element of T is the same whether computed in R or T; similarly with right projections.*

Proof. We must show that the ring R constructed in the preceding portion is $*$-regular. (*We already have an embedding $T \to R$ extending the involution.*)

(1) Choose $(E', f, E) \in R$. There exists a projection p in T such that $f^{rQ} = pQ_H(T)$. As $T \subset R \subset Q_H(T)$ and $p \in T$, it follows that $f^{rR} = pR$. Now $E \cap pT$ is essential in pT and, similarly, $E \cap (1-p)T$ is essential in $(1-p)T$, so $G = (E \cap pT) \oplus E \cap (1-p)T$ is essential and ecg. Also $f(G) = f(E \cap (1-p)T)$ and ecg. There exists a projection q in T with $f(G)$ essential in qT (find $1-q$ in T with $(f(G))^{lR} = R(1-q)$, as in the corresponding result with p). Then $f(G) \oplus (1-q)T$ is essential and ecg. We define an element $(g, f(G) \oplus (1-q)T)$ in $Q_H(T)$ by

$$g = \begin{cases} 0 & \text{on} \quad (1-q)T \\ f^{-1} & \text{on} \quad f(E \cap (1-p)T) \quad (f \text{ is one-to-one on } E \cap (1-p)T). \end{cases}$$

As in the argument of Lemma 2.7, $g \in Q_H(T)$ and $fgf = f$. Now we show $T:g = \{t \in T \mid tg \in T\}$ is essential and ecg, so that $g \in R$.

Consider the element fg of $Q_H(T)$. On $f(E \cap (1-p)T) \ominus (1-q)T$, fg agrees with the projection q and, by essentiality, $fg = q$ as elements of $Q_H(T)$. Thus $((E')f)g \subset T$. As usual, $(E')f$ is an ecg left ideal. However, $g(f(G) \oplus (1-q)T)$ is essential in $(1-p)T$, so their left annihilators are the same, so $pg(f(G) \oplus (1-q)T) = 0$; as $f(G) \oplus (1-q)T$ is essential, $pg = 0$. Setting $H = (E')f + Tp$, we have H is a sum of ecg left ideals, so is ecg; $H^r = ((E')f)^{rT} \cap (1-p)T \subset pT \cap (1-p)T = 0$, so H is essential by Lemma 2.2 with the involution. Finally, $Hg \subset (E')fg + (Tp)g = (E')fg \subset T$, so $H \subset T:g$, showing $T:g$ is essential and ecg, so $g \in R$. Thus R is regular.

Choose f in R nonzero. There exists t in T with $ft \neq 0$ and $ft \in T$. Then $t^*f^*ft = (ft)^*(ft) \neq 0$, so $f^*f \neq 0$. Thus R is $*$-regular.

(2) Let P be a projection of R. There exists a projection q in T such that $(1-P)^{rQ} = qQ_H$; thus $(1-P)^{rR} = qR$. But $(1-P)^{rR} = PR$, so $PR = qR$, and so $P = q$.

(3) The map of lattices is $p \mapsto pR$. As this is clearly one-to-one and, by (2), is onto (since R is $*$-regular), it suffices to show it is a lattice homomorphism. Let p, q be projections of T, and set $h = p \vee q$ (computed in T). There exists a projection j in R such that $jR = pR + qR$. Now $p \leqslant j$, $q \leqslant j$ ($\leqslant$ is equationally determined, e.g., $pj = p$, so does not depend on the ring); by (2), $j \in T$, so $h \leqslant j$, whence $hR \leqslant jR$. On the other hand, $pT + qT$ is contained in hT, so $pR + qR$ is contained in hR, whence $jR \leqslant hR$, so $j \leqslant h$, and thus $h = j$, so the map of lattices $p \mapsto pR$ preserves suprema. Similarly (or apply the orthocomplement), the map preserves infima, and thus is a lattice isomorphism.

(4) From (3), and Corollary 1.1(c).

(5) Choose r in T. Let p be the right projection of r as computed in R. Then p belongs to T and $r^{\mathbf{r}R} = (1-p)R$. But

$$r^{\mathbf{r}T} = r^{\mathbf{r}R} \cap T = (1-p)R \cap T = (1-p)T,$$

so the right projection of r computed in T is also p. Similarly with left projections, as the involution extends to R. ■

3. $\aleph_0$-Continuous Rings Are Unit Regular

We would like to prove that in the $\aleph_0$-continuous ring R constructed in Section 2 ordinary equivalence of projections implies perspectivity. Actually, we prove a more general result by dropping the involution. Throughout this section, R will denote an $\aleph_0$-continuous ring; that is, a regular ring whose lattice of principal right ideals is $\aleph_0$-continuous. From the order-reversing map $eR \mapsto R(1-e)$, the lattice of principal left ideals is also $\aleph_0$-continuous.

A ring A is *unit regular* if for all r in A, there exists a unit u such that $rur = r$.

PROPOSITION 3.1 [11]. *Let A be a regular ring. The following are equivalent*:

(a) *A is unit regular.*

(b) *If e, f are idempotents, $e \sim f$ implies $1 - e \sim 1 - f$.*

(c) *If e, f are idempotents, $e \sim f$ implies eA is perspective to fA in the lattice of principal right ideals.*

(d) *Perspectivity is transitive on the lattice of principal right ideals of the 2×2 matrix ring over A.*

Further, if A is unit regular, then both the ring of $n \times n$ matrices over A, and eAe $(e = e^2)$ are unit regular.

We want to show R is unit regular. Here again we use a Q_H-construction, to employ Proposition 3.1(d). Fortunately, the details of the construction are much simpler. First we observe that in *any* regular ring A, every countably generated right ideal is of the form $\bigoplus_{i \in \mathbb{N}} e_i A$, where the e_i are orthogonal idempotents (e.g., Goodearl [8, Lemma 12]). Thus we have the analog of Proposition 2.1 for free. Let us check the analog of Corollary 1.1(d), (f) for $\aleph_0$-continuous rings.

PROPOSITION 3.2. *Let R be an $\aleph_0$-continuous ring.*

(a) *If $\{r_iR\}$ is a countable collection of principal right ideals, then in the lattice of principal right ideals*

$$\sup\{r_iR\} = (\sum r_iR)^{\mathrm{lr}}, \qquad \inf\{r_iR\} = \bigcap r_iR.$$

(b) *If $\{r_i\}$ is a countable subset of R, there exist idempotents e, f such that*

$$\{r_i\}^{\mathrm{r}} = eR, \qquad \{r_i\}^{\mathrm{l}} = Rf.$$

(c) *If I is an essentially countably generated right (left) ideal, then I is essential in I^{lr} (I^{rl}).*

Proof. (a) There exists $g = g^2$ in R such that $gR = \inf\{r_iR\}$. As $gR \leqslant r_iR$, we have $gR \leqslant \bigcap r_iR$. Conversely, if $t \in \bigcap r_iR$, then $tR \leqslant r_iR$, so by the definition of infimum, $tR \leqslant gR$, so $\bigcap r_iR = gR$.

Set $hR = \sup\{r_iR\}$. Then $r_iR \leqslant hR$, so $\sum r_iR \leqslant hR$; taking left then right annihilators, $(\sum r_iR)^{\mathrm{lr}} \leqslant hR$. It thus suffices to show the second part of (b).

(b) We may assume the r_i are all idempotents. Then $\{r_i\}^{\mathrm{l}} = \bigcap r_i^{\mathrm{l}} = \bigcap R(1 - r_i)$. By the antiautomorphism between right and left ideals, the inf result of (a) applies to give us an idempotent f such that $\bigcap R(1 - r_i) = Rf$. The other part of (b) comes from the right–left duality.

(c) Routine applications of $\aleph_0$-continuity as in the *proof* of Corollary 1.1(f). ■

Now in this situation, the "analog" of Lemma 2.4 holds, so we may define the subring $Q_J(R)$ of the maximal ring of right quotients of R, just as in Section 2, by considering

$$\{(f, E) \mid E \text{ essential, ecg, and } f: E \to R \text{ a module homomorphism}\}.$$

Replacing "projection" by "idempotent" in Lemma 2.6, that result carries through, and the proof of Lemma 2.7 carries through to show $Q_J(R)$ is regular.

We may also define the ring $[Q_J(R)]_L$, by considering *left* essential and ecg ideals and left module homomorphisms. We shall show (eventually) that $Q_J(R) = [Q_J(R)]_L$.

If $A \to B$ is an inclusion of rings, the map is said to be (right) *intimate* [12] if every principal right ideal is generated by an element of the image of A.

LEMMA 3.1. *The map $R \to Q_J(R)$ is right intimate and all idempotents of $Q_J(R)$ lie in R.*

Proof. Choose $(f, E) \in Q_J(R)$. As $Q_J(R)$ is regular, to prove intimacy, we may assume f is idempotent. By the method of proof of Lemma 2.7, there exists an idempotent e in R with such that

$$(1 - f)^{\mathrm{r}Q} = eQ_J(R).$$

However, $(1-f)^{r\varrho} = fQ_J(R)$, so $fQ_J(R) = eQ_J(R)$, from which intimacy is established. By [12, Corollary 1.6], the last statement holds. ■

We have a strong form of orthogonalization available here that is not generally available in regular rings.

LEMMA 3.2. *For the $\aleph_0$-continuous ring R, given a countable direct sum $\bigoplus e_iR$ of principal right ideals in R, there exist* orthogonal *idempotents $\{f_i\}$ in R such that*

$$f_iR = e_iR \qquad \text{for all} \quad i.$$

Proof. Set $eR = \sup\{e_iR\}$, where e is an idempotent. Then $E = (1-e)R \oplus (\bigoplus e_iR)$ is essential (its left annihilator is zero). Define elements (F_j, E) in Q_J by

$$F_j \text{ is} \begin{cases} 0 & \text{on} \quad (1-e)R \oplus \left(\bigoplus\limits_{i \neq j} e_iR\right), \\ \text{identity} & \text{on} \quad e_jR. \end{cases}$$

Then F_j are clearly idempotents, so by the preceding result, $F_j \in R$. Also $e_jR = F_jE$ is essential in F_jR. Taking left and then right annihilators, we obtain $e_jR = F_jR$. If $i \neq j$, then $F_jF_iE = 0$; as E is essential, $F_jF_i = 0$. Set $f_j = F_j$. ■

THEOREM 3.1. *If R is an $\aleph_0$-continuous ring, then $Q_J(R) = [Q_J(R)]_L$.*

Proof. As the hypothesis on R is right–left symmetric, it suffices to show $Q_J(R) \subseteq [Q_J(R)]_L$. Choose $(s, E) \in Q_J(R)$. We show the left R-ideal, $R: s = \{r \in R \mid rs \in R\}$, is essential and ecg, which will complete the proof.

There exists, by Lemma 3.1, an idempotent f in R with $s^{r\varrho} = fQ_J(R)$. Then, as usual, $(E \cap fR) \oplus (E \cap (1-f)R)$ is essential and ecg. We may thus find a countable orthogonal set of idempotents $\{f_i\}_{i=0}^{\infty}$ such that $f_0 = f$, and $\bigoplus_{i \in \mathbb{N}} f_iR$ is essential in $E \cap (1-f)R$. In particular, s restricted to $\bigoplus_{i \in \mathbb{N}} f_iR$ is one-to-one. Thus the sum $\sum_{i \geqslant 1} s(f_i)R = \bigoplus_{i \geqslant 1} s(f_i)R$ and, of course, $\bigoplus_{i \geqslant 1} s(f_i)R$ is essential in $s(E)$. There exists an idempotent e such that $\bigoplus s(f_i)R$ is essential in eR (take $eR = \sup\{s(f_i) \cdot R\}$). Thus $(1-e)R \oplus (\bigoplus_{i \geqslant 1} s(f_i)R)$ is essential. By the preceding lemma, there exists an orthogonal set of idempotents $\{g_i\}_{i=0}^{\infty}$ such that $g_0R = (1-e)R$ and $g_iR = s(f_i)R$ for $i \geqslant 1$. Thus (multiplying in Q_J), $g_isf_i = sf_i$ for $i \geqslant 1$.

We claim $g_is = g_isf_i$ for $i \geqslant 1$.

Consider

$$(g_is - g_isf_i)f_j = \begin{cases} g_isf_j = g_ig_jsf_j = 0 & \text{if} \quad i \neq j,\ i \geqslant 1. \\ 0 & \text{if} \quad i \geqslant 1,\ i = j. \end{cases}$$

Finally, $sf_0 = sf_i f_0 = 0$, so $(g_i s - g_i s f_i)(\ominus_{j\geqslant 0} f_j R) = 0$. As $\bigoplus f_j R$ is essential, $g_i s = g_i s f_i$, so $g_i s \in R$ for $i \geqslant 1$.

Also, $s(\bigoplus_{i=0}^{\infty} f_i R) = \bigoplus_1^{\infty} g_i R$ and, as $\bigoplus_0^{\infty} f_i R$ is essential, $\bigoplus_1^{\infty} g_i R$ is essential in the image of s in R. There exists an idempotent h in R with $\{g_i\}^{\mathrm{l}Q} = Q_J(R)\cdot h$ and, of course ($\{g_i\} \subset R$!), $\{g_i\}^{\mathrm{l}R} = Rh$, and it is clear that $hs(\bigoplus_0^{\infty} f_i R) = \sum hg_i R = 0$; as $\bigoplus f_i R$ is essential, $hs = 0$ (multiplication in Q_J). Now consider $K = Rh + (\bigoplus_{i\geqslant 1} Rg_i)$; $K^{\mathrm{r}} = h^{\mathrm{r}} \cap \{g_i\}_{i\geqslant 1}^{\mathrm{r}} = \{g_i\}^{\mathrm{lr}} \cap \{g_i\}^{\mathrm{r}} = (0)$, so, by Proposition 3.2(c), K is left essential. Also K is ecg, and since $g_i s \in R$ for all $i \geqslant 1$ and $hs = 0$, $K \subset R : s$, concluding the proof. ■

COROLLARY 3.1. *Suppose the $\aleph_0$-continuous ring R possesses an involution $*$. Then $*$ may be extended uniquely to an involution on $Q_J(R)$. If for any integer n, R satisfies:*

$$\text{if} \quad \{a_i\}_{i=1}^n \subset R, \qquad \text{then} \quad \sum a_i a_i^* = 0 \quad \text{implies} \quad a_i = 0, \tag{3}$$

then $Q_J(R)$ also satisfies (3).

Proof. Pick $(f, E) \in Q_J(R)$ (i.e., E is an ecg essential right ideal of R, $f: E \to R$ is a right R-module homomorphism). Now define the element (f^*, E^*) in $[Q_J(R)]_L$ by $(e^*)f^* = (f(e))^*$. (Observe that if E is essential or ecg as a right ideal, E^* has the same property as a left ideal.) Since $[Q_J(R)]_L = Q_J(R)$, it is readily seen to be a ring involution, agreeing with $*$ on R.

Suppose R satisfies (3). Let $\{q_i\}_{i=1}^n$ be elements of $Q_J(R)$ such that $\sum q_i q_i^* = 0$, but $q_i \neq 0$ for all i. As $Q_J(R)$ is an essential extension of R, there exists $a_1^* \in R$ such that $q_1^* a_1^*$ is nonzero and belongs to R. Then

$$\sum (a_1 q_i)(a_1 q_i)^* = 0$$

and $a_1 q_i \in T$. If $a_1 q_2 \notin R$, we may find a_2 such that $a_2 a_1 q_1 \in R$ and is nonzero. This process may obviously be continued by induction to yield a contradiction to (3). ■

A ring A is right (left) σ-injective if for any countably generated right (left) ideal K and any module homomorphism $f: K \to A$ ($A \leftarrow K: f$) there is an element a in A such that $f(k) = ak$ [$ka = (k)f$] for all k in K.

LEMMA 3.3. *If R is $\aleph_0$-continuous, $Q_J(R)$ is right and left σ-injective, as well as $\aleph_0$-continuous. The inclusion $R \subset Q_J(R)$ induces an isomorphism between the principal right R- and Q_J-ideal lattices.*

Proof. As $Q_J(R) = [Q_J(R)]_L$, it suffices to show that $Q_J(R)$ is right σ-injective. Let K be a countably generated right ideal; since $Q_J(R)$ is regular, we may find orthogonal idempotents $\{e_i\}_{i\in\mathbb{N}}$ such that $K = \bigoplus e_i Q_H$. By

Lemma 3.1, $\{e_i\} \subset R$, and so $\bigoplus e_iR$ is an essential submodule of $K \cap R$. Let $f: K \to Q_J$ be a Q_J-module homomorphism. We have $f(K) \cap R$ is R-essential in $f(K)$.

Now $f(K)$ is ecg, and as all idempotents of $Q_J(R)$ lie in R, $f(K) \cap R$ is ecg as a right R-ideal. Define

$$L = f^{-1}(f(K) \cap R) = \{r \in K \cap R \mid f(r) \in f(K) \cap R\}$$

(considering f as an R-module homomorphism).

The usual arguments show L is an ecg right R-ideal and L is essential in K. Then $f: L \to R$ is a right R-module homomorphism, so there is an element q in $Q_J(R)$ such that $(f - q)L = (0)$. As L is essential in k, $(f - q)K = 0$, yielding the first result. To see $Q_J(R)$ is $\aleph_0$-continuous, it suffices to prove the second sentence.

Since R and $Q_J(R)$ are regular rings, the lattice operations are just addition and intersection. Thus the map $eR \mapsto eQ_J$ is a lattice homomorphism. It is clear that if e, f are idempotents, $eR \subseteq fR$ if and only if $eQ \subseteq fQ$. Also, by Lemma 3.1, the map is onto. Finally if $eR \neq fR$, there exist orthogonal idempotents g, h, k with $eR \cap fR = kR$, $(g + k)R = eR$, $(h + k)R = fR$. It follows that $eQ_J \neq fQ_J$. Thus the map is an isomorphism of partially ordered sets; any such map between lattices is easily seen to be a lattice isomorphism. ∎

This result allows us to prove $\aleph_0$-continuous rings are unit regular.

THEOREM 3.2. *An $\aleph_0$-continuous ring is unit regular.*

Proof. We first show that $Q_J(R)$ is unit regular. To begin we show M_2Q_J, its 2×2 ring of matrices, is $\aleph_0$-continuous.

In [26] and [25, p. 209] are defined right factor correspondences: These are pairs $(u; v)$ such that left multiplication by u on vR and left multiplication by v on uR are inverse module isomorphisms; equivalently $uvu = u$ and $vuv = v$. Factor correspondences may be partially ordered by saying $(u, v) \leqslant (u', v')$ if $uR \subset u'R$ and $v'u = vu$ and $u'v = uv$; it immediately follows that $vR \subset v'R$, and the restriction of left multiplication by v' on uR agrees with that by v, and similarly u' restricts on vR to agree with u. Left factor correspondences may be defined and ordered similarly. Now [26, Theorem 2] asserts that if suprema of countable chains of right and of left factor correspondences exist, then M_2Q_J is $\aleph_0$-continuous. We show right σ-injectivity and upper $\aleph_0$-continuity forces suprema of right factor correspondences to exist, and since Q_H is also left σ-injective and lower $\aleph_0$-continuous, the dual result allows us to use [26, Theorem 2].

So let $\{(x_i; y_i)\}_{i \in \mathbb{N}}$ be an increasing chain of right factor correspondences in Q_J. Then we have isomorphisms

$$\bigcup x_iQ \simeq \bigcup y_iQ \quad \text{and} \quad \bigcup y_iQ \simeq \bigcup x_iQ,$$
$$x_i \mapsto y_ix_i \quad \text{and} \quad y_i \mapsto x_iy_i.$$

Set $eQ = \sup\{x_iQ\}$, $fQ = \sup\{y_iQ\}$. Then by Proposition 3.2(a, b), $\bigcup x_iQ \oplus (1-e)Q$ and $\bigcup y_iQ \oplus (1-f)Q$ are essential and countably generated. By right σ-injectivity, there exist y, $x \in Q_J$ with $yx_i = y_ix_i$, $y(1-e) = 0$, and $xy_i = x_iy_i$ and $x(1-f) = 0$. Then $x(\bigcup y_iQ \oplus (1-f)Q) = \bigcup x_iQ$ is essential in xQ, so $xQ = eQ$, and similarly for $yQ = fQ$. For the same reasons $xyQ = xQ$, $yxQ = yQ$, and it easily follows that $(x; y)$ is a factor correspondence; it is clear that $(x; y)$ is the least upper bound of $(x_i; y_i)$. Left σ-injectivity proves the corresponding result on the left, whence by [26, Theorem 2], M_2Q_J is $\aleph_0$-continuous.

By [10, Theorem 5.1], perspectivity is transitive on an $\aleph_0$-continuous lattice, so by Proposition 3.1(d), Q_J is unit regular.

Suppose e, f are idempotents in R and $e \sim f$ in R. Then $e \sim f$ in Q_J (by definition of equivalence), so, by Proposition 3.1(c), eQ is perspective to fQ. But perspectivity is a lattice invariant so eR is perspective to fR. Thus we have verified the criterion of Proposition 3.1(c) for unit regularity of R. ■

There must be a shorter proof of this result.

Corollary 3.2. *Let R be an $\aleph_0$-continuous ring and M a maximal two-sided ideal of R. Then R/M is a right and left self-injective (hence continuous) ring. Further, either both R/M and $Q_J/\langle M\rangle$ are division rings, where $\langle M\rangle$ is the Q_J ideal generated by M, or $R/M = Q_J/\langle M\rangle$.*

Proof. Let us first show that $\langle M\rangle \neq Q_J$. Consider the set of idempotents in M, $\tilde{M}$. Then $\tilde{M}$ satisfies the following properties: Letting e be an idempotent:

(a) If e, f are orthogonal in $\tilde{M}$, then $e + f \in \tilde{M}$.
(b) If $e \sim g$, $g \in \tilde{M}$, then $e \in \tilde{M}$.
(c) If $e = ef$ or $fe = e$, and $f \in \tilde{M}$, then $e \in \tilde{M}$.

Now observe that if e, g are idempotents, then $e \sim g$ in R if and only if $e \sim g$ in Q_J (because eQ_J is perspective to gQ_J). Let us consider $T = \sum_{e \in \tilde{M}} eQ_J$. Clearly T is a right ideal containing $\tilde{M}$. Now if $q', q \in Q_J$ and g is an idempotent such that $gQ_J = qeq'Q_J$, then $g \lesssim e$ in Q_J, hence in R, so $g \in \tilde{M}$. Thus for all $q \in Q_J$, $qT \subset T$, so T is a two-sided ideal of Q_J. If

$T = Q_J$, $Q_J = \sum e_j Q_J$ for some finite set $\{e_j\}$, so $\{e_j\}^{1_Q} = (0)$, whence $\{e_j\}^{1_R} = 0$ and so $R = \sum e_j R$ (since R is regular, $\sum e_j R$ is principal), and thus $M = R$, a contradiction. We have $\langle M \rangle \subset T \neq Q_J$. But it is clear that $T = MQ$, so $\langle M \rangle = T$ is a proper ideal. Now any ideal of a regular ring is generated as a right ideal by its idempotents, so if $\langle M \rangle$ is not maximal, say $\langle M \rangle \subset K$, $\tilde{K}$ would properly contain $\tilde{M}$, and so $M \subsetneq K \cap R$, again a contradiction. Thus $\langle M \rangle$ is maximal, so $Q_J/\langle M \rangle$ is a simple unit-regular ring, and thus itself has no uncountable direct sums of principal right ideals [9, 3.4(6)].

Set $V = Q_J/\langle M \rangle$. We prove, by a proof of Goodearl, that V is right and left self-injective. Let E be right ideal of V, and $f: E \to V$ a homomorphism. In any ring, there is a collection of principal right ideals $\{v_i V\}$ such that $\bigoplus v_i V$ is essential in E. Here, the collection $\{v_i V\}$ must be countable. If $f(v_i) = v v_i$ for all i, some $v \in V$, then $(f - v)(\bigoplus v_i V) = 0$, so $(f - v)E = 0$, i.e., f agrees with left multiplication on E. Now we may find $y_i \in Q_J$ such that $\bar{y}_i = v_i$. Set $K = \sum y_i Q_J$. As K is countably generated there exist orthogonal idempotents $\{e_i\}$ in Q_J such that $\bigoplus e_i Q_J = \sum y_i Q_J$, whence $\bigoplus \bar{e}_i V = \sum v_i V$. For each i, $f(\bar{e}_i) = \bar{x}_i \bar{e}_i$, some $x_i \in Q_J$ ($f(\bar{e}_i) = f(\bar{e}_i{}^2) = f(\bar{e}_i)\bar{e}_i$), and we obtain a map $\bigoplus e_i Q_J \to Q_J$ by $e_i \mapsto x_i e_i$. There exists, by right σ-injectivity of Q_J, x in Q_J with $x_i e_i = x e_i$, so $(f - \bar{x})e_i = 0$; i.e., set $v = \bar{x}$. Dually, left injectivity holds. So $V = Q_J/\langle M \rangle$ is right and left self-injective and simple so, by [23, Corollary 1], is a continuous ring.

There is a natural inclusion $R/M \to Q_J/\langle M \rangle$. Every idempotent of $Q_J/\langle M \rangle$ is an image of an idempotent from Q_J (true from regularity) so every idempotent of $Q_J/\langle M \rangle$ lies in R/M. However, if $Q_J/\langle M \rangle$ is not a division ring, $Q_J/\langle M \rangle$ is generated by its idempotents (by [21]), so $Q_J/\langle M \rangle = R/M$. Otherwise $Q_J/\langle M \rangle$ is a division ring, so R/M must also be a division ring. ■

COROLLARY 3.3. *Let R be an $\aleph_0$-continuous ring. If R is generated by its idempotents (as a ring), then R is right and left σ-injective. If R is σ-injective, then all matrix rings over R are $\aleph_0$-continuous and σ-injective.*

Proof. By Lemma 3.1, if R is generated by its idempotents, then $Q_J(R) = R$, so Lemma 3.3 yields the first statement. It is implicit in the first part of the proof of Theorem 3.2 that if R is $\aleph_0$-continuous and σ-injective, then $M_2 R$ is $\aleph_0$-continuous. Now any 2×2 matrix ring over *any* ring is generated by its idempotents, so $M_2 R$ is σ-injective. By induction $M_{2^n} R$ is $\aleph_0$-continuous. Now there exists $e = e^2 \in M_{2^n} R$ so that $M_n R \simeq e M_{2^n} R e$, and it easily follows that $M_n R$ is $\aleph_0$-continuous. ■

So the combined property "$\aleph_0$-continuous and σ-injective" is a Morita invariant. Goodearl has shown that any $\aleph_0$-continuous ring with no division rings as homomorphic images is generated by its idempotents, so all its matrix rings would be $\aleph_0$-continuous.

Much of the construction in this section is possible under weaker hypotheses; for instance, in Corollary 3.3, "$\aleph_0$-continuous" can be replaced by "upper $\aleph_0$-continuous and directly finite," if σ-injective is weakened to right σ-injective.

4. Main Results

THEOREM 4.1. *Let T be a finite Rickart $*$ ring, satisfying* (1)–(3) *of Lemma* 1.2. *Then*:

(a) *If p and q are* (*ordinary*) *equivalent projections, then there exists a unitary u in T such that $upu^* = q$.*

(b) *All matrix rings over T are directly finite* ($xy = 1$ *implies* $yx = 1$), *so, in particular*, are finite.

(c) *If $t \in T$, the left and right projections of t are unitarily equivalent.*

Proof (a) If $p \sim q$ in T, then in the $\aleph_0$-continuous overring R obtained in Theorem 2.1, $p \sim q$ as well. By Theorem 3.2, R is unit regular, so pR is perspective to qR in the lattice of principal right ideals of R. By Theorem 2.1, the inclusion $T \subset R$ induces an isomorphism between the projection lattice of T and the principal right R-ideal lattice, so the lattice-invariant perspectivity is inherited, i.e., p is perspective to q in T. By Corollary 1.1(g), $p \overset{*}{\sim} q$ in T. Now, since $p \sim q$ in R, $1 - p \sim 1 - q$ in R by Proposition 3.1(b) and the unit regularity of R. The same process leads to $1 - p \overset{*}{\sim} 1 - q$ in T and, of course, unitary equivalence is the same as $p \overset{*}{\sim} q$ and $1 - p \overset{*}{\sim} 1 - q$.

(b) We have $T \subset R$, so $M_n T \subset M_n R$. As R is unit regular, all matrix rings over R are unit regular (Proposition 3.1); it is an easy exercise to see unit regularity implies direct finiteness, so all $M_n R$ are directly finite. Any subring of a directly finite ring is trivially directly finite.

(c) Set $p = \mathrm{LP}(t)$, $q = \mathrm{RP}(t)$. By Theorem 2.1, "left projection" and "right projection" are unambiguous, whether computed in T or R. Since R is a $*$-regular ring, however, $pR = rR$, and $Rq = Rr$. As right R-modules,

$$pR = rR \simeq R/r^{\mathrm{r}_R} = R/(1 - q)R \simeq qR.$$

Thus $p \sim q$ in R. The lattice argument of (a) shows $p \sim q$ in T, whence by (a) p is unitarily equivalent to q. ∎

So LP $\sim$ RP holds in finite Rickart C^* algebras, real or complex.

THEOREM 4.2. *Let T be a finite Rickart $*$-ring satisfying* (1)–(3) *of Lemma* 1.2. *$*$-equivalence is countably additive, that is, if $\{p_i\}_{i\in\mathbb{N}}$, $\{q_i\}_{i\in\mathbb{N}}$ are sets of orthogonal projections such that $p_i \overset{*}{\sim} q_i$, then $\sup p_i \overset{*}{\sim} \sup q_i$.*

Proof. We have three rings T, R, and $Q_J(R)$ (in fact $Q_H(T) = Q_J(R)$, but we do not need this). As $p_i \overset{*}{\sim} q_i$ in T, $p_i \sim q_i$ in $Q_J(R)$. Now $Q_J(R)$ is σ-injective, and it easily follows that

$$\sup_Q(p_iQ_j) \simeq \sup(q_iQ_J).$$

By unit regularity, the two objects are perspective. By the lattice isomorphisms between projections of T, principals of R, and principals of Q_J, it is seen that $(\sup_T p_i)Q_J = \sup_Q\{p_iQ_J\}$ and so $\sup_T\{p_i\}$ is perspective to $\sup_T\{q_i\}$, whence by Corollary 1.1(g), they are $*$-equivalent. ■

THEOREM 4.3. *Let T be a finite Rickart C^* algebra and M a maximal two-sided ideal of T. Then T/M is a finite AW^* algebra.*

Proof. This result is the analog of [2, Theorem 1, p 206]. We observe there are several steps in the proof ot that result, and show that these may be carried through here.

(1) Propositions 1 and 2 of Section 23 of [2] carry through now that we know LP $\sim$ RP holds. Thus if $\tilde{M}$ is the collection of projections of M and $I = T\tilde{M}T$, then T/I is a Rickart $*$ ring, and by [2, Proposition 5, p. 140], the closure of I is M.

(2) Now Lemma 2 of Berberian [2, p. 203] requires only that the bicommutant obtained in the proof be a Rickart C^* algebra, so we have the result that if u is a projection of T/M, $u = \bar{e}$ for some projection e in T. Also Lemmas 4 and 5 [2, p. 204] hold, so countable orthogonal sets of projections in T/M may be lifted to T.

(3) Next, we need a dimension function on T/I. Now every ideal in R is generated by its projections, and the same argument as in the proof of Corollary 3.2 shows RIR is a maximal ideal of R, and $RIR \cap T = I$ (because I is a maximal restricted ideal). But by Corollary 3.2, R/RIR is simple self-injective, so by [9, Theorem 2.2, Theorem 3.17], has a dimension function. This dimension function restricts to T/I, and, as in Definition 1 of [2, p. 204], this allows us to define a function D_M on the projections of T/M.

(4) Now the proofs of Lemmas 7–12 and Theorem 1 of [2, Section 45] carry through verbatim to conclude the proof. ■

One would hope to be able to obtain, analogous to the AW^* case, a decomposition of an arbitrary Rickart C^* algebra along the lines of $T = T_1 \times T_2$, where T_2 is ("properly") infinite and T_1 is finite, and to further write T_1 as a product of an Abelian algebra and a C^* algebra possessing no central Abelian projections. In fact such hopes are dashed by the following examples based on an example of Goodearl in a related situation.

Let T be an AW^* algebra and S an AW^* subalgebra. Set C to be the C^* sum of an uncountable number of copies of T. It is well known and easy to verify that C is itself an AW^* algebra. Define D to be the subalgebra of C consisting of the sequences (t_i) such that all but a countable number of coordinates lie in S. It is routine to verify that $D = D(T, S)$ is a Rickart C^* algebra, and it will be finite if and only if T is finite.

EXAMPLE 1. No finite by properly infinite decomposition.

Let T be a type III factor and S a II_f subfactor. Then $D(T, S)$ has no faithful properly infinite projections, but is not finite; nor does it admit any such decompositions.

EXAMPLE 2. Finite but no Abelian by properly non-Abelian decompositions.

Let T be $M_2\mathbb{C}$ and $S = \mathbb{C} \times \mathbb{C}$ embedded diagonally in T. Then $D(T, S)$ has plenty of Abelian central projections, but no largest one. This example also fails to satisfy central comparability (consider the projections $(\ldots, \left[\begin{smallmatrix}1&0\\0&0\end{smallmatrix}\right], \ldots)$ and $(\ldots, \left[\begin{smallmatrix}0&0\\0&1\end{smallmatrix}\right], \ldots)$), and was the original example due to Goodearl.

EXAMPLE 3. Finte Rickart C^* algebra, with AW^* centre, that is not AW^*.

Set T to be a II_f factor and S a II_f subfactor. Then $D = D(T, S)$ has centre $C(\beta I)$ (I a discrete uncountable set, βI its Stone–Cech compactification), as the centres of C and D agree in this case. Yet D is not AW^* because if J is an uncountable and co-uncountable subset of I, and p is a projection of T not in S, then

$$\mathrm{Sup}\{(\ldots, 0, 0, p, 0, 0, \ldots) \mid p \text{ in the } j\text{th position}, j \in J\}$$

does not exist in D.

Observe that $C(T, S)$ is a sort of AW^* completion of $D(T, S)$. Perhaps there is a structure theorem for finite Rickart C^* algebras along the lines of a $D(T, S)$ construction. The first step in establishing such a result would be to prove, for example, the intersection of the maximal (two-sided) ideals is zero. This would follow if we could prove the following result:

If p, q are projections in the finite Rickart C^* algebra T and

$$\begin{bmatrix} p & 0 \\ 0 & p \end{bmatrix} \overset{*}{\precsim} \begin{bmatrix} q & 0 \\ 0 & q \end{bmatrix} \quad \text{in} \quad M_2 T,$$

does it follow that $p \precsim q$? This will be shown by ordered group techniques in a forthcoming paper by J. Lawrence, D. Higgs, and the author.

5. General Comments

In our construction of the regular ring R in Section 2, we took a subring of the maximal ring of quotients, where the filter varied over the essentially countably generated essential right ideals. *Sometimes* we can use only the essentially finitely generated (efg) right ideals, and the computations are much easier.

Recall that a ring is (right) *semihereditary* if every finitely generated right ideal is projective as a right module. Any finitely generated submodule of a projective module is projective for these rings [18; Example 6, p. 86].

Proposition 5.1 (Essentially in Chase [5, Theorem 2.2]). *Let A be a right semihereditary ring, let I, J be finitely generated right ideals, and $f: I \to A$ a module homomorphism.*

(a) *$I \cap J$ is finitely generated.*
(b) *$f^{-1}0 = \{r \in I \mid f(r) = 0\}$ is finitely generated.*
(c) *$f^{-1}J = \{r \in I \mid f(r) \in J\}$ is finitely generated.*

Proof. (a) The diagonal map $\Delta: I \cap J \to I \times J$ (direct product), $\Delta(i) = (i, i)$, makes the following sequence exact

$$(0) \to I \cap J \xrightarrow{\Delta} I \times J \to I + J \to (0), \qquad (i,j) \mapsto i - j.$$

But $I + J$ is finitely generated, thus projective, and the sequence splits. Therefore $I \cap J$ is a direct summand of the finitely generated module $I \times J$, and so is finitely generated.

(b) The sequence

$$(0) \to f^{-1}0 \to I \xrightarrow{f} f(I) \to 0$$

splits as $f(I)$ is finitely generated and projective.

(c) The sequence

$$(0) \quad f^{-1}0 \to f^{-1}J \xrightarrow{f} f(I) \cap J \to (0)$$

splits and, by (a), (b), $f^{-1}0, f(I) \cap J$ are finitely generated. As $f^{-1}J \simeq f^{-1}0 \oplus (f(I) \cap J)$, we have the result. ■

It easily follows from Proposition 5.1 that for a semihereditary ring (or any nonsingular ring satisfying Proposition 5.1(c)) A, the collection of pairs of the form (f, E), E essential and efg $f: E \to A$, forms a subring $Q_F(A)$ of the maximal ring of quotients of A. If we knew in Section 2 that Rickart C^* algebras were semihereditary, we could have used this much simpler construction for R.

For general semihereditary rings, $Q_F(A)$ need not be regular: in fact, $Q_F(A)$ is regular if and only if for all finite subsets $\{a_i\} \subset A$,

$$\{a_i\}^l = 0 \quad \text{implies} \quad \sum a_i A \text{ is essential.}$$

When $Q_F(A)$ is regular, $A \to Q_F(A)$ is a flat epimorphism of rings. If, however, A is either a free algebra on two or more variables or a W^* factor of type III (both classes of semihereditary rings), then $Q_F(A) = A$.

For Rickart $*$ rings, it is probably true that (4) is equivalent to the formally weaker condition

$$r \in A, \qquad r^l = 0 \quad \text{implies} \quad rA \quad \text{is essential,}$$

but a Rickart $*$ ring need not be semihereditary.

References

1. I. Amemiyah and I. Halperin, Complemented modular lattices, *Canad. J. Math.* **2** (1959), 481–520.
2. S. K. Berberian, "Baer $*$ Rings," Grundlehren Band 195, Springer-Verlag, Berlin and New York, 1972.
3. G. M. Bergman, Hereditary commutative rings and centres of hereditary rings, *Proc. London Math. Soc.* **23** (3) (1971), 214–236.
4. V. C. Cateforis, Flat regular quotient rings, *Trans. Amer. Math. Soc.* **138** (1969), 241–249.
5. S. U. Chase, Direct products of modules, *Trans. Amer. Math. Soc.* **97** (1960), 457–473.
6. N. Fine, L. Gillman, and J. Lambek, "Rings of Quotients of Rings of Continuous Functions," McGill Univ. Press, Montreal, 1965.
7. L. Gillman and M. Jerison, "Rings of Continuous Functions," Princeton Univ. Press, Princeton, New Jersey, 1960.
8. K. R. Goodearl, Simple regular rings and rank functions, *Math. Ann.* **214** (1975), 267–287.
9. K. R. Goodearl and D. Handelman, Simple self-injective rings, *Comm. in Algebra* **3**(9) (1975), 797–834.
10. I. Halperin, On the transitivity of perspectivity in continuous geometries, *Trans. Amer. Math. Soc.* **44** (1938), 537–562.
11. D. Handelman, Perspectivity and cancellation in regular rings, *J. Algebra* **48** (1977), 1–16.
12. D. Handelman, Coordinatization applied to finite Baer* rings, *Trans. Amer. Math. Soc.* **235** (1978), 1–34.
13. D. Handelman and J. Lawrence, Strongly prime rings, *Trans. Amer. Math. Soc.* **211** (1975), 209–223.
14. B. Jònsson, Representations of complemented modular lattices, *Trans. Amer. Math. Soc.* **97** (1960), 64–94.
15. I. Kaplansky, Projections in Banach algebras, *Ann. of Math.* **53**(2) (1951), 235–249.
16. I. Kaplansky, Any orthocomplemented complete modular lattice is a continuous geometry, *Ann. of Math.* **61** (1955), 524–541.

17. I. Kaplansky, "Rings of Operators," Benjamin, New York, 1968.
18. J. Lambek, "Lectures on Rings and Modules" (rev.), Chelsea, New York, 1976.
19. S. Maeda, On $*$-rings satisfying the square root axiom, *Proc. Amer. Math. Soc.*
20. S. Maeda and S. S. Holland, Jr., Equivalence of projections in Baer $*$ rings, *J. Algebra*
20a. G. Renault, Anneaux réguliers autoinjectifs à droite, *Bull. Soc. Math. Fr.* **101** (1973).
21. W. Stephenson, Rings generated by their nilpotents, idempotents and units, to appear.
22. H. Storrer, Epimorphismen von kommutativen Ringen, *Comment. Math. Helv.* **43** (1968), 378–401.
23. Y. Utumi, "On a theorem on modular lattices," *Proc. Japan Acad.* **35** (1959), 16–21.
24. I. Vidav, On some *-regular rings, *Acad. Serbe. des Sci. Publ. Inst. Math.* (*Beograd*) **13** (1959), 73–80.
25. J. von Neumann, "Continuous Geometry," Princeton Univ. Press, Princeton, New Jersey, 1960.
26. M. Wonenberger, Matrix $\aleph$-rings, *Proc. Amer. Math. Soc.* **14** (1963), 211–215.
27. J. D. Maitland Wright, Regular $\aleph$-completions of C^* algebras, *J. London Math. Soc.* **12** (1976), 299–309.

AMS (MOS) subject classifications: primary, 46L05, 16A30; secondary, 46K99, 16A08, 16A52, 06A30.

STUDIES IN ANALYSIS
ADVANCES IN MATHEMATICS SUPPLEMENTARY STUDIES, VOL. 4

Approximations of Differentiable Functions

JOSÉ G. LLAVONA

Facultad de Ciencias Matemáticas
Universidad Complutense
Madrid, Spain

Contents

INTRODUCTION

The contents of this chapter belong to the theory of approximation of differentiable functions.

In a more precise way: Let E, F be two real Banach spaces. (Throughout the chapter every vector space will be supposed to be defined over the field of real numbers. Let $\dim(E) > 0$, $C^n(E, F)$ the vector space of n times continuously differentiable functions from E into F, $n \geqslant 1$, with the usual topology, i.e., with the topology of uniform convergence of functions and their derivatives up to order n over the compact sets in E. If $F = R$ and if A is a subalgebra of $C^n(E, R) = C^n(E)$, we may state the following problem:

Under which conditions may we assure density of A in $C^n(E)$?

Nachbin, in 1949 [20] proved the result that follows:

If $\dim(E) < \infty$, A is dense in $C^n(E)$ if and only if:

(1) For each $x \in E$, there is an $a \in A$ such that $a(x) \neq 0$.
(2) For each $x, y \in E$, with $x \neq y$, there is an $a \in A$ such that $a(x) \neq a(y)$.
(3) For each $x, v \in E$, with $v \neq 0$, there is an $a \in A$ such that $Da(x)v \neq 0$.

ISBN 0-12-599150-9

We shall call these conditions "the Nachbin conditions."

Nachbin himself stated the interest of the study of that problem, in the case where the dimension of E is infinite, and the possible extension of his result.

Until recently (we shall remark later on the contributions of Lesmes and Prolla) no other results were reached about this question. The reason is, in a certain way, due to the fact that if the extension of Nachbin's result is not suitably conjectured when the dimension of E is infinite, then it is false. Indeed, by trying to extend the result to that case, we found a counter example that allows us to prove the nondensity of the polynomials of finite type $P_f(E)$ which are evidently a fundamental example of a $C^n(E)$ subalgebra satisfying the Nachbin conditions, when $n \geqslant 2$ and when E is a Hilbert space of infinite dimension.

In a more precise way: If E is a Hilbert space in the preceding condition, the function $p: E \to R$ defined as $p(x) = \|x\|^2 = (x \mid x)$ cannot be approximated in $C^2(E)$ by polynomials of finite type.

Lesmes [17] found this counterexample independently; and also verified the validity of Nachbin's result for a Hilbert space E, when $n = 1$, at which point, by extending Nachbin's result with the same formulation structure, research may be directed in the following way. It consists of noting the preceeding counterexample and finding topologies in the space $C^n(E, F)$ that allow us to extend the result.

By thinking that the usual topology of the space $C^n(E, F)$ is not suitable for the study of such density problems and by approaching the problem in a more general way, we have chosen this direction. In that sense, we have found certain topologies that we shall call differential approximations, induced by bornologies of approximation in the initial space E, which play, as we will see in the next sections, a fundamental role in the theory of approximation of differentiable functions.

We must remark that Prolla has obtained some new contributions in this direction. In a recent visit to the University of Madrid, he communicated to me a theorem that extends Nachbin's result to the case of Hilbert spaces (see [23]). This theorem constitutes a special case of the results presented in Section IV.

We shall now summarize the contents of this chapter. We start with definitions and notations, followed by Section 0, in which we state the problem of the extension of Nachbin's result, formulating this last in a precise way and stating the counterexample mentioned previously.

In Section I, we introduce topologies of differential approximation in the space $C^n(E, F)$. We prove that a necessary and sufficient condition for the density of $C^n(E) \otimes F$ in $C^n(E, F)$ is to consider in $C^n(E, F)$ a topology of differential approximation. We also formulate a characterization of Grothendieck's approximation property in terms of the approximation of differentiable functions.

We prove, in Section II, a theorem of the "Nachbin type" for any Banach space E. In particular, we prove that, independently of the topology of differential approximation chosen in $C^n(E)$, polynomials of finite type $P_f(E)$ are always dense in $C^n(E)$.

In Section III we prove a result of simultaneous approximation and interpolation. More precisely, we prove that for any compact set $K \subset E$, for any $f \in C^n(E)$ and $\{t_1, \ldots, t_p\}$ a finite number of points belonging to E, and for S a finite-dimensional subspace of E, a polynomial of finite type can always be found that approximate f up to order n over K and that interpolates f over S up to the same order n.

In Section IV we make use of the notion of pseudoalgebras (see [8, Definition 2, Section 3]) to enunciate "Nachbin type" theorems for vector-valued mappings. In particular we prove that independently of the topology of differential approximation chosen in $C^n(E, F)$, polynomials of finite type $P_f(E, F)$ are always dense in $C^n(E, F)$.

The aim of Section V is to prove, making use of the results obtained in Sections II and IV, the following question: If E is a product of Banach spaces, then functions from E to F that are a product of a vector in F and a numerical function of class n defined in a factor of E generate a vector subspace that is dense in $C^n(E, F)$. This result establishs a Dieudonné density theorem for differentiable functions.

In Section VI, the techniques of Sections I, II, and IV are adapted to the class of locally convex spaces. We study particular cases, such as metrizable spaces and spaces with countable Schauder decomposition.

Basic Notation and Definitions

We shall use the letters X and Y to denote locally convex spaces and the letters E, F, and G, for Banach spaces. All the vector spaces that we shall use will be defined over the field of real numbers. We shall suppose also that they are Hausdorff when we deal with locally convex spaces.

We shall denote by $L(E, F)$ [or $L(X, Y)$] the space of continuous linear mappings from E into F [from X into Y] endowed with the norm topology (with the topology of uniform convergence on the bounded sets of X).

We shall deal with the space $C^n(E, F)$, $n \geqslant 1$, of n times continuously differentiable functions from E into F, and the space $C^n(X, Y)$ with a concept of differentiability which will be expressed in Section VI. We shall denote by $C_u^n(E,F)$ and $C_u^n(X,Y)$, respectively, these spaces endowed with the usual topology, i.e., the topology of the uniform convergence of functions and their derivatives up to order n on compact sets of E and X. We shall use the notation $C^n(E)$ and $C^n(X)$ for the spaces $C^n(E,R)$ and $C^n(X,R)$. The vector space of all Y-valued continuous functions on X will be denoted by $C(X,Y)$.

DEFINITION. Let H be a subset of the space $C^n(E,F)$, $n \geqslant 1$. We say that H satisfies *Nachbin's conditions* if:

(1) For each $x \in E$, there is an $h \in H$ such that $h(x) \neq 0$.
(2) For each $x, y \in E$, with $x \neq y$, there is an $h \in H$ such that $h(x) \neq h(y)$.
(3) For each $x, v \in E$, with $v \neq 0$, there is an $h \in H$ such that $Dh(x)v \neq 0$.

DEFINITION. For each integer $n \geqslant 1$, we shall denote by $P_f({}^nX, Y)$, the vector subspace of $C(X,Y)$ generated by the set of all maps of the form:

$$x \to \langle x, x' \rangle^n \cdot y, \tag{1}$$

where $x' \in X'$, the topological dual of X, and $y \in Y$. The elements of $P_f({}^nX, Y)$ are called *continuous n-homogeneous polynomials of finite type* from X into Y. The vector subspace generated by the union all $P_f({}^nX, Y)$, $n \geqslant 1$, and the constants maps is denoted by $P_f(X,Y)$. The elements of $P_f(X,Y)$ are called *continuous polynomials of finite type* from X into Y, similarly for $P_f(E,F)$. When $Y = R$ or $F = R$, $P_f(X,Y) = P_f(X)$ and $P_f(E,F) = P_f(E)$; we define them in an analogous way, provided that we write equation (1) as

$$x \to \langle x, x' \rangle^n.$$

We use the usual concept of bornology, i.e., if H is a set, we call *bornology* on H each family β of subsets of H such that they form a covering of H, are hereditary by inclusion, and stable by finite union. The pair (H, β) consisting of H and β is called a *bornological set*. The elements of β are the bounded sets in H. A basis of the bornology β is each subset β_0 of β such that each element of β is contained in some element of β_0. A map between two bornological sets is called *bounded* provided that it transforms bounded sets in bounded sets. β_1, β_2 are two bornologies on H, we say that β_1 is *finer* than β_2 if $\beta_1 \subset \beta_2$. We shall use bornologies on vector spaces, and in this sense we call a space a *bornological vector space* when each vector space endowed with a vector bornology, i.e, a bornology such that the two canonical mappings of the vector space are bounded. A vector bornology is convex if the convex hulls of the bounded sets are also bounded.

0. Nachbin's Theorem. The Problem of Its Extension

Nachbin proved in 1949 [20], the following theorem.

THEOREM 0.1. *Let E be a real, finite-dimensional, vector space, A a subalgebra of the space $C^n(E)$, $n \geqslant 1$. Then A is dense in $C_u{}^n(E)$ if and only if A verifies Nachbin's conditions.*

Remark. Nachbin proved the same result for n times continuously differentiable functions over real finite-dimensional manifolds.

COROLLARY 0.1. *Let E be a finite-dimensional vector space. Then $P_f(E)$ is dense in $C_u{}^n(E)$.*

For the purpose of obtaining theorems of the "Nachbin type" when E is an infinite-dimensional Banach space, we point out the following counterexample.

COUNTER EXAMPLE. Let E be a real, infinite-dimensional Hilbert space. Then $P_f(E)$ is not dense in $C_u{}^2(E)$.

Proof. We consider the function $p: E \to R$ defined by $p(x) = \|x\|^2 = (x \mid x)$. For each $x \in E$, one has that $D^2p(x) = 2j$, where $j: E \to E'$ is the canonical isomorphism. If $h \in P_f(E)$, then $\operatorname{Im}(Dh) \subset S$, where S is a finite-dimensional subspace of E'. So, for each $x \in E$, $D^2h(x) \in L(E, S)$, i.e, for each $x \in E$, $D^2h(x)$ does not belong to the set of the continuous isomorphisms of E into E'. As this set is open in $L(E, E')$, we conclude that p cannot be approximated in $C_u{}^2(E)$ by continuous polynomials of finite type, over any compact $\{x\}$ reduced to a point.

Remark. This counterexample was obtained simultaneously by Lesmes (see [17]).

I. Topologies of Differential Approximation in Banach Spaces. The Approximation Property

Let E be a real Banach space, $L(E)$ the vector space of the continuous endomorphisms of E, and $E' \otimes E$ the space of the finite-rank continuous endomorphisms of E. In [14] one introduces the concept of *topologies of approximation* in $L(E)$ and the concept of *bornologies of approximation* in E. More precisely one calls approximation topologies in $L(E)$ those locally convex topologies that make $E' \otimes E$ dense in $L(E)$. When one of these topologies is the uniform convergence over all the balanced convex compacts of E, one says that E has the approximation property. Enflo [7] proves the existence of Banach spaces such that the approximation property is not verified. Usually the locally convex topologies in $L(E)$ are β-topologies, where β is a convex bornology in E. So, we say that β is an approximation bornology in E, if β is convex, finer than βc (bornology of the balanced convex compacts), and the β-topology in $L(E)$ (topology of uniform convergence over the elements of β) is an approximation topology. In [14] one can see various examples of approximation bornologies in E. As an example, other then the finite-dimensional (whose basis is formed by the finite-dimensional balanced convex compacts), we shall detail the *rapidly decreasing bornology*. A sequence (x_n) in E is rapidly decreasing in E if for each integer $p \geqslant 0$, the sequence $(n^p \cdot x_n)$ is bounded in E. Then $A \subset E$ is rapidly decreasing if it is contained in the balanced, closed, convex hull of a rapidly decreasing sequence. The rapidly decreasing bornology in E is formed by the subsets A rapidly decreasing in E. In [14] it is proved that this is an approximation bornology in E. In the same way one arrives at the existence of maximal approximation bornologies with respect to inclusion.

Now we introduce what we call *topologies of differential approximation*, which play a fundamental role in the spaces of differentiable functions.

DEFINITION 1.1. Let E and F be two nonzero Banach spaces, β a convex bornology over E finer than βc. We denote by $C_\beta{}^n(E, F)$ the space $C^n(E, F)$, $n \geqslant 1$, endowed with the topology given by the family of seminorms

$$\|f\|_{k,n} = \operatorname*{Max}_{0 \leqslant j \leqslant n} [\sup\{\|D^j f(x) y\| : x \in K, y \in K^j\}],$$

K running over the balanced convex compacts of a basis of β, K_β.

Remark. If E is infinite-dimensional and has the approximation property, then the topology of the space $C_u{}^n(E, F)$ is strictly finer than that of $C^n_{\beta c}(E, F)$. Of course, the topology of the space $C_u{}^n(E, F)$ induces in $L(E, F)$ the space $L_b(E, F)$ (norm convergence); the topology of the $C^n_{\beta c}(E, F)$ induces in $L(E, F)$ the space $L_c(E, F)$ (uniform convergence over compacts sets). If E verifies the approximation property, the identity in E is adherent to $E' \otimes E$ in $L_c(E, E)$, but on the other hand, it is clear that the identity in E is not adherent to $E' \otimes E$ in $L_b(E, E)$.

If E is finite-dimensional, both topologies are the same, because in this case E is locally compact, and we can restrict ourselves to a basis K_β of β formed by simple compacts (balls). So, the family $\{p_{K,n}\}$, $K \subset E$ compact, where

$$p_{K,n}(f) = \underset{0 \leqslant j \leqslant n}{\text{Max}} \left[\sup\{\|D^j f(x)\| : x \in K\}\right]$$

of seminorms which define the topology of $C_u{}^n(E, F)$, is equivalent to the family that defines the topology of $C^n_{\beta c}(E, F)$.

DEFINITION 1.2. We shall refer to the *topology of differential approximation* as each topology is introduced in $C^n(E, F)$ by the approximation bornologies β over E, i.e, the topology of the space $C_\beta{}^n(E, F)$ is of differential approximation if and only if β is an approximation bornology over E.

DEFINITION 1.3. Let E be a Banach space, β a convex bornology over E finer than β_c, $n \geqslant 1$.

(1) We say that $\mathscr{F}$ is a differentiable *β-approximation of the identity of finite type on E*, if

(a) $\mathscr{F} \subset C^n(E) \otimes E$, and
(b) the identity in E can be approximated in $C_\beta{}^n(E, E)$ by functions of $\mathscr{F}$.

(2) We say that $\mathscr{F}$ is a *linear β-approximation of the identity of finite type on E*, if

(a) $\mathscr{F} \subset E' \otimes E$, and
(b) the identity in E can be approximated in $C_\beta{}^n(E, E)$ by functions of $\mathscr{F}$.

(3) We say that $\mathscr{F}$ is a *differentiable approximation of the identity of finite type* (respectively; a *linear approximation of the identity of finite type*) on E if $\mathscr{F}$ is a differentiable βc-approximation of the identity of finite type (respectively, a linear βc-approximation of the identity of finite type) on E.

THEOREM 1.1. *Let E be a Banach space, β a convex bornology over E finer than βc, $n \geqslant 1$. Then the following statements are equivalent*:

(1) *$C^n(E) \otimes F$ is dense in $C_\beta{}^n(E, F)$ for each Banach space F.*
(2) *There exists a differentiable β-approximation of the identity of finite type on E.*
(3) *There exists a linear β-approximation of the identity of finite type on E.*
(4) *β is an approximation bornology over E.*

Proof. (1) $\Rightarrow$ (2). If $F = E$, then $\mathscr{F} = C^n(E) \otimes E$ is a differentiable β-approximation of the identity of finite type on E.

(2) $\Rightarrow$ (3). If $\mathscr{F}$ is a differentiable β-approximation of the identity of finite type on E, then $\{Du(y) : u \in F, y \in E\}$ is a linear β-approximation of the identity of finite type on E.

(3) $\Rightarrow$ (4). Trivial.

Before proving that (4) $\Rightarrow$ (1), we give some preliminary results.

LEMMA 1.1. *Let E and F be Banach spaces, β an approximation bornology over E. If $f \in C^n(E, F)$ and $A_f = \{f_0 u : u \in E' \otimes E\}$, then f is adherent to A_f in $C_\beta{}^n(E, F)$.*

Proof. β is an approximation bornology over E and so $E' \otimes E$ is dense in $L_\beta(E)$ (the vector space of the continuous endomorphisms of E, endowed with the topology of uniform convergence over the elements of β). For each $K \in K_\beta$, there is (u_p), a sequence contained in $E' \otimes E$, such that

$$\|I_E - u_p\|_k \to 0, \qquad p \to \infty$$

Then:

(i) $\|f(x) - f(u_p(x))\| \to 0$, $p \to \infty$, uniformly in $x \in K$, because f is uniformly continuous over k.

(ii) $D^j f$ being continuous, $1 \leqslant j \leqslant n$, it follows that $\{\|D^j f(x)\| : x \in K, 1 \leqslant j \leqslant n\}$ is bounded. On the other hand, it is clear that $\{\|u_p(x)\| : x \in K, p \in N\}$ is bounded. Moreover if $x, x_1, x_2, \ldots, x_n \in K$, we have that:

$$\begin{aligned}
&\|D^j f(x)x_1, \ldots, x_j - D^j(f_0 u_p)(x)x_1, \ldots, x_j\| \\
&\quad \leqslant \|D^j f(x)x_1, \ldots, x_j - D^j f(x)u_p(x_1), \ldots, u_p(x_j)\| \\
&\qquad + \|D^j f(x)u_p(x_1), \ldots, u_p(x_j) - D^j f(u_p(x))u_p(x_1), \ldots, u_p(x_j)\| \\
&\quad \leqslant \|D^j f(x) - D^j f(u_p(x))\| \, \|u_p(x_1)\| \cdots \|u_p(x_j)\| \\
&\qquad + \|D^j f(x)\| \, \|x_1 - u_p x_1\| \, \|x_2\| \cdots \|x_j\| + \cdots \\
&\qquad + \|D^j f(x)\| \, \|u_p x_1\| \cdots \|u_p x_{j-1}\| \, \|x_j - u_p x_j\|.
\end{aligned}$$

It results that $\|f - f_0 u_p\|_{k,n} \to 0, p \to \infty$.

LEMMA 1.2. *Let E be a finite-dimensional vector space and F a Banach space. Then*

$$C^n(E) \otimes F \quad \textit{is dense in} \quad C^n_{\beta c}(E, F)$$

Proof. See [29, p. 448]. Now, we prove that (4) $\Rightarrow$ (1) in Theorem 1.1. Let F be a Banach space, $f \in C^n(E, F)$, K a compact set in K_β, $\epsilon > 0$. Because of Lemma 1.1., there is $u \in E' \otimes E$ such that

$$\|f - f_0 u\|_{k,n} < \epsilon/2.$$

If $S = u(E)$ and $g = f|S$, we have $g \in C^n(S, F)$ and so, given the compact set $u(K)$ in S, by Lemma 1.2, we conclude the existence of $b \in C^n(S) \otimes F$ such that

$$\|g - b\|_{u(K),n} < \epsilon/2.$$

As $D^j f(x)|S^j = D^j g(x)$, for every $x \in S$, $1 \leqslant j \leqslant n$, it follows that

$$\|f_0 u - b_0 u\|_{k,n} \leqslant \|g - b\|_{u(K),n} < \epsilon/2.$$

Since $b_0 u \in C^n(E) \otimes F$, the prove is finished.

COROLLARY 1.1. *Let E be a Banach space, $n \geqslant 1$. The following statements are equivalent.*

(1) *$C^n(E) \otimes F$ is dense in $C^n_{\beta c}(E, F)$ for each Banach space F.*
(2) *There exists a differentiable approximation of the identity of finite type on E.*
(3) *There exists a linear approximation of the identity of finite type on E.*
(4) *E satisfies the approximation property.*

Proof. We remark that E satisfies the approximation property if and only if βc is an approximation bornology over E; then Theorem 1.1 gives the required conclusion.

II. NACHBIN'S THEOREM FOR INFINITE-DIMENSIONAL SPACES

We have seen (Introduction and Section 0), a counterexample indicating that we must adopt a suitable approach to generalize Nachbin's theorem for differentiable functions defined in infinite-dimensional Banach spaces. In this section we study results analogous to this theorem for the topologies of differential approximation.

DEFINITION 2.1. If $H \subset C^n(E, F)$ and $\mathscr{F} \subset C^n(E, E)$, we say that H is *$\mathscr{F}$-stable* if, for each $h \in H$, $f \in \mathscr{F}$, one has

$$h_0 f \in H.$$

THEOREM 2.1. *Let E be a Banach space, β an approximation bornology on E, $A \subset C^n(E)$ a subalgebra $\mathscr{F}$-stable for some $\mathscr{F}$, a differentiable β-approximation of the identity of finite type on E. Then A is dense in $C_\beta{}^n(E)$ if and only if A satisfies Nachbin's conditions.*

Before proving the theorem, we give some preliminary lemmas.

LEMMA 2.1. *Let E be a Banach space, $K \subset E$ compact set, and (u_p) a sequence contained in $C(E,F)$ such that* $\lim_p \|I_E - u_p\|_k = 0$. *Then*

$$\bar{K} = K \cup \{u_p(x) : p \in N, x \in K\}$$

is a compact subset of E.

Proof. We observe that the mapping

$$e: C(K,E) \times E \to E, \qquad e(f,x) = f(x)$$

is continuous, when we take in $C(K,E)$ the topology given by the norm of supremum.

LEMMA 2.2. *Let $E_1, \ldots, E_n$ be Banach spaces, $L: E_1 \times \cdots \times E_n \to E$ an n-linear continuous mapping. Then L is hypocontinuous, i.e, for each neighborhood V of zero in E and for each $B_1 \times B_2 \times \cdots \times B_{j-1} \times B_{j+1} \times \cdots \times B_n$ contained in $E_1 \times E_2 \times \cdots \times E_{j-1} \times E_{j+1} \times \cdots \times E_n$, B_i bounded in E_i, $i \neq j$, there exists a neighborhood V_j of zero in E_j such that*

$$L(B_1 \times B_2 \times \cdots \times B_{j-1} \times V_j \times B_{j+1} \times \cdots \times B_n) \subset V.$$

Proof. Trivial.

LEMMA 2.3. *Let E be a Banach space, β an approximation bornology over E, $\mathscr{F}$ a differentiable β-approximation of the identity of finite type on E. If $f \in C^n(E)$ and $A_f = \{f_0 u : u \in \mathscr{F}\}$, then f is adherent to A_f in $C_\beta{}^n(E)$.*

Proof. *If $K \in K_\beta$ and $u \in C^n(E,E)$, then* $\bar{K} = \bigcup_{j=0}^n D^j u(K) K^j$ is compact in E. This is so because, for each $j = 0, 1, 2, \ldots, n$, the mapping $L_j: K^{j+1} \to E$ defined by $L_j(x_0, x_1, \ldots, x_j) = D^j u(x_0) x_1 \cdots x_j$ is continuous.

If $f \in C^n(E,F)$ and $u \in C^n(E,E)$, then

$$D^n(f_0 u)(a) x_1 \cdots x_n = \sum_{p=1}^n \sum D^p f(u(a)) D^{j_1} u(a) x_{1,j_1} \cdots x_{j_1,j_1} \cdots D^{j_p} u(a) x_{1,j_p} \cdots x_{j_p,j_p},$$

where the second summation is taken over all the partitions of the set $\{x_1, \ldots, x_n\}$ in p disjoint sets $\{x_{1,j_1}, \ldots, x_{j_1,j_1}\}, \ldots, \{x_{1,j_p}, \ldots, x_{j_p,j_p}\}$, such that each one of these sets is expressed in increasing order of the index and $j_1 + \cdots + j_p = n$. Now to obtain a bound for $\|f - f_0 u\|_{K,n}$ we must give a bound for

$$|D^r f(x) x_1 \cdots x_r - D^r(f_0 u)(x) x_1 \cdots x_r|, \qquad 0 \leqslant r \leqslant n,$$

and $x, x_1, \ldots, x_r, \ldots, x_n \in K$, so, we need only to find a bound for

(i) $|f(x) - f(u(x))|$, $x \in K$,
(ii) $|Df(x)y - D(f_0 u)(x)y|$, $x, y \in K$, and
(iii) $|D^j f(x)x_1 \cdots x_j - D^j f(u(x))Du(x)x_1 \cdots Du(x)x_j|$, $2 \leqslant j \leqslant n$, because the other terms in the summation are:

$$D^p f(u(x))D^{j_1}u(x)x_{1,j_1} \cdots x_{j_1,j_1} \cdots Du(x)x_{1,j_p} \cdots x_{j_p,j_p},$$

$j_1 + \cdots + j_p = r$, and some of the $j_1, \ldots, j_p$ is $\geqslant 2$, i.e., some of the r arguments tend to zero and the rest are bounded. By Lemma 2.2, it follows that one must estimate only terms (i)–(iii).

If (u_p) is a sequence contained in $\mathscr{F}$ such that $\|I_E - u_p\|_{K,n} \to 0$, $p \to \infty$, then:

(i) $|f(x) - (f_0 u_p)(x)| \to 0$, $p \to \infty$, uniformly in $x \in K$.
(ii)

$$\begin{aligned}|Df(x)y - Df(u_p(x))Du_p(x)y| &\leqslant |Df(x)y - Df(x)Du_p(x)y| \\ &\quad + |Df(x)Du_p(x)y - Df(u_p(x))Du_p(x)y| \\ &\leqslant \|Df(x)\| \, \|y - Du_p(x)y\| \\ &\quad + \|Df(x) - Df(u_p(x))\| \, \|Du_p(x)y\| \to 0, \; p \to \infty,\end{aligned}$$

uniformly in $x, y \in K$.
(iii)

$$\begin{aligned}&|D^j f(x)x_1 \cdots x_j - D^j f(u_p(x))Du_p(x)x_1 \cdots Du_p(x)x_j| \\ &\quad \leqslant |D^j f(x)x_1 \cdots x_j - D^j f(x)Du_p(x)x_1 \cdots Du_p(x)x_j| \\ &\qquad + |D^j f(x)Du_p(x)x_1 \cdots Du_p(x)x_j - D^j f(u_p(x))Du_p(x)x_1 \cdots Du_p(x)x_j| \\ &\quad \leqslant M\|D^j f(x)\| \, |\|x_1 - Du_p(x)x\| + \cdots + \|x_j - Du_p(x)x_j\|| \\ &\qquad + \|D^j f(x) - D^j f(u_p(x))\| \, \|Du_p(x)x_1\| \cdots \|Du_p(x)x_j\| \to 0, \quad p \to \infty,\end{aligned}$$

uniformly in $x, x_1, x_2, \ldots, x_j \in K$.

Now, we prove Theorem 2.1.

Proof. The necessity of the conditions are easily verified.

Conversely, assume that subalgebra A satisfies Nachbin's conditions. Let $f \in C^n(E)$, $K \in K_\beta$ and $\epsilon > 0$. There exists, by Lemma 2.3, $u \in C^n(E) \otimes E$, $u \in \mathscr{F}$, such that $\|f - f_0 u\|_{K,n} < \epsilon/2$. Let S be the finite-dimensional vector subspace of E final for the map u. If $g = f|S$, it is clear that $g \in C^n(S)$. If $A|S = \{a|S : a \in A\}$, we obtain $A|S \subset C^n(S)$, and it is easily proved that $A|S$ is a subalgebra of $C^n(S)$ satisfying Nachbin's conditions. Furthermore, as

S is finite-dimensional, $C_u{}^n(S) = C^n_{\beta c}(S)$; then for a given compact $\bar{K} = \bigcup_{j=0}^n D^j u(K)K^j$ (first remark in the proof of Lemma 2.3) in S, it follows (Theorem 0.1) that the existence of $a \in A$ such that if $b = a|S$, $\|g - b\|_{\bar{K},n} < \epsilon/2$ is verified. As $D^j g(x) = D^j f(x)|S^j$ and $D^j b(x) = D^j a(x)|S^j$ for every $x \in S$, $1 \leqslant j \leqslant n$, it follows that

$$\|f_0 u - a_0 u\|_{K,n} \leqslant \|g - b\|_{\bar{K},n} < \epsilon/2.$$

Since $a_0 u \in A$, the proof is finished.

COROLLARY 2.1. *Let E be a Banach space, satisfying the approximation property*: $A \subset C^n(E)$ *a subalgebra* $\mathscr{F}$*-stable for some* $\mathscr{F}$*-differentiable approximation of the identity of finite type on E. Then A is dense in* $C^n_{\beta c}(E)$ *if and only if A verifies Nachbin's conditions.*

COROLLARY 2.2. *Let E be a Banach space,* β *any approximation bornology on E. Then* $P_f(E)$ *is dense in* $C_\beta{}^n(E)$.

Proof. There exists, by Theorem 1.1, a linear β-approximation to the identity of finite type on E, $\mathscr{F} \subset C^n(E, E)$. Then $P_f(E)$ is an $\mathscr{F}$-stable subalgebra, for any $\mathscr{F}$-linear β-approximation of the identity of finite type on E.

COROLLARY 2.3. *Let E be a Banach space with the approximation property. Then* $P_f(E)$ *is dense in* $C^n_{\beta c}(E)$.

We remark that if we take a maximal approximation bornology for the inclusion over E (see [14]), Theorem 2.1 is an extension of Nachbin's theorem (Theorem 0.1) for topologies of differential maximal approximation in $C^n(E)$. That is, even if βc is not an approximation bornology, results are obtained for topologies of differential maximal approximation.

III. SOME RESULTS ON SIMULTANEOUS APPROXIMATION AND INTERPOLATION IN DIFFERENTIABLE FUNCTION SPACES

In this section we shall study classical interpolation problems, which are at the same time approximation problems. In the space $C^n(E)$ a classical interpolation problem consists of finding sufficient conditions for a subset H of $C^n(E)$ to interpolate each function in a finite number of given points $\{a_1, \ldots, a_p\} \subset E$; i.e., in finding sufficient conditions so that when any f is given, we might assure the existence of $h \in H$ such that $h(a_i) = f(a_i)$, $i = 1, \ldots, p$, and $D^j h(a_i) = D^j f(a_i)$, $i = 1, \ldots, p$; $j = 1, \ldots, n$.

We obtain results on simultaneous approximation interpolation, i.e., we find sufficient conditions for H to be not only an interpolation set in the preceding sense, but also for H to be dense.

This section starts with Yamabe's theorem, which may be seen in [31]; it will play a fundamental role throughout the exposition.

THEOREM 3.1. *Let E be a normed real space, $H \subset E$ a dense subspace, and $\{x_1', \ldots, x_n'\} \subset E'$. Then for every $\epsilon > 0$ and for every $x \in E$, there exists $y \in H$ such that $\|x - y\| < \epsilon$ and $x_i'(x) = x_i'(y)$, $i = 1, \ldots, n$.*

COROLLARY 3.1. *Let (E, p) be a seminormed real space. $H \subset E$ a dense subspace, and $\{x_1', \ldots, x_n'\} \subset L(E, R^m)$. Then for every $\epsilon > 0$ and for every $x \in E$, there exists $y \in H$ such that $p(x - y) < \epsilon$ and $x_i'(x) = x_i'(y)$, $i = 1, \ldots, n$.*

We enunciate now the main theorem of this section, which will allow us to obtain interesting results on the simultaneous approximation and interpolation problem.

THEOREM 3.2. *Let E be a real Banach space, β an approximation bornology on E, $A \subset C^n(E)$ a subalgebra $\mathscr{F}$-stable for some $\mathscr{F}$-differentiable β-approximation of the identity of finite type on E. If A verifies Nachbin's conditions, then given a compact K belonging to the basis of β, $\epsilon > 0$, $f \in C^n(E)$, $\{a_1, \ldots, a_p\} \subset E$, and $S \subset E$ finite-dimensional subspace, the existence of $g \in A$ follows such that $\|f - g\|_{K,n} < \epsilon$ and $D^j f(a_i) | S^j = D^j g(a_i) | S^j$, $j = 0, 1, \ldots, n$; $i = 1, \ldots, p$.*

It will be very useful to us, before proving the theorem, to see some preliminary lemmas. If $K \subset E$ is compact, we consider the space

$$\prod_{j=0}^{n} C(K, L^j(E, R))$$

endowed with the topology given by the seminorm

$$\bar{p}[(f_i)_{i=0}^n] = \max_{0 \leqslant j \leqslant n} [\sup\{|f_j(x)y| : x \in K, y \in K^j\}].$$

LEMMA 3.1. *Let $K \subset E$ be a compact set, $K \in K_\beta$. Then the mapping*

$$\pi : C^n(E) \to \prod_{j=0}^{n} C(K, L^j(E, R)),$$

defined by $\pi(f) = (D^j f \,|\, K)_{j=0}^n$ is linear and continuous for $\bar{p}$ and $\| \ \|_{K,n}$.

Proof. It is clear because $\bar{p}(\pi(f)) = \|f\|_{K,n}$.

LEMMA 3.2. *Let E be a real Banach space, $H \subset C^n(E)$ a subset, and β an approximation bornology on E. Then the following statements are equivalent*:

(1) *For any $f \in C^n(E)$, any compact set $K \subset E$, $K \in K_\beta$, $\{a_1, \ldots, a_p\} \subset E$, $\epsilon > 0$, and S a finite-dimensional subspace of E, the existence of $g \in H$ follows such that $\|f - g\|_{K,n} < \epsilon$ and $D^j f(a_i)|S^j = D^j g(a_i)|S^j$, $i = 1, \ldots, p$; $j = 0, 1, \ldots, n$.*

(2) *For any $f \in C^n(E)$, $K \in K_\beta$, $\{a_1, \ldots, a_p\} \subset E$, $\epsilon > 0$, and $S \subset E$ a finite-dimensional subspace such that $B^*(0,1) \cap S \subset K$, the existence of $g \in H$ follows such that $\|f - g\|_{K,n} < \epsilon$ and $D^j f(a_i)|S^j = D^j g(a_i)|S^j$, $i = 1, \ldots, p$; $j = 0, 1, \ldots, n$.*

Remark. $B^*(0,1)$ is the closed unit ball in E.

Proof. We have only to note that $K_1 = K \cup B_0{}^*(0,1)$ is compact and belongs to β. Furthermore $\|f - g\|_{K,n} \leqslant \|f - g\|_{K_1,n}$. [We have also noted that $B_0{}^*(0,1) = B^*(0,1) \cap S$.]

We now prove Theorem 3.2.

Proof. We have already noted that we may assume $\{a_1, \ldots, a_p\} \subset K$. Let $f \in C^n(E)$, $\epsilon > 0$, and $\{b_1, \ldots, b_q\}$ a normal basis in S. By Lemma 3.2, we may suppose $\{b_1, \ldots, b_q\} \subset K$. From Lemma 3.1 and Theorem 2.1, it follows that $\pi(A)$ is dense in $\mathrm{Im}(\pi)$. For each i, $i = 1, \ldots, p$, we considered the mapping $a_i{}' : \mathrm{Im}(\pi) \to R^m$ defined as:

$a_i{}'[(D^j f|K)_{j=0}^n] = (f(a_i),\ (Df(a_i)b_s), \ldots,\ (D^n f(a_i)c_r));\ s = 1, \ldots, q; \ldots;$ $r = 1, \ldots, q^n$; where $(b_s)_{s=1}^q$ is the basis in S; ... ; $(c_r)_{r=1}^{q^n}$ is the canonical basis in S^n obtained from $\{b_1, \ldots, b_q\}$. Since for each i, $i = 1, \ldots, p$,

$$\|a_i{}'[(D^j f|K)_{j=0}^n]\| \leqslant \bar{p}[(D^j f|K)_{j=0}^n],$$

it follows that $\{a_i{}', \ldots, a_p{}'\} \subset L(\mathrm{Im}(\pi), R^m)$. It follows, from Corollary 3.1, that there exists $g \in A$ such that

$$\bar{p}[(D^j f|K)_{j=0}^n - (D^j g|K)_{j=0}^n] < \epsilon$$

and

$$a_i{}'[(D^j f|K)_{j=0}^n] = a_i{}'[(D^j g|K)_{j=0}^n],$$

i.e., $\|f - g\|_{K,n} < \epsilon$ and $D^j f(a_i)|S^j = D^j g(a_i)|S^j$, $j = 0, 1, \ldots, n$; $i = 1, \ldots, p$.

COROLLARY 3.2. *Let E be a Banach space with the approximation property: $A \subset C^n(E)$ a subalgebra $\mathcal{F}$-stable for some $\mathcal{F}$-differentiable approximation of the identity of finite type on E. If A satisfies Nachbin's conditions, then given $K \subset E$ a compact set, $\epsilon > 0$, $\{a_1, \ldots, a_p\} \subset E$, $f \in C^n(E)$, and $S \subset E$*

finite-dimensional, the existence of $g \in A$ *follows such that* $\|f - g\|_{K,n} < \epsilon$ *and* $D^j f(a_i)|S^j = D^j g(a_i)|S^j$, $i = 1, \ldots, p$; $j = 0, 1, \ldots, n$.

Proof. It is an immediate consequence of Theorem 3.2 and Corollary 1.1.

COROLLARY 3.3. *Let* E *be a Banach space,* β *an approximation bornology over* E. *Then for any* $K \in K_\beta$, $f \in C^n(E)$, $\epsilon > 0$, $\{a_1, \ldots, a_p\} \subset E$, *and* $S \subset E$ *a finite-dimensional subspace, the existence of* $g \in P_f(E)$ *follows such that* $\|f - g\|_{K,n} < \epsilon$ *and* $D^j f(a_i)|S^j = D^j g(a_i)|S^j$, $i = 1, \ldots, p$; $j = 0, 1, \ldots, n$.

IV. NACHBIN'S THEOREM FOR VECTOR-VALUED MAPPINGS. PSEUDOALGEBRAS

The first difficulty in stating Nachbin's theorem for vector-valued differentiable functions is to define a concept similar to that of algebra in the case of real functions.

In this section we use the concept of pseudoalgebras, which play an essential role in the study of density properties.

DEFINITION 4.1. Let E, F be Banach spaces. We say that A is a *pseudoalgebra* of $C(E, F)$ if:

(1) A is a vector subspace of $C(E, F)$.

(2) $u_0 g \in A$ for each $g \in A$ and for each linear continuous map $u: F \to F$ of finite rank.

(3) There exists a bilinear, continuous, nonzero, function $p: F \times F \to F$ such that for every $g, h \in A$, the function $x \in E \to p(g(x), h(x)) \in F$ belongs to A too. (See [8, Def. 2; Section 3].)

LEMMA 4.1. *Let* $A \subset C(E, F)$ *be a vector subspace. The following statements are equivalent.*

(1) $T(h_1, \ldots, h_n) \in A$ *for every* $h_1, \ldots, h_n \in A$, *and for every* $T \in L_f(^nE, F)$, $n \geqslant 1$, *where* $L_f(^nE, F)$ *is the set of* n *linear continuous maps of finite type from* E^n *into* F, *i.e., it is the vector space generated by all the maps of the form*

$$(x_1, \ldots, x_n) \to \langle x_1, x_1' \rangle \cdots \langle x_n, x_n' \rangle \cdot f,$$

where $x_i' \in E'$, $f \in F$.

(2) $p_0 g \in A$ *for every* $g \in A$ *and for every* $p \in P_f(F, F)$ *with* $p(0) = 0$.

(3) $M = \{y'(h) : y' \in F', h \in A\}$ *is an algebra contained in* $C(E)$ *such that* $M \otimes F \subset A$.

(4) A *is a pseudoalgebra of* $C(E, F)$.

Proof. See [25, Lemma 2.2, p. 249].

We may mention that $P_f(E, F)$ is an example of a pseudoalgebra.

The next theorem will allow us to obtain results on the density in $C^n(E, F)$ using the results obtained in Section II.

THEOREM 4.1. *Let E be a Banach space, β an approximation bornology on E, $n \geqslant 1$. The following statements are equivalent.*

(1) *Every subalagebra $M \subset C^n(E)$ satisfying Nachbin's conditions is dense in $C_\beta{}^n(E)$.*

(2) *Every pseudoalgebra $A \subset C^n(E, F)$ satisfying Nachbin's conditions is dense in $C_\beta{}^n(E, F)$, where F is any Banach space.*

To show this theorem, a preliminary lemma is needed.

LEMMA 4.2. *If $A \subset C^n(E, F)$ is a pseudoalgebra satisfying Nachbin's conditions, then $M = \{u'(a) : a \in A,\ u' \in F'\}$ is an algebra of $C^n(E)$ that also satisfies Nachbin's conditions.*

Proof. M is a subalgebra of $C^n(E)$ such that $M \otimes F \subset A$ by Lemma 4.1(3). Then M satisfies Nachbin's conditions trivially.

Proof of Theorem 4.1. (1) $\Rightarrow$ (2) Let A be a pseudoalgebra of $C^n(E, F)$ satisfying Nachbin's conditions. Then $M = \{u'(a) : a \in A, u' \in F'\}$, by Lemmas 4.2 and 4.1 (3), is a subalgebra of $C^n(E)$ which satisfies Nachbin's conditions and $M \otimes F \subset A$. Then M is dense in $C_\beta{}^n(E)$. So the adherence of $M \otimes F$ in $C_\beta{}^c(E, F)$ is precisely $C^n(E) \otimes F$. As $C^n(E) \otimes F$ is dense in $C_\beta{}^n(E, F)$, by Theorem 1.1, it follows that A is dense in $C_\beta{}^n(E, F)$.

(2) $\Rightarrow$ (1) Trivial.

THEOREM 4.2. *Let E, F be Banach spaces, β an approximation bornology on E, $A \subset C^n(E, F)$ a pseudoalgebra, $\mathscr{F}$-stable for some $\mathscr{F}$-differentiable β-approximation of the identity of finite type on E. Then A is dense in $C_\beta{}^n(E, F)$ if and only if A satisfies Nachbin's conditions.*

Proof. It is clear that if A is dense in $C_\beta{}^n(E, F)$, then A satisfies Nachbin's conditions. If A satisfies Nachbin's conditions, then, by Lemma 4.2, $M = \{u'(a) :\ a \in A,\ u' \in F'\}$ satisfies these conditions too. Evidently, M is $\mathscr{F}$-stable for some differentiable β-approximation of the identity of type on E, i.e., M is dense in $C_\beta{}^n(E)$ by Theorem 2.1. The proof holds following step-by-step what we did in the proof of (1) $\Rightarrow$ (2), in Theorem 4.1.

COROLLARY 4.1. *Let E, F be Banach spaces, E with the approximation property, $A \subset C^n(E, F)$ a pseudoalgebra, $\mathscr{F}$-stable for some $\mathscr{F}$-differentiable β-approximation of the identity of finite type on E. Then A is dense in $C_\beta{}^n(E, F)$ if and only A satisfies Nachbin's conditions.*

COROLLARY 4.2. *Let E, F be Banach spaces, β an approximation bornology on E. Then $P_f(E, F)$ is dense in $C_\beta{}^n(E, F)$.*

Proof. By Theorem 2.1 there exists a linear β-approximation of the identity of finite type on E. Then $P_f(E, F)$ is a pseudoalgebra $\mathscr{F}$-stable for any $\mathscr{F}$-linear β-approximation of the identity of finite type on E.

COROLLARY 4.3. *Let E be a Banach space with the approximation property, F any Banach space. Then $P_f(E, F)$ is dense in $C^n_{\beta c}(E, F)$.*

Remark. We may note the fact that if we consider on E an maximal approximation bornology for the inclusion, then Theorem 4.2 is a generalization of Nachbin's theorem to the general case $C^n(E, F)$, assuming in $C^n(E, F)$ a topology of differential maximal approximation, i.e., even if E does not verify the approximation property, we obtain results for differential maximal topologies.

V. THE DIEUDONNÉ DENSITY THEOREM FOR DIFFERENTIABLE FUNCTIONS

The Dieudonne density theorem establishes the fact that any numerical continuous function defined in the product $X_1 \times X_2$ of two compact spaces can be approximated, uniformly, by product sums $g_1 \cdot g_2$, with $g_i \in C(X_i)$, $i = 1, 2$.

In this, section we obtain two extensions of this theorem for differentiable functions.

THEOREM 5.1. *Let E_1, E_2 be two Banach spaces, β_1 and β_2, two approximation bornologies on E_1 and E_2, respectively. Then the set of functions $g: E_1 \times E_2 \to R$, in the form*

$$x = (x_1, x_2) \to g_1(x_1) \cdot g_2(x_2), \qquad \text{where} \quad g_i \in C^n(E_i), \quad i = 1, 2,$$

generate a vector subspace A dense in $C_\beta{}^n(E_1 \times E_2)$, where β is the product bornology.

Proof. If $E = E_1 \times E_2$, with β the product approximation bornology on E, we have only to observe that the algebra A is in the condition of Theorem 2.1. By Theorem 1.1(3), there is $\mathscr{F}_1 \subset E_i' \otimes E_1$ ($\mathscr{F}_2 \subset E_2' \otimes E$), a linear β_1-approximation of the identity of finite type on E_1 (respectively, a linear β_2-approximation of the identity of finite type on E_2).

If $\lambda \in \mathscr{F}_1$ and $\beta \in \mathscr{F}_2$, we define $\lambda \otimes \beta$ as the map $x = (x_1, x_2) \to (\lambda(x_1), \beta(x_2))$. Then $\mathscr{F} = \mathscr{F}_1 \otimes \mathscr{F}_2$, contained in $E' \otimes E$, is a linear β-approximation of the identity of finite type on E. If

$$\phi = \sum_{i=1}^{n} f_i^1 g_i^2 \in A, \qquad f_i^1 \in C^n(E_1), \quad g_i^2 \in C^n(E_2), \quad \lambda \otimes \beta \in \mathscr{F},$$

then

$$\phi \circ (\lambda \otimes \beta) = \sum_{i=1}^{n} (f_i^1 \circ \lambda)(g_i^2 \circ \beta) \in A,$$

i.e., A is $\mathscr{F}$-stable. Nachbin's conditions are trivially satisfied.

COROLLARY 5.1. *Let E_1 and E_2 be two Banach spaces satisfying the approximation property. Then the set of functions g defined in $E_1 \times E_2 \to R$ of the form*

$$x = (x_1, x_2) \to g_1(x_1) g_2(x_2), \qquad \text{where} \quad g_i \in C^n(E_i), \quad i = 1, 2,$$

generate a vector subspace A, dense in $C^n_{\beta c}(E_1 \times E_2)$.

Remark. If E_1 and E_2 are finite-dimensional, Corollary 5.1 is the Dieudonné density theorem obtained by Nachbin (see [20]).

Making use of the instruments of Section IV, different versions of the Dieudonné density theorem are obtained for the case of vector-valued functions. In this sense we enunciate and prove the following theorems.

THEOREM 5.2. *Let E_1 and E_2 be two Banach spaces, β_1 and β_2 two approximation bornologies on E_1 and E_2, respectively. Then, if F is any Banach space, the set of functions*

$$g: (x_1, x_2) \to g_1(x_1) g_2(x_2) y, \qquad \textit{with} \quad g_i \in C^n(E_i), \quad i = 1, 2,$$

$y \in F$, generate a vector subspace A dense in $C_\beta{}^n(E_1 \times E_2, F)$, where β is the product bornology.

Proof. If $u: F \to F$ is a linear and continuous map, then $(u_0 g)(x_1, x_2) = g_1(x_1) g_2(x_2) u(y)$, so $u_0 g \in A$. Let $p: F \times F \to F$ be a map defined as $p(y_1, y_2) =$

$\langle y_1, e' \rangle \langle y_2, e' \rangle e$, where $e \neq 0$, $e \in F$, $e' \in F'$ with $\langle e, e' \rangle = 1$, and p a bilinear, continuous, nonzero map. If $h(x_1, x_2) = h_1(x_1)h_2(x_2)y_1$, $g(x_1, x_2) = g_1(x_1)g_2(x_2)y_2$, it follows that

$$p(g(x), h(x)) = g_1(x_1)g_2(x_2)h_1(x_1)h_2(x_2)p(y_1, y_2),$$

then A is a pseudoalgebra. If we construct $\mathscr{F} = \mathscr{F}_1 \oplus \mathscr{F}_2$ as in Theorem 5.1, we obtain that A is $\mathscr{F}$-stable, and trivially at last that A satisfies Nachbin's conditions.

COROLLARY 5.2. *Let E_1, E_2 be two Banach spaces satisfying the approximation property, where F is any Banach space. Then the set of functions $g: E_1 \times E_2 \to F$ in the form*

$$g(x_1, x_2) = g_1(x_1)g_2(x_2)y, \qquad g_i \in C^n(E_i), \quad i = 1, 2,$$

$y \in F$, generate a vector subspace A, dense in $C^n_{\beta c}(E_1 \times E_2, F)$.

VI. APPROXIMATION PROPERTY IN LOCALLY CONVEX SPACES. ITS RELATION TO DENSITY PROPERTIES IN THE FRAMEWORK OF THE DIFFERENTIABLE FUNCTIONS

In Section I we introduced the concept of topologies of approximation and bornologies in the framework of Banach spaces. This introduction can be translated in a similar way to the case of locally convex spaces. More detailed study of these concepts can be found in [14].

DEFINITION 6.1. We define $L^n(X, Y)$, $n > 1$, *to be* $L(X, L^{n-1}(X, Y))$. If $f \in L^n(X, Y)$ we write: $f(x_1, \ldots, x_n) = f(x_1)(x_2) \cdots (x_n)$; $x_1, \ldots, x_n \in X$ and in this way we identify $L^n(X, Y)$ with a subspace of the n-linear mappings from X into Y.

Remark. With this definition, $L^n(X, Y)$ is not the space of the continuous n-linear mappings from X into Y, but a bigger space. Precisely, u belongs to $L^n(X, Y)$ if and only if u is n-linear from X into Y, and for each V neighborhood of o in Y, and for each $B_1, \ldots, B_{n-1}$ bounded in X, there exists U neighborhood of o in X, such that $u(U \times B_1 \times B_2 \times \cdots \times B_{n-1}) \subset V'$, i.e., the elements of $L^n(X, Y)$ are the n-linear mappings from X into Y *hypocontinuous* with respect to the last $(n-1)$ variables. So, an equicontinuous family $\mathscr{F} \subset L(X, L^{n-1}(X, Y))$ is identified with a family $\mathscr{F} \subset L^n(X, Y)$ equihypocontinuous with respect to the last $(n-1)$ variables, i.e., for each V neighborhood from o in Y, and for $B_1, \ldots, B_{n-1}$ bounded in X, there exists

U, a neighborhood of o in X, such that $f \in \mathscr{F}$ implies $f(U \times B_1 \times \cdots \times B_{n-1}) \subset V$.

DEFINITION 6.2. Let $C^n(X, Y)$ be the space of *n-times continuously differentiable functions from X into Y* with respect to a notion of differentiability that satisfies:

(1) The chain rule for differentiable functions is valid.

(2) The composition of a function of class n and a continuous linear map is a function of class n.

(3) If $f \in C^n(X, Y)$, then for every p, $0 \leqslant p \leqslant n$, $D^p f : X \to L^p(X, Y)$ is continuous and $D^p f(x)$ is p-linear symmetric.

(4) If $u \in L(X, Y)$, then $u \in C^n(X, Y)$ and $Du(x) = u$ for every $x \in X$, $D^p u(x) = 0$ if $p > 1$.

(5) If $\dim(X) < \infty$, then we have the usual notion of differentiability.

Remark. The usual notions of differentiability verify these properties. See, for example Yamamuro [32].

DEFINITION 6.3. Let β be an approximation bornology on X. We say that β is *quasi-bounded* if for each $B \in \beta$, there exists a net $(u_i)_{i \in I}$ contained in $X' \otimes X$, such that $(u_i) \to I_X$ uniformly in B and $\bigcup_{i \in I} u_i(B)$ is bounded.

Remark 1. If X is metrizable, each approximation bornology on X is quasi-bounded. 2. As an example of a nonmetrizable space with quasi-bounded approximation bornologies, we notice the locally convex barreled spaces, with a denumerable Schauder basis. (For the properties of this kind of spaces, see Dieudonné [5]).

DEFINITION 6.4. Let X, Y be locally convex spaces, β a convex bornology on X finer than βc. We denote by $C_\beta{}^n(X, Y)$, $n > 1$, the space $C^n(X, Y)$, endowed with the topology defined by the family of seminorms:

$$\|f\|_{k,p}^n = \max_{0 \leqslant j \leqslant n} [\sup\{p(D^j f(x)y) : x \in K, y \in K^j\}]$$

$K \in K_\beta$, p runs over a saturated family of seminorms which defines the topology in Y.

THEOREM 6.1. *Let X be a barreled locally convex space, Y locally convex space, β a quasi-bounded approximation bornology in X. If $f \in C^n(X, Y)$ and $A_f = \{f \circ u : u \in X' \otimes X\}$, then f belongs to the closure in $C_\beta{}^n(X, Y)$ of A_f.*

Proof. Let $(u_i)_{i \in I}$ be a net contained in $X' \otimes X$ such that $(u_i) \to I_X$ uniformly on K, and $B = K \cup \{u_i(K) : i \in I\} \subset X$ is bounded. As f is uniformly continuous on K, we have the following.

(i) There exists U_1, neighborhood of 0 in X, such that:

$$x \in K, \quad y \in X, \quad x - y \in U_1 \qquad \text{implies} \qquad p(f(x) - f(y)) < \epsilon.$$

Let $1 \leqslant j \leqslant n$. Then $D^j f : X \to L^j(X, Y)$ is continuous. So $\{D^j f(x) : x \in K\}$ is compact in $L^j(X, Y)$. From the fact that X is barreled, $\{D^j f(x) : x \in K\}$ is equicontinuous in $L(X, L^{j-1}(X, Y))$, i.e., equihypocontinuous in $L^j(X, Y)$ (remember the identification made in remarks of Definition 5.1). Now, we have that:

(ii) There exists U_2, neighborhood of 0 in X, such that

$$\begin{aligned} p(D^j f(x)(U_2 \times B \times \cdots \times B)) &= p(D^j f(x)(B \times U_2 \times \cdots \times B)) \\ &= p(D^j f(x)(B \times B \times \cdots \times B \times U_2)) < \epsilon/2j, \end{aligned}$$

for each $x \in K$.

The mapping $D^j f : X \to L^j(X, Y)$ is continuous. From that, we have:

(iii) There exists U_3, a neighborhood of 0 in X, such that $x \in K$, $y \in X$, $x - y \in U_3$ imply

$$p(D^j f(x)z - D^j f(y)z) < \epsilon/2 \qquad \text{if} \quad z \in B \times B \times \overset{\langle j \rangle}{\overbrace{\cdots}} \times B$$

Let $U = U_1 \cap U_2 \cap U_3$ and let $u = u_i$ with $i \geqslant i_0$, where $i_0 \in I$ satisfies $x \in K$ and $i \geqslant i_0$ implies

$$x - u_i(x) \in U.$$

If $x, x_1, x_2, \ldots, x_n \in K$, it follows that

(a) $p(f(x) - f(ux)) < \epsilon$ by (i).
(b) $p(D^j f(x)x_1 \cdots x_j - D^j f(ux)ux_1 \cdots ux_j)$
$\leqslant p(D^j f(x)x_1 \cdots x_j - D^j f(x)ux_1 \cdots ux_j)$
$+ p(D^j f(x)ux_1 \cdots ux_j - D^j f(ux)_1 \cdots ux_j)$
$\leqslant j\epsilon/2j + \epsilon/2 = \epsilon$ by (ii) and (iii).

From (a) and (b) we conclude the existence of $u \in X' \otimes X$ such that

$$\|f - f_0 u\|^n_{K,p} < \epsilon.$$

THEOREM 6.2. *Let X be a barreled, locally convex space, β a convex bornology on X, finer than βc, $n \geqslant 1$. Then the following hold.*

(1) *$C^n(X) \otimes Y$ is dense in $C_\beta{}^n(X, Y)$, for each locally convex space Y.*

(2) *The identity in X can be approximated in $C_\beta{}^n(X, X)$ by mappings from $C^n(X) \otimes X$.*

(3) *The identity in X can be approximated in $C_\beta{}^n(X, X)$ by mappings from $X' \otimes X$.*

(4) *β is an approximation bornology on X.*

(5) *β is a quasi-bounded approximation bornology on X.*

These are related by: (1) ⇒ (2); (2) ⇒ (3); (3) ⇒ (4); and (5) ⇒ (1).

Proof. (1) ⇒ (2). Take $X = Y$.

(2) ⇒ (3). Suppose $\mathscr{F} \subset C^n(X) \otimes X$ such that the identity on X can be approximated in $C_\beta{}^n(X, X)$ by elements of $\mathscr{F}$. If $\mathscr{F}_1 = \{Du(y) : u \in \mathscr{F}, y \in X\}$, it follows that the identity on X can be approximated in $C_\beta{}^n(X, X)$ by mappings of $\mathscr{F}_1$ and, clearly, $\mathscr{F}_1 \subset X' \otimes X$.

(3) ⇒ (4) Trivial.

(5) ⇒ (1). First, observe that if $\dim(X) < \infty$, then $C^n(X) \otimes Y$ is dense in $C^n_{\beta c}(X, Y)$, for each locally convex space Y (see Treves [29, p. 44]).

Then if $f \in C^n(X, Y)$, $K \in K_\beta$, $\epsilon > 0$, and p is a continuous seminorm in Y, by Theorem 6.1 we conclude the existence of $u \in X' \otimes X$ such that $\|f - f_0 u\|^n_{k,p} < \epsilon/2$. Let $S = u(X)$ and $g = f\,|\,S$. By Definition 6.2, $g \in C^n(S, Y)$, and so, when the compact set $u(K)$ is given in S, there must exist $b \in C^n(S) \otimes Y$ such that $\|g - b\|^n_{u(K),p} < \epsilon/2$. As $\|g - b\|^n_{u(K),p} \geqslant \|f_0 u - b_0 u\|^n_{K,p}$, it follows that $\|f - b_0 u\|^n_{K,p} < \epsilon$. As $b_0 u \in C^n(X) \otimes Y$, the conclusion follows.

COROLLARY 6.1. *Let X be a barreled locally convex space.*

(1) *$C^n(X) \otimes Y$ is dense in $C^n_{\beta c}(X, Y)$ for each locally convex space Y.*

(2) *The identity in X can be approximated in $C^n_{\beta c}(X, X)$ by mappings from $C^n(X) \otimes X$.*

(3) *The identity in X can be approximated in $C^n_{\beta c}(X, X)$ by mappings from $X' \otimes X$.*

(4) *X has the approximation property.*

(5) *βc is a quasi-bounded approximation bornology on X.*

These statements are related by: (1) ⇒ (2); (2) ⇒ (3); (3) ⇒ (4); *and* (5) ⇒ (1).

COROLLARY 6.2. *Let X be a Fréchet space, β a convex bornology on X finer than βc, $n \geqslant 1$. Then the* (1)–(5) *in Theorem 6.2 are equivalent.* (See Remark 1 after Definition 6.3).

COROLLARY 6.3. *Let X be a Fréchet space, $n \geqslant 1$. Then X has the approximation property if and only if for each locally convex space Y, $C^n(X) \otimes Y$ is dense in $C^n_{\beta c}(X, Y)$.*

COROLLARY 6.4. *Let X be a barreled locally convex space with a countable Schauder basis, β a convex bornology on X finer than βc, $n \geqslant 1$. Then the* (1)–(5) *in Theorem* 6.2 *are equivalent.* (See, Remark 2 after Definition 6.3).

COROLLARY 6.5. *If X is a barreled locally convex space with a countable Schauder basis, then $C^n(X) \otimes Y$ is dense in $C^n_{\beta c}(X, Y)$ for each locally convex space Y.*

THEOREM 6.3. *Let X be a barreled locally convex space and β a quasi-bounded approximation bornology on X. Then $P_f(X)$ is dense $C_\beta{}^n(X)$.*

Proof. Let $K \in K_\beta$, $\epsilon > 0$, $f \in C^n(X)$. By Theorem 6.2 there is $u \in X' \otimes X$ such that $\|f - f_0 u\|^n_K < \epsilon/2$. If $S = u(X)$ and $g = f \mid S \in C^n(S)$, it follows (see Corollary 0.1), that there is $b \in P_f(S)$ such that $\|g - b\|^n_{u(K)} < \epsilon/2$. It follows that $b_0 u \in P_f(X)$ and $\|f_0 u - b_0 u\|^n_K < \epsilon/2$, so the proof is complete.

COROLLARY 6.6. *If X is a Fréchet space and β an approximation bornology on X, then $P_f(X)$ is dense in $C_\beta{}^n(X)$.*

COROLLARY 6.7. *Let X be a Fréchet space. Then X verifies the approximation property if and only if $P_f(X)$ is dense in $C^n_{\beta c}(X)$.*

COROLLARY 6.8. *If X is a barreled space with a countable Schauder basis, then $P_f(X)$ is dense in $C^n_{\beta c}(X)$.*

The vector-valued version of these results is easily obtained.

THEOREM 6.4. *Let X be a barreled locally convex space, Y a locally convex space, β a quasi-bounded approximation bornology on X, $n \geqslant 1$. Then $P_f(X, Y)$ is dense in $C_\beta{}^n(X, Y)$.*

Proof. By Theorem 6.3, $P_f(X)$ is dense in $C_\beta{}^n(X)$. So the adherence of $P_f(X) \otimes Y$ in $C_\beta{}^n(X, Y)$ is precisely $C^n(X) \otimes Y$. By Theorem 6.2 (implication $(5) \Rightarrow (1)$), $P_f(X, Y) = P_f(X) \otimes Y$ is dense in $C_\beta{}^n(X, Y)$.

Remark. With Theorem 6.4 we can express Corollaries 6.6–6.8 in a vectorial way.

REFERENCES

1. R. M. ARON AND M. SCHOTTENLOHER, Compact holomorphic mappings on Banach spaces and the approximation property, *Bull. Amer. Math. Soc.* **80**(6) (1974).

2. H. CARTAN, "Calcul differentiel," Collection Méthodes, Hermann, Paris, 1967.
3. E. W. CHENEY "Introduction to Approximation Theory," McGraw–Hill, New York, 1966.
4. D. DACUNHA-CASTELLE, "Contre-exemple à la propiété d'approximation uniforme dans les espaces de Banach (d'aprés Enflo et Davie)," Séminaire Bourbaki, 25e-année no. 443 (1972–1973).
5. J. DIEUDONNÉ, On biorthogonal systems, *Michigan Math. J.* **2** (1954), 7–20.
6. J. DIEUDONNÉ, Sur les fonctions continues numériques definies dans un produit de deux espaces compacts, *C. R. Acad. Sci. Paris* **205** (1937), 40–58.
7. P. A. ENFLO, Counterexample to the approximation problem, *Acta. Math.* **130** (1973), 309–317.
8. A. DE LA. FUENTE ANTUNEZ, Algunos resultados sobre Aproximación de funciones vectoriales tipo Teorema Weierstrass–Stone, Ph.D. Thesis, 1973, Univ. of Madrid.
9. G. GLAESER, Algebras Taylorianas en R^n, *J. Analyse Math.* **6**, 1958.
10. G. GLAESER, Algèbres et sous-algèbres de fonctions différentiables, *An. Acad. Brasil. Ci.* **37** (1965), 395–406.
11. V. GOODMAN, Quasi-differentiable functions on Banach spaces, *Proc. Amer. Math. Soc.* **30**(2) (1971), 367–371.
12. L. M. GRAVES, Some general approximation theorems, *Ann. of Math.* **42** (1941), 281–293.
13. A. GROTHENDIECK, Produits tensoriels topologiques et espaces nucléaires, *Mem. Amer. Math. Soc.* **16** (1966).
14. H. HOGBE-NLEND, Techniques de bornologie en théorie des espaces vectoriels topologiques, "Summer School on Topological Vector Spaces," Lecture Notes in Math. No. 331, p. 84, Springer-Verlag. Berlin and New York, 1970.
15. J. HORVATH, "Topological Vector Spaces and Distributions," Addison–Wesley, Reading, Massachusetts, 1959.
16. J. R. ISBELL AND Z. SEMADENI, Projection constants and spaces of continuous functions, *Trans. Amer. Math. Soc.* **107** (1963), 38–48.
17. J. LESMES, On the approximation of continuously differentiable functions in Hilbert Spaces, *Rev. Colombiana Mat.* **8** (1974), 217–223.
18. J. T. MARTI, "Introduction to the Theory of Bases," Springer-Verlag, Berlin and New York, 1969.
19. N. MOULIS, Approximation de fonctions différentiables sur certains espaces de Banach, *Ann. Inst. Fourier* **21**, (1971), 293–345.
20. L. NACHBIN, Sur les algébres denses de fonctions différentiables sur une variéte, *C. R. Acad. Sci. Paris* **288**, (1949), 1549–1551.
21. L. NACHBIN, "Elements of Approximation Theory," Van Nostrand Reinhold, Princeton, New Jersey, 1967.
22. P. M. PRENTER, Lagrange and Hermite Interpolation in Banach-Spaces, *J. Approximation Theory* **4** (1971), 419–432.
23. J. B. PROLLA, "On Polynomial Algebras of Continuously Differentiable Functions," Inst. für Angew. Math. und informatik, der Universitat Bonn, 1974.
24. J. B. PROLLA, The weighted Dieudonné theorem for density in tensor products, *Indag. Math.* **33** (1971), 170–175.
25. J. B. PROLLA AND S. MACHADO, Weighted Grothendieck, subspaces, *Trans. Amer. Math. Soc.* **186** (1973), 247–258.
26. G. A. REID, Concepts of differentiability and analyticity on certain classes of topological groups, *Proc. Cambridge Philos. Soc.* **61** (1965), 347–379.
27. G. RESTREPO, An infinite dimensional version of a theorem of Bernstein, *Proc. Amer. Math. Soc.* **23** (1969), 193–198.
28. H. SCHAEFER, "Topological Vector Spaces," Springer-Verlag, Berlin and New York, 1971.

29. FR. TREVES, "Topological Vector Spaces, Distributions, and Kernels," Academic Press, New York, 1967.
30. D. WULBERT, "Approximation by C^k-Functions," Department of Math., Univ. of Washington. Seattle, Washington.
31. H. YAMABE, On an extension of the Helly's theorem, *Osaka Math. J.* **2** (1950), 15–17.
32. S. YAMAMURO, "Differential Calculus in Topological Linear Spaces," Lecture Notes in Math. No. 374, Springer-Verlag, Berlin and New York, 1974.
33. G. I. ZAPATA, Sur le problème de Bernstein et les algèbres de fonctions continûment différentiables, *C. R. Acad. Sci. Paris.* **274**, (1972), 70–72.
34. G. I. ZAPATA, "Aproximação ponderada para funções diferenciáveis," Monografias do Centro Brasileiro de Pesquisas Fisicas, Rio de Janeiro, G. B., Brasil, to appear.

STUDIES IN ANALYSIS
ADVANCES IN MATHEMATICS SUPPLEMENTARY STUDIES, VOL. 4

On the Convergence of Walsh–Fourier Series for $L^2(0, 1)$

JOHN A. GOSSELIN

Department of Mathematics
University of Georgia
Athens, Georgia

The purpose of this chapter is to highlight the new proof of the Carleson–Hunt theorem [2, 5] given by Fefferman [3]. To accomplish this we will prove that the Walsh–Fourier series of an L^2 function converges almost everywhere by adapting Fefferman's techniques to the Walsh system. For a proof of the same result based on Carleson's techniques the reader is referred to Hunt [6] or Gosselin [4]. With the clarity of the Walsh system it is possible to illustrate the main concepts in Fefferman's construction without many of the technical problems encountered in the trigonometric system. We feel this is important for several reasons. First, as Fefferman points out, his techniques are well suited to n-dimensional problems which gives his constructions intrinsic interest. Also, Fefferman's proof is essentially an L^2 proof. This could prove useful in dealing with orthonormal systems on the line where standard L^p $(p \neq 2)$ inequalities are known to fail. For examples of such systems, the reader is referred to [1].

For the definition and basic properties of the Walsh–Paley system see [4, 6]. Let $S_n f$ denote the nth partial sum of the Walsh–Fourier series of f. Then $S_n f = f * D_n$, where D_n denotes the nth Dirichlet kernel and $*$ denotes convolution over the Walsh group $\{0, 1\}^\omega$. Let $S_n{}^* f$ denote the modified nth partial sum. Then $S_n{}^* f = w_n S_n(f w_n) = f * D_n{}^*$, $D_n{}^* = w_n D_n$. If $n = \sum_{j=0}^{\infty} \epsilon_j 2^j$, $\epsilon_j = 0$ or 1, $D_n{}^* = \sum_{j=0}^{\infty} \epsilon_j \delta_j{}^*$, where

$$\delta_j{}^* = \begin{cases} 2^j & 0 < x < 2^{-j-1}, \\ -2^j & 2^{-j-1} < x < 2^{-j}, \\ 0 & \text{otherwise} \end{cases}$$

(see [6] for more details). Let $Mf(x) = \sup_n |S_n f(x)|$. Define the integer-valued function $n'(x)$ as $\inf\{n : |S_n f(x)| > \frac{1}{2} Mf(x)\}$. Fix a large positive integer N and let $n(x) = \min[n'(x), N]$. We will establish the basic inequality $\|S_{n(x)} f(x)\|_p \leqslant C_p \|f\|_2$, $1 < p < 2$, where C_p depends only on p. The a.e. convergence readily follows from this (see Fefferman [3]).

ISBN 0-12-599150-9

We now introduce some notation. I will denote a dyadic interval of $(0, 1)$ and $|I|$ will denote the Lebesque measure of I. ω will denote an interval of the form $[j2^k, (j+1)2^k)$ and ω^* will denote the interval $[j'2^{k+1}, (j'+1)2^{k+1})$ which contains ω. Throughout this work we will use pairs $[\omega, I]$ where ω is an interval as above of length $|I|^{-1}$. p will denote a pair $[\omega, I]$. We say $p \ni x$ if $x \in I$.

Given a pair $p = [\omega, I]$, $|I| = 2^{-k}$, $\omega = [j2^k, (j+1)2^k)$ we define an operator T_p as follows: Let $E(p) = \{x \in I : n(x) \in \omega\}$ and let $j2^k = \sum_{s=k}^{\infty} \epsilon_s 2^s$. Then

$$T_p f(x) = \left(\epsilon_k \int_0^1 f(t) w_{j2^k}(x-t)\, \delta_k{}^*(x-t)\, dt \right) \chi_{E(p)}(x)$$

where $\chi_{E(p)}(\cdot)$ is the characteristic function of $E(p)$. Note that $T_p f(\cdot)$ vanishes outside I. For $x \in (0, 1)$ let $I_k(x)$, $0 \leqslant k \leqslant N$, denote the dyadic interval of length 2^{-k} containing x and let $n(x) \in \omega_k(x) = [j2^k, (j+1)2^k)$. Then

$$\begin{aligned}
\sum_{p \ni x} T_p f(x) &= \sum_{k=0}^{\infty} T_{[\omega_k(x), I_k(x)]} f(x) \\
&= \sum_{k=0}^{\infty} \epsilon_k(j2^k) \int_0^1 f(t) w_{j2^k}(x-t)\, \delta_k{}^*(x-t)\, dt \\
&= \sum_{k=0}^{\infty} \epsilon_k(n(x)) \int_0^1 f(t) w_{n(x)}(x-t)\, \delta_k{}^*(x-t)\, dt \\
&= \int_0^1 f(t) w_{n(x)}(x-t) \sum_{k=0}^{\infty} \epsilon_k(n(x))\, \delta_k{}^*(x-t)\, dt \\
&= \int_0^1 f(t) w_{n(x)}(x-t) D_{n(x)}^*(x-t)\, dt \\
&= \int_0^1 f(t) D_{n(x)}(x-t)\, dt = S_{n(x)} f(x)
\end{aligned}$$

We set $Tf(x) = \sum_{p \ni x} T_p f(x) = \sum_p T_p f(x)$ and our object is to show $\|Tf\|_r \leqslant C_r \|f\|_2$, $1 < r < 2$.

The basic idea is to estimate $\|T^{\mathscr{P}}\|_2$, where $T^{\mathscr{P}} = \sum_{p \in \mathscr{P}} T_p$ and $\mathscr{P}$ is a collection of pairs with certain properties. We begin with a basic estimate for $\|T_p\|_2$, $p = [\omega, I]$, $|I| = 2^{-k}$.

$$\begin{aligned}
|T_p f(x)| &\leqslant \left| \int_0^1 f(t) w_{j2^k}(x-t)\, \delta_k{}^*(x-t)\, dt \right| \chi_{E(p)}(x) \\
&\leqslant \left(\int_0^1 |f(t)|\, |\delta_k{}^*(x-t)|\, dt \right) \chi_{E(p)}(x) \\
&= \left(1/|I| \int_I |f(t)|\, dt \right) \chi_{E(p)}(x)
\end{aligned}$$

Thus

$$\begin{aligned}\|T_p f\|_2 &\leqslant \left(1/|I| \int_I |f(t)|\,dt\right)|E(p)|^{1/2}\\ &\leqslant (1/|I|)\left(\int_I |f(t)|^2\,dt\right)^{1/2} |I|^{1/2}|E(p)|^{1/2}\\ &= \left(\int_I |f(t)|^2\,dt\right)^{1/2} (|E(p)|/|I|)^{1/2}\end{aligned}$$

Since T_p depends only on the restriction of f to I, it follows that $\|T_p\|_2 \leqslant (|E(p)|/|I|)^{1/2}$. Because of this, the quantity $|E(p)|/|I|$ is important in analyzing pairs. Given a pair $p = [\omega, I]$ we define

$$A(p) = \sup_{\substack{p' = [w', I'] \\ I' \supset I,\ \omega' \subset \omega^*}} (|E(p')|/|I'|)$$

Let $\mathscr{B}$ denote the collection of all pairs $[\omega, I]$. The following partial ordering $<$ on $\mathscr{B}$ will also be useful in analyzing pairs. Given two pairs $p_1 = [\omega_1, I_1]$ and $p_2 = [\omega_2, I_2]$ we say $p_1 < p_2$ if $I_1 \subset I_2$ and $\omega_1 \supset \omega_2$. We define a tree with top p_0 as a collection of pairs $\mathscr{P}$ satisfying

(i) $p_1 < p_2 < p_3$ and $p_1, p_3 \in \mathscr{P}$ implies $p_2 \in \mathscr{P}$;
(ii) $p < p_0$ for every $p \in \mathscr{P}$.

We now analyze $T^{\mathscr{P}} = \sum_{p \in \mathscr{P}} T_p$ where $\mathscr{P}$ is a tree with top $p_0 = [\omega_0, I_0]$. Let $x \in I_0$ and let $\mathscr{P}(x) = \{p \in \mathscr{P} : p \ni x\}$. Suppose $|I_0| = 2^{-k_0}$. For $k \geqslant k_0$ we define $\epsilon_k(\mathscr{P})$ as follows: Let $\omega_0 \subset [\alpha_k 2^k, (\alpha_k + 1)2^k)$ and let $\alpha_k 2^k = \sum_{s=k}^{\infty} \epsilon_s(\alpha_k 2^k) 2^s$. Let $\epsilon_k(\mathscr{P}) = \epsilon_k(\alpha_k 2^k)$. Let $n(\mathscr{P}) = \sum_{k \geqslant k_0} \epsilon_k(\mathscr{P}) 2^k$. Assume $x \in E(p)$ for some $p \in \mathscr{P}(x)$. Let $p_1(x) = [\omega_1, I_1]$ denote that pair with $x \in E(p_1(x))$ and $|I_1|$ maximal. Then $\mathscr{P}(x)$ is a finite chain $p_1(x) > p_2(x) > \cdots > p_m(x) = [\omega_m, I_m]$. Suppose $|I_1| = 2^{-k_1(x)}$ and $|I_m| = 2^{-k_m(x)}$. For the sake of notation we call $w_{\alpha_{k_0} 2^{k_0}}$ simply $w_{\mathscr{P}}$. Then

$$\begin{aligned}T^{\mathscr{P}} f(x) &= \sum_{i=1}^{m} T_{p_i(x)} f(x)\\ &= \sum_{v=k_1(x)}^{k_m(x)} \epsilon_v(\alpha_v 2^v) \int_0^1 f(t) w_{\alpha_v 2^v}(x - t)\, \delta_v{}^*(x - t)\,dt\\ &= \sum_{v=k_1(x)}^{k_m(x)} \epsilon_v(\mathscr{P}) \int_0^1 f(t) w_{\mathscr{P}}(x - t)\, \delta_v{}^*(x - t)\,dt\\ &= w_{\mathscr{P}}(x) \int_0^1 f(t) w_{\mathscr{P}}(t) \left(\sum_{v=k_1(x)}^{k_m(x)} \epsilon_v(\mathscr{P})\, \delta_v{}^*(x - t) \right) dt\\ &= w_{\mathscr{P}}(x) \left(f w_{\mathscr{P}} * \sum_{v=k_1(x)}^{k_m(x)} \epsilon_v(\mathscr{P})\, \delta_v{}^* \right)(x)\\ &= w_{\mathscr{P}}(x) [S_{2^{k_m(x)+1}} - S_{2^{k_1(x)}}] (S^*_{n(\mathscr{P})}(f w_{\mathscr{P}}))(x).\end{aligned}$$

Thus

$$|T^{\mathscr{P}}f(x)| \leqslant |S_{2^{k_m(x)+1}}(S^*_{n(\mathscr{P})}(fw_{\mathscr{P}}))(x)| + |S_{2^{k_1(x)}}(S^*_{n(\mathscr{P})}(fw_{\mathscr{P}}))(x)|$$
$$\leqslant 2 \sup_{\substack{I \ni x \\ |I| \geqslant 2^{-k_m(x)-1}}} (1/|I|) \int |S^*_{n(\mathscr{P})}(fw_{\mathscr{P}})(x)|\, dx$$

since $S_{2^j}f(x)$ is just the average of f over the dyadic interval of length 2^{-j} containing x. Now suppose $\mathscr{P}$ is not only a tree with top p_0, but also that $A(p) \leqslant \delta$ for every $p \in \mathscr{P}$. Let $\{J_s\}$ be a partition of I_0 into maximal dyadic intervals, $|J_s| \geqslant 2^{-N}$, such that $A(\omega, J_s) > \delta$ where ω is the appropriate interval of length $|J_s|^{-1}$ containing ω_0. If $[\omega, I] \in \mathscr{P}$ and $I \cap J_s \neq \varnothing$, then $I \supsetneqq J_s$. Otherwise $(\omega, J_s) \in \mathscr{P}$ and contradicts $A(p) \leqslant \delta$ for all $p \in \mathscr{P}$. Let $J_s{}^*$ denote the dyadic interval containing J_s with $|J_s{}^*| = 2|J_s|$. Let $E_s{}^* = E(\omega, J_s{}^*)$, where $\omega \supset \omega_0$ and set $E_s = E_s{}^* \cap J_s$. Then $|E_s| \leqslant |E_s{}^*| \leqslant \delta |J_s{}^*| \leqslant 2\delta |J_s|$. Suppose $x \in \bigcup_{p \in \mathscr{P}} E(p)$ and J_s is the partition element containing x. Then if $\omega \supset \omega_0$ and $|\omega| = |J_s{}^*|^{-1}$, $(\omega, J_s{}^*) < p$ for every $p \in \mathscr{P}(x)$. Thus $x \in E(p)$ implies $x \in E(\omega, J_s{}^*) = E_s{}^*$. Thus

$$\bigcup_p E(p) = \bigcup_s \left(\bigcup_p E(p) \cap J_s \right) \subset \bigcup_s (E_s{}^* \cap J_s) = \bigcup_s E_s.$$

Now $T^{\mathscr{P}}f$ has support in $\bigcup_p E(p) \subset \bigcup_s E_s$ and, by the previous estimate, we have

$$|T^{\mathscr{P}}f(x)| \leqslant 2 \sup_{I \supset J_s} (1/|I|) \int_I |S_{n(\mathscr{P})}(fw_{\mathscr{P}})(t)|\, dt,$$

where J_s is the partition element containing x. Letting Mf denote the dyadic maximal function over I_0, we have

$$\int_{I_0} |T^{\mathscr{P}}f(x)|^2\, dx = \sum_s \int_{E_s} |T^{\mathscr{P}}f(x)|^2\, dx$$
$$\leqslant 2 \sum_s \left(\sup_{I \supset J_s} (1/|I|) \int_I |S^*_{n(\mathscr{P})}(fw_{\mathscr{P}})(t)|\, dt \right)^2 |E_s|$$
$$\leqslant 4\delta \sum_s \left(\sup_{I \supset J_s} (1/|I|) \int_I |S^*_{n(\mathscr{P})}(fw_{\mathscr{P}})(t)|\, dt \right)^2 |J_s|$$
$$\leqslant 4\delta \sum_s \int_{J_s} (M(S^*_{n(\mathscr{P})}(fw_{\mathscr{P}}))(x))^2\, dx$$
$$= 4\delta \int_{I_0} (M(S^*_{n(\mathscr{P})}(fw_{\mathscr{P}}))(x))^2\, dx$$
$$\leqslant 4c\, \delta \|f\|_2^2$$

since $f \to Mf$ and $f \to S^*_{n(\mathscr{P})}f$ are bounded (sub) linear operators on L^2. Since $T^{\mathscr{P}}$ has its support continued in I_0, it follows that $\|T^{\mathscr{P}}\|_2 \leqslant C\,\delta^{1/2}$ whenever $\mathscr{P}$ is a tree with top p_0 and $A(p) < \delta$ for every $p \in \mathscr{P}$. This is the basic estimate for trees.

The following estimates will be used later to deal with some error terms. Let $p_0 = [\omega_0, I_0]$ be fixed. Suppose $\mathscr{P}$ is a collection of pairs $[\omega, I]$ with $I \subset I_0$ and $\omega^* \supset \omega_0{}^*$. Also suppose $p_0 \in \mathscr{P}$, no pairs in $\mathscr{P}$ are comparable, and $A(p) \leqslant \delta$ for every $p \in \mathscr{P}$. Then if $T^{\mathscr{P}} = \sum_{p \in \mathscr{P}} T_p$ and $r > 1$, we have $\|T^{\mathscr{P}}\|_r \leqslant C_r\,\delta^{1/r}$. We partition I_0 into maximal dyadic intervals $\{J_s\}$ such that $|E(p_s)| > \delta |J_s|$ where $p_s = [\omega_s, J_s]$, $\omega_s \supset \omega_0$. If $p = [\omega, I] \in \mathscr{P}$ and $I \cap J_s \neq \varnothing$, then $J_s \subset I$. To see this suppose $I \subset J_s$. Then ω_s and ω^* both contain $\omega_0{}^*$ and $|\omega_s| \leqslant |\omega^*|$ so $\omega^* \supset \omega_s$. Thus $[\omega_s, J_s]$ is considered when computing $A(p)$. Thus $A(p) \leqslant \delta$ implies $|E(p_s)| \leqslant \delta |J_s|$, a contradiction to the construction of our partition. Thus $J_s \subset I$. For J_s in the partition we define E_s exactly as in our basic estimate for trees. Recall $|E_s| \leqslant 2\delta|J_s|$. Since no two pairs in $\mathscr{P}$ are comparable, the sets $E(p)$ are mutually disjoint as p ranges over $\mathscr{P}$. Thus for $x \in I_0$, $T^{\mathscr{P}}f(x)$ has at most one nonzero term, say $T_{p(x)}f(x)$, $p(x) = [\omega_x, I_x]$. By our basic estimate for a single pair $T_p(x)f(x) \leqslant (1/|I_x|) \int_{I_x} |f(t)|\,dt$. If we set $Mf(x) = \sup_{I_0 \supset I \supset J_s \ni x} (1/|I|) \int_I |f(t)|\,dt$, it follows that $T^{\mathscr{P}}f$ is dominated pointwise by $Mf(x)$. But $T^{\mathscr{P}}f$ has support in $\bigcup_p E(p) \subset \bigcup_s E_s$. Also $Mf(\cdot)$ is constant over each J_s. Let $Mf(J_s)$ denote its value over J_s. Then

$$\begin{aligned}
\int_{I_0} |T^{\mathscr{P}}f(x)|^r\,dx &= \sum_s \int_{E_s} |T^{\mathscr{P}}f(x)|^r\,dx \leqslant \sum_s \int_{E_s} |Mf(x)|^r\,dx \\
&= \sum_s (Mf(J_s))^r |E_s| \leqslant 2\delta \sum_s (Mf(J_s))^r |J_s| \\
&= 2\delta \sum_s \int_{J_s} (Mf(x))^r\,dx = 2\delta \int_{I_0} (Mf(x))^r\,dx \\
&\leqslant 2c_r\,\delta \int_{I_0} |f(x)|^r\,dx
\end{aligned}$$

Thus $\|T^{\mathscr{P}}\|_r \leqslant C_r\,\delta^{1/r}$ for any $r > 1$.

We now establish the basic estimate for noncomparable pairs. Let $\mathscr{P}$ be a collection of pairs no two of which are comparable and assume each $p \in \mathscr{P}$ satisfies $A(p) \leqslant \delta \leqslant 1$. Then $\|T^{\mathscr{P}}\|_2 \leqslant C\,\delta^{1/3}$. We need the adjoint of T_p, $T_p{}^*$. Let $p = [\omega, I]$, $\omega = [j2^k, (j+1)2^k)$. Then

$$T_p{}^*f(x) = \epsilon_k(j2^k)(f\chi_{E(p)} * w_{j2^k}\,\delta_k{}^*)(x).$$

Now

$$w_{j2^k}\,\delta_k{}^* = w_{j2^k} \sum_{s=2^k}^{2^{k+1}-1} w_s = w_{j2^k} w_{2^k} \sum_{s=0}^{2^k-1} w_s = \sum_{s \in \omega^* \setminus \omega} w_s.$$

Thus $T_p{}^*$ projects $f\chi_{E(p)}$ onto a subspace spanned by a subset of w_n, $n \in \omega^*$. Thus if $p_1 = [\omega_1, I_1]$ and $p_2 = [\omega_2, I_2]$ satisfy $\omega_1{}^* \cap \omega_2{}^* = \varnothing$, $\langle T_{p_1}^* f, T_{p_2}^* f\rangle = 0$. Also note that $T_p{}^*$ has support in I so that $\langle T_{p_1}^* f, T_{p_2}^* f\rangle = 0$ if $I_1 \cap I_2 = \varnothing$. We estimate

$$\|T^{\mathscr{P}*}f\|_2^2 = \sum_{p,p' \in \mathscr{P}} \langle T_p{}^*f, T_{p'}^* f\rangle$$

$$= \sum_p \sum_{\substack{p'=[\omega',I'] \\ |I'| \geqslant |I|}} \langle T_p{}^*f, T_{p'}^* f\rangle + \sum_p \sum_{\substack{p'=[\omega',I'] \\ |I'| < |I|}} \langle T_p{}^*f, T_{p'}^* f\rangle = S_1 + S_2.$$

We first consider S_1.

$$S_1 = \sum_{[\omega',I']} \left\langle \sum_{\substack{[\omega,I] \\ I \subset I'}} T_p{}^*f, T_{p'}^* f \right\rangle = \sum_{[\omega'I']} \left\langle \sum_{\substack{[\omega,I] \\ I \subset I' \\ \omega^* \supset \omega'^*}} T_p{}^*f, T_{p'}^* f \right\rangle$$

Fix $p' = [\omega', I']$. Using the estimate $|T_p{}^*f(x)| \leqslant (1/|I|) \int_{E(p)} |f(t)|\, dt$ (which stems from the convolution $f\chi_{E(p)} * w_{j2^k}\delta_k{}^*$) we have

$$\left|\left\langle \sum_{\substack{[\omega,I] \\ I \subset I' \\ \omega^* \supset \omega'^*}} T_p{}^*f, T_{p'}^* f \right\rangle\right| = \left| \int_{I'} T_{p'}^* f(x) \left(\sum_{\substack{[\omega,I] \\ I \subset I' \\ \omega^* \supset \omega'^*}} T_p{}^*f(x) \right) dx \right|$$

$$\leqslant \left(1/|I'| \int_{E(p')} |f(t)|\, dt \right) \left(\int_{I'} \left| \sum_{\substack{[\omega,I] \\ I \subset I' \\ \omega^* \supset \omega'^*}} T_p{}^*f(x) \right| dx \right)$$

$$\leqslant \left(\int_{E(p')} |f(t)|\, dt \right) \left(1/|I'| \int_{I'} \left| \sum_{\substack{[\omega,I] \\ I \subset I' \\ \omega^* \supset \omega'^*}} T_p{}^*f(x) \right|^r dx \right)^{1/r}$$

Let $\mathscr{P}(I') = \{[\omega, I] : I \subset I',\ \omega^* \supset \omega'^*\}$. By our first estimate on noncomparable pairs we have $\|T^{\mathscr{P}(I')}\|_r = \|T^{\mathscr{P}(I')*}\|_r \leqslant C_r \delta^{1/r}$, $r > 1$. Thus

$$\left|\left\langle \sum_{\substack{[\omega,I] \\ I \subset I' \\ \omega^* \supset \omega'^*}} T_p{}^*f, T_{p'}^* f \right\rangle\right| \leqslant C_r \delta^{1/r} \left(\int_{E(p')} |f(t)|\, dt \right) \left(1/|I'| \int_{I'} |f(x)|^r\, dx \right)^{1/r}$$

$$\leqslant C_r \delta^{1/r} \int_{E(p')} |f(t)|\, M_r f(t)\, dt$$

where $M_r f(x) = \sup_{I \ni x} (1/|I| \int_I |f(t)|^r\, dt)^{1/r}$, $1 < r < 2$. Summing over the p's in S_1 and recalling that the sets $E(p')$ are mutually disjoint, we have

$$|S_1| \leqslant C_r \delta^{1/r} \int_0^1 |f(t)| M_r f(t)\, dt \leqslant C_r \delta^{1/r} \|f\|_2 \|M_r f\|_2 \leqslant C_r{}' \delta^{1/r} \|f\|_2^2$$

since $f \to M_r f$ is bounded on L^2 for $1 < r < 2$. The same estimate holds for S_2 by simply interchanging the roles of I and I'. Thus we have

$$\|T^{\mathscr{P}*}f\|_2^2 \leqslant C_r' \delta^{1/r} \|f\|_2^2, \qquad 1 < r < 2$$

Thus $\|T^{\mathscr{P}*}\|_2 \leqslant C_r \delta^{1/2r}$. The basic estimate for noncomparable pairs follows by setting $r = \frac{3}{2}$.

The operators $T^{\mathscr{P}}$, $\mathscr{P}$ a tree, are the basic building blocks that we shall use to construct $T = \sum_{p \in \mathscr{B}} T_p$. To put trees together we use the notion of orthogonal operators. Let $\{A_i\}$ be a sequence of linear operators on H, a Hilbert space. Then $\{A_i\}$ is said to be *orthogonal* if $i \neq j$ implies $A_i A_j^* = 0$ and $A_i^* A_j = 0$. Then if $A = \sum_i A_i$, $\{A_i\}$ an orthogonal sequence with $\|A_i\| \leqslant M$ for all i, it follows that $\|A\| \leqslant M$.

Let $\mathscr{P}_1$ and $\mathscr{P}_2$ be trees with tops $p_1 = [\omega_1, I_1]$ and $p_2 = [\omega_2, I_2]$. We say $\mathscr{P}_1$ and $\mathscr{P}_2$ are *separated* if $I_1 \cap I_2 = \varnothing$ or if $[\omega, I] \in \mathscr{P}_i$ and $I \subset I_j$ implies $\omega^* \cap \omega_j^* = \varnothing$, $i, j = 0$ or 1, $i \neq j$. Suppose $\mathscr{P}_1$ and $\mathscr{P}_2$ are separated. It follows that $\bigcup_{p \in \mathscr{P}_1} E(p)$ and $\bigcup_{p \in \mathscr{P}_2} E(p)$ are disjoint. Thus if $f, g \in L^2(0,1)$, $\langle T^{\mathscr{P}_1}f, T^{\mathscr{P}_2}g\rangle = 0$ since $T^{\mathscr{P}_1}f$ and $T^{\mathscr{P}_2}f$ have disjoint supports. If $I_1 \cap I_2 = \varnothing$, $\langle T^{\mathscr{P}_1*}f, T^{\mathscr{P}_2*}g\rangle = 0$ since $T^{\mathscr{P}_1*}$ and $T^{\mathscr{P}_2*}$ have disjoint supports. Finally if $I_1 \cap I_2 \neq \varnothing$ we write $\langle T^{\mathscr{P}_1*}f, T^{\mathscr{P}_2*}g\rangle = \sum_{p \in \mathscr{P}_1} \sum_{p' \in \mathscr{P}_2} \langle T_p^* f, T_{p'}^* g\rangle$. Fix $p_0 = [\omega_0, I_0] \in \mathscr{P}_1$. Then

$$\sum_{p' \in \mathscr{P}_2} \langle T_{p_0}^* f, T_{p'}^* g\rangle = \sum_{\substack{[\omega', I'] \in \mathscr{P}_2 \\ I' \supset I_0}} \langle T_{p_0}^* f, T_{p'}^* g\rangle + \sum_{\substack{[\omega', I'] \in \mathscr{P}_2 \\ I_0 \supsetneqq I'}} \langle T_{p_0}^* f, T_{p'}^* g\rangle$$

We consider the first term, the second being similar. Fix $[\omega', I'] \in \mathscr{P}_2$. Then $I_0 \subset I' \subset I_2$ implies $\omega_0^* \cap \omega_2^* = \varnothing$ since $\mathscr{P}_1$ and $\mathscr{P}_2$ are separated. Since $p' < p_2$, $\omega'^* \supset \omega_2^*$. Since $I_0 \subset I'$, $|\omega_0^*| \geqslant |\omega'^*|$. Thus if $\omega_0^* \cap \omega'^* \neq \varnothing$, we would have $\omega_0^* \supset \omega'^* \supset \omega_2^*$, a contradiction. Thus $\omega_0^* \cap \omega'^* = \varnothing$, so $\langle T_{p_0}^* f, T_{p'}^* g\rangle = 0$. Summing first over $p' \in \mathscr{P}_2$ and then $p_0 \in \mathscr{P}_1$, we finally obtain $\langle T^{\mathscr{P}_1*}f, T^{\mathscr{P}_2*}g\rangle = 0$. Summarizing, we have shown that if $\mathscr{P}_1$ and $\mathscr{P}_2$ are two trees that are separated, then $T^{\mathscr{P}_1}$ and $T^{\mathscr{P}_2}$ are orthogonal operators.

Let $\{\mathscr{P}_j\}$ be a family of trees with tops $[\omega_j, I_j]$. In addition assume

(i) $[\omega_j, I_j] \in \mathscr{P}_j$ for any j;
(ii) $A(p) \leqslant \delta \leqslant 1$ for any $p \in \bigcup_j \mathscr{P}_j$;
(iii) $i \neq j$, $p \in \mathscr{P}_i$, $p' \in \mathscr{P}_j$ implies $p \not< p'$ for any i, j.

Such a family of trees is called a *forest*. Let $\mathscr{P} = \bigcup_j \mathscr{P}_j$ be a forest. We define $\mathscr{P}'$ as the collection of pairs $p \in \mathscr{P}$ such that there is at least one pair $\hat{p} \in \mathscr{P}$ that satisfies $p \gneqq \hat{p}$. Let $\mathscr{P}'' = \mathscr{P} \setminus \mathscr{P}'$. We first consider $\mathscr{P}''$. The pairs in $\mathscr{P}''$ are mutually noncomparable under $<$ and satisfy $A(p) \leqslant \delta$. Thus, by the basic estimate for noncomparable pairs, $\|T^{\mathscr{P}''}\|_2 \leqslant c\,\delta^{1/3}$. We now write

$\mathscr{P}'$ as $\bigcup_j (\mathscr{P}_j \cap \mathscr{P}')$. Each $\mathscr{P}_j \cap \mathscr{P}'$ is still a tree, and the family of trees $\{\mathscr{P}_j \cap \mathscr{P}'\}$ are separated. To see this assume $p_1 = [\omega_1, I_1]$ is the top of $\mathscr{P}_1 \cap \mathscr{P}'$, $p_2 = [\omega_2, I_2] \in \mathscr{P}_2 \cap \mathscr{P}'$, and $I_2 \subset I_1$. Then there exists $p_3 = [\omega_3, I_3] \in \mathscr{P}_2$ with $p_2 \gneqq p_3$. By property (iii) $p_3 \nless p_1$ so $\omega_3 \not\supset \omega_1$. But $|\omega_3| \geqslant 2|\omega_1|$ implies $\omega_3 \cap \omega_1{}^* = \varnothing$. Since $p_2 > p_3$ we have $\omega_3 \supset \omega_2{}^*$. Hence $\omega_1{}^* \cap \omega_2{}^* = \varnothing$. It follows that the trees $\{\mathscr{P}_j \cap \mathscr{P}'\}$ are separated and thus the operators $T^{\mathscr{P}_j \cap \mathscr{P}'}$ are mutually orthogonal. Thus

$$\begin{aligned}\|T^{\mathscr{P}}\|_2 &= \|T^{\mathscr{P}'} + T^{\mathscr{P}''}\|_2 \leqslant \|T^{\mathscr{P}'}\|_2 + \|T^{\mathscr{P}''}\|_2 \\ &\leqslant \Big\|\sum_j T^{\mathscr{P}' \cap \mathscr{P}_j}\Big\|_2 + \|T^{\mathscr{P}''}\|_2 \\ &\leqslant \max_j \|T^{\mathscr{P}' \cap \mathscr{P}_j}\|_2 + \|T^{\mathscr{P}''}\|_2 \leqslant C\,\delta^{1/2} + C'\,\delta^{1/3} \leqslant C''\,\delta^{1/3}\end{aligned}$$

Since for each j, $\mathscr{P}' \cap \mathscr{P}_j$ is a tree with $A(p) \leqslant \delta$, the basic estimate for trees applies.

We are now ready to assemble the various parts of this proof to establish the inequality $\|Tf\|_r \leqslant C_r\|f\|_2$, $1 \leqslant r < 2$. We write $\mathscr{B} = \bigcup_n \mathscr{B}_n$ where $\mathscr{B}_n = \{p : 2^{-n-1} \leqslant A(p) < 2^{-n}\}$. Of course $T = \sum_n T^{\mathscr{B}_n}$. Let $\{p_j\}$ denote the maximal pairs that satisfy $|E(p_j)|/|I_j| \geqslant 2^{-n-1}$, i.e., if $p_j < p$, $|E(p)|/|I| < 2^{-n-1}$. We write $\mathscr{B}_n = \mathscr{B}_n{}' \cup \mathscr{B}_n''$, where $\mathscr{B}_n{}'$ is the set of pairs $p \in \mathscr{B}_n$ such that $p < \hat{p}$ for some $\hat{p} \in \mathscr{B}_n$ and $\mathscr{B}_n'' = \mathscr{B}_n \backslash \mathscr{B}_n{}'$. Then no two pairs in $\mathscr{B}_n''$ are comparable, and by the basic estimate for noncomparable pairs we have $\|T^{\mathscr{B}_n''}\|_2 \leqslant C(2^{-n})^{1/3} = C2^{-n/3}$. We now show that if $p \in \mathscr{B}_n{}'$, then $p < p_j$ for some j. It suffices to show there exists p' with $p < p'$ and $|E(p')|/|I'| \geqslant 2^{-n-1}$. Since $p \in \mathscr{B}_n{}'$, there exists $p'' = [\omega'', I''] \in \mathscr{B}_n$ with $p < p''$. Since $A(p'') \geqslant 2^{-n-1}$ there exists $p' = [\omega', I']$ with $|E(p')|/|I'| \geqslant 2^{-n-1}$, $I' \supset I''$, and $\omega' \subset \omega''^*$. Since $p < p''$, we have $I \subset I''$ and $\omega \supset \omega'^* \supset \omega'$. Thus $p < p'$. Thus every $p \in \mathscr{B}_n{}'$ satisfies $p < p_j$ for some j.

Fix a large number K. Since no two of the p_js are comparable, the sets $E(p_j)$ are mutually disjoint and so $\sum_j |E(p_j)| \leqslant 1$. Thus $\sum_j |I_j| \leqslant 2^{n+1} \sum_j |E_j| \leqslant 2^{n+1}$. Define G_n as the set of points x belonging to more than $K2^{2n-1}$ of the I_js. Then it follows that $|G_n| \leqslant 4/2^n K$. Now write $\mathscr{B}_n{}' = \tilde{\mathscr{B}}_n \cup \tilde{\tilde{\mathscr{B}}}_n$, where $\tilde{\mathscr{B}}_n = \{p = [\omega, I] \in \mathscr{B}_n{}' : I \not\subset G_n\}$ and $\tilde{\tilde{\mathscr{B}}}_n = \mathscr{B}_n{}' \backslash \tilde{\mathscr{B}}_n$. It follows that if $x \in {}^c G_n$, $T^{\mathscr{B}_n'} f(x) = T^{\tilde{\mathscr{B}}_n} f(x)$ since $T_p f(x)$ has support in I for any $p = [\omega, I]$. After deleting those p_js for which $I_j \not\subset G_n$, we have that no point of $(0, 1)$ belongs to more than $K2^{2n-1}$ of the I_js, and every $p \in \tilde{\tilde{\mathscr{B}}}_n$ satisfies $p < p_j = [\omega_j, I_j]$ with $I_j \subset G_n$ for some j. For $p \in \tilde{\mathscr{B}}_n$, let $M(p)$ denote the number of p_js with $p < p_j$. Then $1 \leqslant M(p) \leqslant K2^{2n-1}$. Let $\mathscr{B}_{ns} = \{p \in \tilde{\mathscr{B}}_n : 2^s \leqslant M(p) < 2^{s+1}\}$. There are less than $2n \log K$ $\mathscr{B}_{ns}$'s. We now show that each $\mathscr{B}_{ns}$ is a forest with $\delta = 2^{-n}$. Let $\{p_{ns}^k\}_k$ denote the maximal pairs in $\mathscr{B}_{ns}$. Let $\mathscr{P}_{ns}^k = \{p \in \mathscr{B}_{ns} : p < p_{ns}^k\}$. Each $\mathscr{P}_{ns}^k$ is a tree with top p_{ns}^k. (It is routine to check that each time we reduce our set of pairs we preserve property (i) in the definition of a tree.) Clearly $p_{ns}^k \in \mathscr{P}_{ns}^k$ for all k. By construction $A(p) < 2^{-n}$ for every $p \in \mathscr{B}_{ns}$. We must finally check the last property of a forest. Suppose $p \in \mathscr{P}_{ns}^k$ and $p' \in \mathscr{P}_{ns}^{k'}$, $k \neq k'$, and $p < p'$. Then no single maximal pair p_j is simultaneously $> p_{ns}^k$ and $> p_{ns}^{k'}$ for otherwise we would have

$$p < p_{ns}^k < p_j, \qquad p < p_{ns}^{k'} < p_j$$

which implies p_{ns}^k and $p_{ns}^{k'}$ are comparable, a contradiction to their being maximal in $\mathscr{B}_{ns}$. Thus assuming $p < p'$ implies $M(p) \geqslant M(p_{ns}^k) + M(p_{ns}^{k'}) \geqslant 2^{s-1} + 2^{s-1} = 2^s$ which contradicts $p \in \mathscr{B}_{ns}$. Thus we have shown that $\mathscr{B}_{ns}$ is a forest with $\delta = 2^{-n}$. Therefore our basic estimate for forests implies $\|T^{\mathscr{B}_{ns}}\|_2 \leqslant C2^{-n/3}$. Summing over the $\mathscr{B}_{ns}$'s we obtain $\|T^{\mathscr{B}_n}\|_2 \leqslant \sum_s \|T^{\mathscr{B}_{ns}}\|_2 \leqslant C(n \log K)2^{-n/3}$. Finally

$$\begin{aligned}
\|T^{\mathscr{B}_n}f\|_{L^2(^cG_n)} &\leqslant \|T^{\mathscr{B}_n'}f\|_{L^2(^cG_n)} + \|T^{\mathscr{B}_n''}f\|_{L^2(^cG_n)} \\
&\leqslant \|T^{\tilde{\mathscr{B}}_n}f\|_{L^2(^cG_n)} + \|T^{\mathscr{B}_n''}f\|_{L^2(0,1)} \\
&\leqslant \{C(n \log K)2^{-n/3} + C'2^{-n/3}\}\|f\|_{L^2(0,1)} \\
&\leqslant C''(n \log K)2^{-n/3}\|f\|_{L^2(0,1)}
\end{aligned}$$

Let $G = \bigcup_n G_n$. Then $|G| \leqslant \sum_n |G_n| \leqslant (1/K) \sum_n 1/2^{n-2} = C/K$. Summing over n, we obtain

$$\begin{aligned}
\|Tf\|_{L^2(^cG)} &\leqslant \sum_n \|T^{\mathscr{B}_n}f\|_{L^2(^cG_n)} \\
&\leqslant C''(\log K)\left(\sum_{n=1}^{\infty} n2^{-n/3}\right)\|f\|_{L^2(0,1)} \leqslant C'''(\log K)\|f\|_{L^2(0,1)}
\end{aligned}$$

Thus, given $\lambda > 0$, we have

$$\begin{aligned}
|\{x : |Tf(x)| > \lambda\}| &\leqslant \lambda^{-2}\|Tf\|^2_{L^2(^cG)} + |G| \\
&\leqslant C'''^2(\log K)^2\|f\|_2^2\lambda^{-2} + (C/K)
\end{aligned}$$

with C and C''' independent of K. Choosing K to minimize the right-hand side, we obtain

$$|\{x : |Tf(x)| > \lambda\} \leqslant C_\epsilon(\|f\|_2/\lambda)^{2-\epsilon}$$

for any $\epsilon > 0$ which implies that if $1 \leqslant r < 2$, $\|Tf\|_r \leqslant C_r\|f\|_2$. This completes the proof.

References

1. D. L. Burkholder, Maximal inequalities as necessary conditions for almost everywhere convergence, *Z. Wahrscheinlichkeitstheorie und Verw. Gebiete* **3** (1964), 75–88.

2. L. Carleson, On convergence and growth of partial sums of Fourier series, *Acta Math.* **116** (1966), 135–157.
3. C. Fefferman, Pointwise convergence of Fourier series, *Ann. of Math.* **98** (1973), 551–571.
4. J. A. Gosselin, Almost everywhere convergence of Vilenkin–Fourier series, *Trans. Amer. Math. Soc.* **185** (1973), 345–370.
5. R. A. Hunt, On the convergence of Fourier series, *in* "Orthogonal Expansions and Their Continuous Analogues," pp. 345–370, S.I.U. Press, Carbondale, Illinois, 1968.
6. R. A. Hunt, Almost everywhere convergence of Walsh–Fourier series of L^2 functions, *Proc. Internat. Congr. Math., Nice, 1970*, pp. 655–661.

AMS (MOS) subject classifications: 42A20, 43A70.

STUDIES IN ANALYSIS
ADVANCES IN MATHEMATICS SUPPLEMENTARY STUDIES, VOL. 4

On the Kuranishi Space of a Holomorphic Principal Bundle over a Compact Complex Manifold

D. SUNDARARAMAN

The Ramanujan Institute
University of Madras
Madras, India

1. INTRODUCTION

Let X be a compact complex manifold and G be a connected complex Lie group. Let P be a holomorphic principal bundle over X with structure group G. We denote this bundle by $G \to P \to X$. Without changing G, we may deform this bundle by varying the bundle structure P and keeping the complex structure of X fixed or we may vary both P and X. Here, for simplicity, we consider only holomorphic principal bundles over a fixed compact complex manifold with a fixed structure group G. By a *holomorphic family of principal bundles* over X parametrized by a connected complex space M, we mean a holomorphic principal bundle $\mathbb{P}$ over $X \times M$. For $t \in M$, let P_t be the restriction of $\mathbb{P}$ to $X \times \{t\}$. Thus we have a holomorphic family of principal bundles $\{G \to P_t \to X\}_{t \in M}$, which we denote by $\mathbb{P} = \{P_t\}_{t \in M}$. Given a bundle $G \to P \to X$, if there exists a family $\mathbb{P} = \{P_t\}_{t \in M}$, such that $G \to P \to X$ is isomorphic with $G \to P_{t_0} \to X$ for some $t_0 \in M$, we say that the family is a *holomorphic family of deformations* of the given bundle. A holomorphic family $\mathbb{P} = \{P_t\}_{t \in M}$ is said to be *complete* at $t_0 \in M$ if for any holomorphic family $\mathbb{R} = \{R_s\}_{s \in N}$ of deformations of $P_{t_0} \approx R_{s_0}$, $s_0 \in N$, there exist an open neighborhood U of s_0 in N and a holomorphic map $f: U \to M$ such that $\mathbb{P}$ induces $\mathbb{R}$ over $X \times U$ by the map $X \times U \to X \times M$ given by $(x, r) \mapsto (x, f(r))$.

It is known that for any holomorphic principal bundle $G \to P \to X$, X compact, there exists a locally complete family of deformations. As Griffiths remarks [1], the construction of locally complete family of deformations of a principal bundle over a compact complex manifold is analogous to the construction of the locally complete family of deformations of a compact complex manifold given by Kuranishi [3]. As our main result is based on this construction, we give an explicit account of this construction and refer

ISBN 0-12-599150-9

to the locally complete family thus constructed as the *Kuranishi family* of the given bundle. Let the Kuranishi family of the given bundle be denoted by $\mathbb{P} = \{P_t\}_{t \in K}$, where K is the analytic space parametrizing the Kuranishi family. We call K the *Kuranishi space* of the given bundle. Let $\mathbb{R} = \{R_s\}_{s \in N}$ be any family of deformations of the given bundle. Then $\mathbb{P}$ induces $\mathbb{R}$ by a holomorphic map $f\colon N \to K$. We note that f may not be unique. Let $0 \to L(P) \to Q \to T \to 0$ be the Atiyah exact sequence of vector bundles over X of the given bundle. Let the corresponding sheaf sequence be $0 \to \mathbf{L(P)} \to \mathbf{Q} \to \mathbf{T} \to 0$. It follows from the construction of the Kuranishi family that the inducing map f is unique if the dimension of $H^0(X, \mathbf{L(P)})$ is zero. We prove the following stronger result.

THEOREM 1.1. *Let $\mathbb{P} = \{P_t\}_{t \in K}$ be the Kuranishi family of the given bundle $P \approx P_{t_0}$, $t_0 \in K$. Let the dimension of $H^0(X, \mathbf{L}(\mathbf{P}_t))$ be independent of t in a neighborhood of t_0. Then we can find a neighborhood V of the identity in the group of bundle diffeomorphisms of P and a neighborhood K' of t_0 in K such that for any $f \in V$, $t, t' \in K'$, f is not a holomorphic map of P_t to $P_{t'}$.*

The proof of this theorem is analogous to the proof of Theorem 1 in [5].

2. Almost Complex Principal Bundles

Associated with the given bundle, we have the Atiyah sequence

$$0 \to L(P) \to Q \to T \to 0 \tag{2.1}$$

Here T is the holomorphic tangent bundle of X, Q the bundle of tangent vectors to P defined along one of its fibers and invariant under G, and $L(P) = P \times g$ the vector bundle with fiber the Lie algebra g of G, associated with P by the adjoint representation of G. Let the exact sheaf sequence corresponding to the previous sequence be

$$0 \to \mathbf{L(P)} \to \mathbf{Q} \to \mathbf{T} \to 0 \tag{2.2}$$

Let $A^{0,q}$ be the sheaf of germs of scalar-valued $(0, q)$ forms on X and θ be the sheaf of germs of holomorphic functions on X. Let

$$A^q = \mathbf{L(P)} \bigotimes_\theta A^{0,q}, \qquad B^q = \mathbf{Q} \bigotimes_\theta A^{0,q}, \qquad C^q = \mathbf{T} \bigotimes_\theta A^{0,q} \tag{2.3}$$

$$A = \sum_{q \geqslant 0} A^q, \qquad B = \sum_{q \geqslant 0} B^q, \qquad C = \sum_{q \geqslant 0} C^q \tag{2.4}$$

The exterior differentiation operator $\bar{\partial}$ and the bracket operator [,] can be extended to act on A, B, C and we denote the extended operators also by the same symbols $\bar{\partial}$ and [,]. Then, corresponding to the sheaf sequence

(2.2), we get an exact sequence of Lie algebra complex

$$0 \to A \to B \xrightarrow{\tau} C \to 0 \tag{2.5}$$

Let **X** be the underlying differentiable manifold of X and $CT\mathbf{X}$ be the complexified tangent bundle of **X**.

DEFINITION 2.1. An almost complex structure on **X** is defined to be a *C^∞-vector subbundle over* $\mathbb{C}$, say T'', of $CT\mathbf{X}$ such that we have a direct sum decomposition $CT\mathbf{X} = T'' \oplus \overline{T''}$.

Any complex structure on **X** induces canonically an almost complex structure on **X**. Let J_0' be the underlying almost complex structure tensor of X. An almost complex structure is said to be *integrable* if it is induced by a complex structure. Thus J_0' is integrable. Let J_0 be the underlying almost complex structure of P.

DEFINITION 2.2. An *almost complex fiber bundle structure* $G \to P \to X$ is defined to be a pair (J, J'), where J (respectively, J') is an almost complex structure on **P** (respectively on **X**) such that

(i) J is G-invariant,
(ii) the almost complex structure on P/G induced by J is J' and
(iii) J restricted to a fiber gives the integrable almost complex structure on **G**.

As above the notion of integrable almost complex fiber bundle structures are defined.

DEFINITION 2.3. An *almost complex principal bundle structure* on $G \to P \to X$ is an almost complex structure J on **P** such that

(i) J is G-invariant,
(ii) the almost complex structure on P/G induced by J is J_0' and
(iii) J restricted to a fiber gives the integrable almost complex structure on **G**.

In deformation theory, the spaces A^1, B^1, C^1 play important roles in view of the following propositions [1].

PROPOSITION 2.1. *There is a one-to-one correspondence between almost complex structures J' on* **X** *which are sufficiently close to J_0' and elements $\varphi \in C^1$ which are near zero. The integrability condition is*

$$\bar{\partial}\varphi - \tfrac{1}{2}[\varphi, \varphi] = 0 \tag{2.6}$$

PROPOSITION 2.2. *There is a one-to-one correspondence between almost complex fiber bundle structures* (J, J') *on* $G \to P \to X$, *which are sufficiently close to* (J_0, J_0') *and elements* $\psi \in B^1$ *which are close to zero and which satisfy* $\pi(\psi) = \varphi$, *where* φ *corresponds to* J'. *The integrability condition is*

$$\bar{\partial}\psi - \tfrac{1}{2}[\psi, \psi] = 0 \tag{2.7}$$

PROPOSITION 2.3. *There is a one-to-one correspondence between almost complex principal bundle structures* J *on* $G \to P \to X$, *which are sufficiently close to* J_0 *and elements* $\theta \in A^1$ *which are near zero. The integrability condition is*

$$\bar{\partial}\theta - \tfrac{1}{2}[\theta, \theta] = 0 \tag{2.8}$$

3. KURANISHI SPACE

Fix a Hermitian scalar product (θ, φ) for $\theta, \varphi \in A^p$ and δ the formal adjoint of $\bar{\partial}$. Let $\Box = \bar{\partial}\delta + \delta\bar{\partial}$. A form θ is said to be *harmonic* if $\Box\theta = 0$ or equivalently if $\bar{\partial}\theta = 0 = \bar{\partial}\theta$. Let H^p be the space of harmonic forms in A^p. It is well known that H^p is finite-dimensional. Let $H_{\bar{\partial}}^{0,p}(A)$ be the p-dimensional cohomology group of complex A with respect to $\bar{\partial}$ and $H^p(X, \mathbf{L}(\mathbf{P}))$ be the p-dimensional cohomology group of X with coefficients in $\mathbf{L}(\mathbf{P})$. Then it is well known that $H^p \approx H_{\bar{\partial}}^{0,p}(A) \approx H^p(X, \mathbf{L}(\mathbf{P}))$. Also there exist the harmonic projection operator H and the corresponding Green's operator G, yielding the Hodge–Kodaira decomposition: For any $\theta \in A^p$,

$$\theta = H\theta + \bar{\partial}\delta G\theta + \delta G\bar{\partial}\theta \tag{3.1}$$

Two types of norms that are usually introduced in these spaces are the Holder and the Sobolov norms. Let α_r be an orthonormal base of $H^1(X, \mathbf{L}(\mathbf{P}))$. Let $\theta_1(t) = \sum_{r=1}^{m} \alpha_r t_r$, where m is the dimension of $H^1(X, \mathbf{L}(\mathbf{P}))$, be a fixed element of $H^1(X, \mathbf{L}(\mathbf{P}))$. Then it is easy to prove the following lemma.

LEMMA 3.1. (i) There is a unique convergent (in any of the two norms) power series solution $\theta(t)$ of the differential equation

$$\theta(t) = \theta_1(t) + \tfrac{1}{2}\delta G[\theta(t), \theta(t)] \tag{3.2}$$

(ii) $\theta(t)$ satisfies the integrability condition if and only if

$$H[\theta(t), \theta(t)] = 0 \tag{3.3}$$

We have now the local completeness theorem of Kuranishi [3].

THEOREM 3.1 (Kuranishi). *Let $G \to P \to X$ be a holomorphic principal bundle over a compact complex manifold X. Let α_r be a base for $H^1(X, \mathbf{L}(\mathbf{P}))$. Let $W = \{\theta_1(t) = \sum_{r=1}^{m} \alpha_r t_r, |t| < \epsilon\}$, $K = \{t \in W \mid H[\theta(t), \theta(t)] = 0\}$. Then*

(i) *There is a one-to-one correspondence between $\theta_1 \in W$ and solutions θ of equation* (3.2) *for sufficiently small ϵ.*

(ii) *The family $\{\theta(t)\}_{t \in K}$ represents a holomorphic family $\mathbb{P} = \{P_t\}_{t \in K}$ of deformations of the given bundle.*

(iii) *The family $\mathbb{P} = \{P_t\}_{t \in K}$ is locally complete at the reference point.*

This theorem can be reformulated in the following form.

THEOREM 3.2 (Kuranishi). *There exist a neighborhood W of the origin in $H^1(X, \mathbf{L}(\mathbf{P}))$ and a complex analytic injective map $\Phi: W \to A^1$ such that if $K = \{t \in W \mid H[\Phi(t), \Phi(t)] = 0\}$, then $\{\Phi(t)\}_{t \in k}$ represents a locally complete family of deformations of the given bundle.*

The mapping Φ is of great importance and we call it the *universal canonical map*.

4. PROOF OF THEOREM 1.1

Let $w \in A^1$ represent an almost complex principal bundle. Define $\bar{\partial}_w$: $A^p \to A^{p+1}$ by

$$\bar{\partial}_w(\theta) = \bar{\partial}\theta - [w, \theta] \tag{4.1}$$

It is easy to check that $\bar{\partial}_w$ is exact if and only if w is integrable. Even when $\bar{\partial}_w$ is not exact, we can define its formal adjoint δ_w and the Laplace operator $\square_w = \bar{\partial}_w \delta_w + \delta_w \bar{\partial}_w$. Let $A_w{}^p$ be the space of $(0, p)$ forms with values in $L(P_w)$. Then if w is sufficiently close to zero, there is an isomorphism Γ_w: $A^0 \to A_w{}^0$ such that Γ_w corresponds to $\bar{\partial}$ with respect to the almost complex principal bundle structure defined by w, and Γ_w depends differentiably on w. Let $\xi \in A^0$ be near zero. Then we can define a bundle diffeomorphism $e_w(\xi)$ as follows:

$$e_w(\xi) = \exp(\Gamma_w(\xi) + \overline{\Gamma_w(\xi)}) \tag{4.2}$$

LEMMA 4.1. *If the dimension of $H^0(x, \mathbf{L}(\mathbf{P}_t))$ is independent of t in a neighborhood of t_0 and if w is integrable, then $e_w(\xi)$ covers a neighborhood of the identity in the group of holomorphic bundle automorphisms of P.*

Proof. It is well known that the eigenvalues of $\Box_w$ form a discrete subset of $\mathbb{R}$ and that they are all nonnegative. For $\alpha > 0$, let

$$A^p(\alpha) = \{\theta \in A^p \,|\, \Box_w \theta = \alpha\theta\}.$$

$A^p(\alpha)$ is finite-dimensional and $A^p(\alpha)$ and $A^p(\beta)$ are orthogonal for $\alpha \neq \beta$. Let $A(\alpha) = \sum_{p \geqslant 0} A^p(\alpha)$. Then $A(\alpha)$ is a subcomplex of A and its injection induces an isomorphism of cohomology groups [4, Proposition 9.4]. Let λ be the smallest eigenvalue of $\Box$ on A^0. Let $H_w{}^{\#} = \sum_{\alpha \leqslant \lambda} A^0(\alpha)$. This is a finite sum and $H_w{}^{\#}$ is finite-dimensional. In a neighborhood of zero, $H_w{}^{\#}$ depends differentiably on w if w is restricted to a finite-dimensional submanifold of A^1 [2]. If w is integrable and if the dimension of $H^0(X, \mathbf{L}(\mathbf{P}_t))$ is independent of t in a neighborhood of t_0, $H_w{}^{\#} = H_w{}^0$, the space of holomorphic vector fields of A^0 with respect to w. Hence $\Gamma_w(H_w{}^{\#})$ is the space of holomorphic vector fields on P_w, and $e_w(\xi)$ covers a neighborhood of the identity in the group of bundle automorphisms of P_w.

Let B_{diff} denote the group of bundle diffeomorphisms of P. Then we have the following:

LEMMA 4.2. *There exist an open subset W of A^0 in C^1-topology and an injective map $E: W \to B_{\text{diff}}(P)$ such that $E(W)$ is an open subset of $B_{\text{diff}}(P)$ in C^0-topology.*

The proof of this lemma is exactly similar to that of Proposition 8.7 in [4]. We omit the proof. Let $f \in B_{\text{diff}}(P)$. Let J_1, J_2 be two almost complex principal bundle structures on $G \to P \to X$. We say that f is an isomorphism of J_1 to J_2 if its differential df sends J_1 into J_2. Let $w \in A^1$ be sufficiently small so that P_w is defined. Then there is a unique almost complex principal bundle structure J such that f is an isomorphism of J to P_w. Hence there must be an element $\theta \in A^1$ which defines J. We set $\theta = w \circ f$. Then θ is defined when f is sufficiently close to the identity in $B_{\text{diff}}(P)$ in the C^1-topology. With these notations, we have:

LEMMA 4.3. *There exist a neighborhood M of the origin in A^1 and a neighborhood N of the origin in ${}^{\perp}A^0$ such that for any w in M, there exists a unique ξ (depending on w) in N such that $\delta(w \circ E(\xi)) = 0$.*

The proof of this lemma is rather involved but it is analogous to the proof of a proposition in [3, p. 152]. We therefore omit the proof. It should be remarked that this lemma is crucial in proving the local completeness theorem of Kuranishi family given in Theorem 3.2. We need one more lemma to prove the main theorem.

LEMMA 4.4. *Let the dimension of $H^0(X, \mathbf{L}(\mathbf{P}_t))$ be independent of t in a neighborhood of t_0. Let W be a neighborhood of the origin in ${}^{\perp}A^0$. Then we can find a neighborhood V of the identity in $B_{\text{diff}}(P)$ and a neighborhood Y of the origin in H^1 such that for any $f \in V$, $t \in Y$ we can write $f = e_{\Phi(t)}(\zeta) \circ E(\xi)$, for suitable ξ in W and ζ in $H^{\#}_{\Phi(t)}$, where Φ is the universal canonical mapping given in Theorem* 3.2.

Proof. $A^0 = H^0 \oplus {}^{\perp}A^0$. But for t sufficiently small, $H^0 = H^{\#}_{\Phi(t)}$ (see the proof of Lemma 4.1). We define a mapping F_t from a neighborhood of the origin in $H^{\#}_{\Phi(t)} \oplus {}^{\perp}A^0$ into A^0 as

$$E(F_t(\zeta, \xi)) = e_{\Phi(t)}(\zeta) \circ E(\xi) \tag{4.3}$$

where F_t is a differentiable map depending differentiably on t. Also it is easy to check that $(dF_0)_0 =$ identity. Hence by applying the implicit mapping theorem, we obtain the conclusion of the lemma.

Proof of Theorem 1.1. We take a neighborhood W of the origin in ${}^{\perp}A^0$ having the property of Lemma 4.3. If possible let $\Phi(t_2) = \Phi(t_1) \circ f$, with $f \in V$. The theorem is proved if we prove $t_2 = t_1$. By Lemma 4.4, we can write $f = e_{\Phi(t_1)}(\zeta) \circ E(\xi)$ for suitable ξ in W and ζ in $H_{\Phi(t_1)}$. By Lemma 4.1, $e_{\Phi(t_1)}(\zeta)$ is holomorphic with respect to $P_{\Phi(t_1)}$. Hence $\Phi(t_2) = \Phi(t_1) \circ E(\xi)$. Applying Lemma 4.3, we conclude that $\xi = 0$. Hence $\Phi(t_2) = \Phi(t_1)$. Since Φ is injective, $t_2 = t_1$. This completes the proof of the theorem.

REFERENCES

1. PH.A. GRIFFITHS, The extension problem for compact submanifolds of complex manifolds, I, *Proc. Conf. Complex Analysis, Minneapolis, 1964*, pp. 113–142, Springer-Verlag, Berlin and New York, 1965.
2. K. KODAIRA AND D. C. SPENCER, On deformations of complex analytic structures III, *Ann. of Math.* **71** (1960), 43–76.
3. M. KURANISHI, New proof for the existence of locally complete families of complex structures, *Proc. Conf. Complex Analysis, Minneapolis, 1964*, pp. 142–154, Springer-Verlag, Berlin and New York, 1965.
4. M. KURANISHI, "Deformations of Compact Complex Manifolds," Univ of Montreal Press, Montreal, 1969.
5. M. KURANISHI, A note on families of complex structures, *in* "Global Analysis, Papers in Honor of K. Kodaira," pp. 309–313, Univ. of Tokyo Press, Japan and Princeton Univ. Press, Princeton, New Jersey, 1969.

AMS (MOS) subject classifications: primary, 32G05, 32G13, 32L05.

STUDIES IN ANALYSIS
ADVANCES IN MATHEMATICS SUPPLEMENTARY STUDIES, VOL. 4

On Bifurcation and Asymptotic Bifurcation for Nondifferentiable Potential Operators and for Systems of the Hammerstein Type

HUGO BEIRÃO DA VEIGA[†]

Instituto de Física e Matemática
Lisbon, Portugal

The greatest positive bifurcation point (cf. Theorem 1.1) and the greatest positive asymptotic bifurcation point (cf. Theorems 1.2 and 1.3) for a class of nondifferentiable potential operators on Hilbert spaces are characterized. These results, announced in [3], are used to study bifurcation for systems of the Hammerstein type (cf. Section 4).

0. INTRODUCTION

Notation

Let H be a real Hilbert space with norm and inner product denoted by $\| \ \|$ and $(\ , \ )$, respectively. The strong convergence in H is denoted by $\to$ and the weak convergence by $\rightharpoonup$.

In the following $\mathbb{K}$ is a closed convex cone in H with vertex at the origin, i.e., $\mathbb{K}$ is a closed subset of H such that $t\mathbb{K} \subset \mathbb{K}$ for all $t \geqslant 0$ and $\mathbb{K} + \mathbb{K} \subset \mathbb{K}$. We assume that $\mathbb{K}$ is nonvoid and different from $\{0\}$. We denote by int $\mathbb{K}$ the interior of $\mathbb{K}$. If $\rho \geqslant 0$, we put

$$S_\rho = \{u \in H : \|u\| = \rho\}, \qquad V_\rho = \{u \in H : \|u\| \leqslant \rho\}, \quad \mathbb{K}_\rho = \mathbb{K} \cap V_\rho.$$

Some Definitions

A real number λ is said to be a *bifurcation point* for an operator $\Gamma : \mathbb{K} \to H$ if for every $\epsilon > 0$ there exist a pair $(\lambda_\epsilon, u_\epsilon) \in \mathbb{R} \times \mathbb{K}$ with $|\lambda - \lambda_\epsilon| < \epsilon$ and $0 < \|u_\epsilon\| < \epsilon$ such that $\Gamma(u_\epsilon) = \lambda_\epsilon u_\epsilon$.

Similarly $+\infty$ is said to be a *bifurcation point* for Γ if the preceding conditions hold with $\lambda_\epsilon > 1/\epsilon$ instead of $|\lambda - \lambda_\epsilon| < \epsilon$; a similar definition holds for $-\infty$.

[†] Present address: Libera Università Degli Studi di Trento, Facoltà di Scienze, Povo (Trento), Italy.

ISBN 0-12-599150-9

If in the preceding definition we replace the condition $0 < \|u_\epsilon\| < \epsilon$ by the condition $\|u_\epsilon\| > 1/\epsilon$, one says that λ (respectively, $+\infty$ or $-\infty$) is an *asymptotic bifurcation point* for the operator Γ.

If S is a nonempty subset of H and Φ is a real functional defined on S, one says that Φ is *weakly continuous on* S if $u_n \in S$ and $u_n \rightharpoonup u_0 \in S$ implies that $\Phi(u_0) = \lim_{n\to+\infty} \Phi(u_n)$. One says that Φ is *weakly upper semicontinuous on* S if under the same conditions one has $\Phi(u_0) \geqslant \lim_{n\to+\infty} \sup \Phi(u_n)$. The *weak lower semicontinuity on* S is defined similarly.

We denoted by $\overline{S}$ the *weak closure* of S in the following sense: $\overline{S} = \{u \in H : \exists u_n \in S,\ n = 1, 2, \ldots, u_n \rightharpoonup u\}$. Obviously $\overline{S} = S$ if S is a convex and (strongly) closed subset of H.

Let $\Phi: S \to \mathbb{R}$. We say that Φ is *weakly continuous on* $\overline{S}$ if there exists an extension $\overline{\Phi}$ of Φ to $\overline{S}$ that is weakly continuous on $\overline{S}$. One gives similar definitions for the *weak (upper or lower) semicontinuity of* Φ *on* $\overline{S}$.

In the following, different positive constants shall be denoted by the same symbol c.

The Classical Result

The following result is well known (cf. [6, §6, Theorem 2.1] and [10, §17, Theorem 17.6]):

THEOREM A. *Let $\Phi: H \to \mathbb{R}$, $\Phi(0) = 0$, be a weakly upper semicontinuous functional. Let Φ be differentiable and put $\Gamma \equiv \nabla\Phi$. Suppose that $\Gamma(0) = 0$ and let Γ be Fréchet differentiable at the origin with $D\Gamma(0) = B$ linear, continuous, self-adjoint, and compact. Then the largest eigenvalue λ_0 of B, if positive, is a bifurcation point for Γ.*

The aim of this chapter is to generalize Theorem A to a class of operators not necessarily differentiable at the origin[1]. The motivation was given by some elementary examples of potential operators on $\mathbb{R}^2$ to which this theorem does not apply. We give these examples now.

Examples with a Nondifferentiable Γ at the Origin

EXAMPLE 1. Consider the following set-up. Let $\phi: [0, +\infty[\to \mathbb{R}$ be a continuously differentiable function which does not vanish identically in any neighborhood of the origin and such that $\phi(0) = \phi'(0) = \phi''(0) = 0$. Let $\psi: [0, 2\pi] \to \mathbb{R}$ be a differentiable function such that $\psi(0) = \psi(2\pi)$, $\psi'(0) = \psi'(2\pi)$, and assume that ψ' is unbounded in $[0, 2\pi]$.

[1] We remark that there is a result of Turner [9, Theorem 2.6] on odd multiplicity bifurcation for nondifferentiable operators verifying $\|(\Gamma - B)(u)\| = O(\|u\|)$ at the origin with B linear and compact.

For any $y \in \mathbb{R}^2$ define a function Φ in polar coordinates (we write $y = (\xi, \eta) = [r, \theta]$, where $\xi = r\cos\theta$, $\eta = r\sin\theta$) by

$$\Phi(y) = \tfrac{1}{2}r^2 + \phi(r)\psi(\theta). \tag{0.1}$$

Then Φ is Fréchet differentiable (F-differentiable) on $\mathbb{R}^2$ and its gradient is given by $\nabla\Phi \equiv \Gamma = I + \omega$, where I is the identity operator on $\mathbb{R}^2$ and ω is given in each point $y = (\xi, \eta) = [r, \theta]$ by $\omega(y) = \omega_r(y) + \omega_t(y)$, where

$$\omega_r(y) = \phi'(r)\psi(\theta)y/r, \qquad \omega_t(y) = (\phi(r)/r)\psi'(\theta)(y^\perp/r), \tag{0.2}$$

and $y^\perp = (-\eta, \xi)$. Note that $\omega_r(y)$ is the radial component of $\omega(y)$ and $\omega_t(y)$ the tangential one.

Since $\Gamma(y) = y + \omega_r(y) + \omega_t(y)$, the equation $\Gamma(y) = \lambda y$ yields, by decomposition in the radial and the tangential components, $y + \omega_r(y) = \lambda y$ and $\omega_t(y) = 0$. Using (0.2) one obtains the system

$$[\phi'(r)/r] + 1 = \lambda, \qquad \psi'(\theta) = 0. \tag{0.3}$$

Obviously the second equation in (0.3) is solvable. Let θ_0 be a fixed solution of $\psi(\theta_0) = 0$. The first equation in (0.3) admits, for each $r > 0$, the solution $\lambda_r = 1 + r^{-1}\phi'(r)\psi(\theta)$. Hence to each solution θ_0 of $\psi'(\theta_0) = 0$ corresponds a branch $y_r, \lambda_r \in \mathbb{R}^2 \times \mathbb{R}$ of solutions of $\Gamma(y_r) = \lambda_r y_r$, with $y_r = [r, \theta_0]$. Obviously $y_r \to 0$ as $r \to 0$; moreover $\lambda_r \to 1$ since $\phi''(0) = 0$. Thus 1 is a bifurcation point (the unique point) for the operator Γ. In order words the greatest bifurcation point for Γ (actually the unique one) is the greatest positive eigenvalue (the unique one) for the linear operator I. However, Theorem A does not apply since Γ is not F-differentiable at the origin. If it were, we would have $D\Gamma(0) = I$ (since this holds in the Gateaux sense) and the remainder $\omega(y)$ would verify

$$\lim_{|y| \to 0} |\omega(y)|/|y| = 0 \tag{0.4}$$

or equivalently, its radial and tangential components should verify

$$\lim_{|y| \to 0} |\omega_r(y)|/|y| = 0 \tag{0.5}$$

and

$$\lim_{|y| \to 0} |\omega_t(y)|/|y| = 0, \tag{0.6}$$

respectively. The first condition holds since $\phi''(0) = 0$, but the second one fails since

$$\sup_{|y| = r} |\omega_t(y)|/|y| = +\infty, \forall r > 0.$$

Our aim is to show that this is not a casual fact but a general one. More precisely, we shall prove in this chapter (see Theorem 1.1) that the thesis of Theorem A holds even if Γ is not F-differentiable at the origin provided that the radial component $\omega_r(u)$ of the remainder $\omega(u) \equiv \Gamma(u) - B(u)$ verifies

$$\lim_{||u||\to 0} \|\omega_r(u)\|/\|u\| = 0, \tag{0.7}$$

or, equivalently,

$$\lim_{||u||\to 0} \|(\omega(u), u)\|/\|u\|^2 = 0. \tag{0.8}$$

We emphasize that, as in the preceding example, we do not exclude the situation

$$\sup_{||u||=r} \|\omega(u)\| = +\infty, \qquad \forall r > 0.$$

EXAMPLE 2. We now give another example of a potential operator $\omega: \mathbb{R}^2 \to \mathbb{R}^2$ such that (0.5) holds but (0.6) fails. This function shall be utilized in Section 4 to show that there exist Hammerstein systems to which our results apply, but Theorem A does not apply; it is an easy exercise to construct other functions on $\mathbb{R}^2$ (or $\mathbb{R}^m$, $m > 2$) for which this holds. We turn now to our example. For the sake of clarity we put $e_1 = (1, 0)$ and $e_2 = (0, 1)$. Let $y = (\xi, \eta) \in \mathbb{R}^2$ and define a function $\Phi: \mathbb{R}^2 \to \mathbb{R}$ by

$$\Phi(y) = \phi(\xi/\eta^2)\xi\eta, \tag{0.9}$$

where $\phi: \mathbb{R} \to \mathbb{R}$, $\phi(0) \neq 0$, is a continuously differentiable function with compact support on $\mathbb{R}$. Then Φ is continuously differentiable on $\mathbb{R}^2$ and its gradient $\nabla\Phi \equiv \omega$ has components given by

$$\begin{aligned} \omega_1(y) &\equiv \partial\Phi/\partial\xi\, e_1 = \{\phi(\xi/\eta^2)\eta + \phi'(\xi/\eta^2)\xi/\eta\}e_1, \\ \omega_2(y) &\equiv \partial\Phi/\partial\eta\, e_2 = \{\phi(\xi/\eta^2)\xi - 2\phi'(\xi/\eta^2)\xi^2/\eta^2\}e_2. \end{aligned} \tag{0.10}$$

By definition Φ and ω vanish when $\eta = 0$. We claim that ω verifies (0.8) but does not verify (0.4). First, one has

$$(\omega(y), y) = [2\phi(\xi/\eta^2) - \phi'(\xi/\eta^2)\xi/\eta^2]\xi\eta. \tag{0.11}$$

On the other hand, it follows from (0.9) that there exists a positive constant c such that

$$|\xi| \geqslant c\eta^2 \quad \Rightarrow \quad \omega(y) = 0 \tag{0.12}$$

since $\phi(\xi/\eta^2) = \phi'(\xi/\eta^2) = 0$ whenever $\xi/\eta^2 \notin \operatorname{supp}\phi$. Consequently one can assume that $|\xi| \leqslant c\eta^2$. Under this assumption (0.11) yields $|(\omega(y), y)| \leqslant$

$c|\xi|\,|\eta| \leqslant c|\eta|^3$ and consequently

$$|(\omega(y), y)|/|y|^2 \leqslant c|y| \to 0 \tag{0.13}$$

as $|y| \to 0$, as desired. However, if one considers the points $y_t = (0, t)$, $t \in \mathbb{R}$, one has

$$|\omega(y_t)|/|y_t| = |\phi(0)| \neq 0, \qquad \forall t \neq 0, \tag{0.14}$$

which contradicts (0.4). Thus our thesis is proved.

As a matter of fact we have proved only that 0 is not the F-derivative of ω at the origin. However, one can easily see that ω Fréchet differentiable at the origin implies $D\omega(0) = 0$.

We take the opportunity to state two estimates that shall be used in Section 4:

$$|\omega_1(y)| \leqslant c|\eta|, \qquad |\omega_2(y)| \leqslant c|\xi|. \tag{0.15}$$

These estimates follow from (0.10) and (0.12).

The two examples just presented show that (0.8) is the natural assumption to be made for the calculus of the largest positive bifurcation point.

Examples with a Nonlinear B

We shall give in this chapter another generalization of the F-differentiability condition requested in Theorem A, whose meaning is shown by elementary examples as, for instance, the function $(y^2/2)\operatorname{sgn} y$, defined in the real line, whose gradient is $|y|$.

Another elementary example (in the plane) is the following one. Using polar coordinates put

$$\Phi(y) = \tfrac{1}{2}r^2 \sin\theta. \tag{0.16}$$

This function is F-differentiable on $\mathbb{R}^2$ with $\nabla\Phi(y) \equiv B(y) = B_r(y) + B_t(y)$, where $B_r(y) = y\sin\theta$ and $B_t(y) = \frac{1}{2}y^\perp \cos\theta$ are, respectively, the radial and the tangential components of $B(y)$. Furthermore, the equation $B(y) = \lambda y$ is equivalent to the system (independent of r) $\lambda = \sin\theta$, $\cos\theta = 0$, which admits the solutions $\theta = \pi/2$, $\lambda = 1$ and $\theta = -\pi/2$, $\lambda = -1$. Hence -1 and 1 are the bifurcation points for B. However, Theorem A (as well as the generalization we have previously mentioned) does not apply since B is not decomposable as a sum $B = A + \omega$ with A linear and ω verifying (0.4) or (0.8). However, the bifurcation points for B, 1 and -1, coincide with

$$M(B) = \sup_{|x|=1} (By, y)/|y|^2 \qquad \text{and} \qquad m(B) = \inf_{|x|=1} (By, y)/|y|^2,$$

respectively. We claim that this is a general fact, provided B is a positively homogeneous operator [i.e., $B(ty) = tB(y)$, $\forall t \geqslant 0$].

In fact, we prove in this chapter that Theorem A holds if B is positively homogeneous and if λ_0 is replaced by $M(B)$. Note that $M(B)$ coincides with the largest eigenvalue λ_0 if B verifies the assumptions of Theorem A.

Operators Γ Defined Only on a Cone $\mathbb{K}$

The third direction in which we generalize Theorem A is to operators Γ defined only on a closed convex cone $\mathbb{K}$ with vertex at the origin. We assume in this case that a suitable geometrical condition holds (see Assumption 1.1). We emphasize that this geometrical condition holds if for each $u \in \mathbb{K}$ there exists $\epsilon = \epsilon(u) > 0$ such that $u + \epsilon\Gamma(u) \in \mathbb{K}$ (or equivalently if the intersection of $\mathbb{K}$ with the half-line $\{v : v = u + t\Gamma(u), t > 0\}$ is not empty) and consequently holds if $\Gamma\mathbb{K} \subset \mathbb{K}$. Note that there are elementary examples for which our condition holds but $\Gamma\mathbb{K} \subset \mathbb{K}$ fails. For instance, let $\mathbb{K}$ be the cone of positive functions (on a domain in $\mathbb{R}^n$) and consider $\Gamma u(x) = \pm \operatorname{sign} u(x)$ or $\Gamma u(x) = -cu(x)$, $c > 0$.

Examples Concerning the Asymptotic Bifurcation

Finally we shall prove also that our results hold for the asymptotic bifurcation case, as stated in Theorem 1.2 (to which we refer the reader). As for the ordinary bifurcation the gradient of function (0.16) gives us a clear motivation for considering in Theorem 1.2 a positively homogeneous B. Similarly, the two first examples of this section show, after slight alterations, that condition (1.2) is the natural one in Theorem 1.2. For the sake of completeness let us write the alterations: in the first example assume that $\lim_{t\to+\infty}[\phi'(t)/t] = \lim_{t\to+\infty}[\phi(t)/t^2] = 0$ instead of $\phi(0) = \phi'(0) = \phi''(0) = 0$. In the second example define

$$\Phi(y) = \phi(\xi^2/\eta)\xi\eta, \tag{0.17}$$

$\Phi(y) = 0$ if $\eta = 0$. This function is continuously differentiable on $\mathbb{R}^2$ and $\omega \equiv \nabla\Phi$ is given by its components

$$\begin{aligned} \omega_1(y) &= [\phi(\xi^2/\eta)\eta + 2\xi^2\phi'(\xi^2/\eta)]e_1, \\ \omega_2(y) &= [\phi(\xi^2/\eta)\xi - (\xi^3/\eta)\phi'(\xi^2/\eta)]e_2, \end{aligned} \tag{0.18}$$

being $\omega(y) = 0$ if $\eta = 0$. Since there exists a positive constant c such that $\omega(y) = 0$ whenever $\xi^2 \geqslant c|\eta|$ it follows easily that

$$|(\omega(y), y)|/|y|^2 \leqslant c/\sqrt{|y|}, \tag{0.19}$$

which implies (1.2). On the other hand, (0.14) holds again and consequently $|\omega(y)|/|y| \to 0$ as $|y| \to +\infty$ fails. Thus ω is not F-differentiable at infinity (note that $D\omega(\infty) = 0$ if ω were F-differentiable at infinity). Finally (0.15) holds again; this shall be used later.

Plan of the Chapter

The plan of the chapter is the following. In Section 1 we state the main results that shall be proved in Section 2. In Section 3 we discuss conditions (1.1) and (1.2) and their relationship with Gateaux differentiability (G-differentiability) and F-differentiability. In Section 4 we apply Theorems 1.1 and 1.2 to the study of bifurcation and asymptotic bifurcation for systems of the Hammerstein type and we establish sufficient conditions in order that (1.1) and (1.2) should hold for these systems. Finally in the Appendix we consider briefly the case $M(B) = 0$ and the case in which B is homogeneous of deg $\alpha > 0$.

1. Main Results

One says that $B: \mathbb{K} \to H$ is *positively homogeneous* if $B(tu) = tB(u)$ for any $u \in \mathbb{K}$ and any $t \geqslant 0$. We define

$$m_{\mathbb{K}}(B) = \inf_{u \in \mathbb{K} \cap S_1} (B(u), u), \qquad M_{\mathbb{K}}(B) = \sup_{u \in \mathbb{K} \cap S_1} (B(u), u).$$

One verifies easily that the following two propositions hold.

PROPOSITION 1.1. *Let $\Gamma = B + \omega$ with $B: \mathbb{K} \to \mathbb{R}$ positively homogeneous and $\omega: \mathbb{K} \to \mathbb{R}$ verifying*

$$\lim_{\substack{u \to 0 \\ 0 \neq u \in \mathbb{K}}} (\omega(u), u)/\|u\|^2 = 0. \tag{1.1}$$

Then the bifurcation points for Γ are contained in the closed interval $[m_{\mathbb{K}}(B), M_{\mathbb{K}}(B)]$.

PROPOSITION 1.2. *Let $\Gamma = B + \omega$ with $B: \mathbb{K} \to \mathbb{R}$ positively homogeneous and $\omega: \mathbb{K} \to \mathbb{R}$ verifying*

$$\lim_{\substack{\|u\| \to +\infty \\ u \in \mathbb{K}}} (\omega(u), u)/\|u\|^2 = 0. \tag{1.2}$$

Then the asymptotic bifurcation points for Γ are contained in the closed interval $[m_{\mathbb{K}}(B), M_{\mathbb{K}}(B)]$.

It is natural to study under what conditions the extreme points of the interval $[m_{\mathbb{K}}(B), M_{\mathbb{K}}(B)]$ are effectively bifurcation points for Γ. First we state some definitions and remarks.

Remark 1.1. Since λ is a bifurcation point (at the origin or asymptotic) for an operator Γ if and only if $-\lambda$ is a bifurcation point for $-\Gamma$, the study concerning the value $m_{\mathbb{K}}(B)$ can be reduced to the study concerning $M_{\mathbb{K}}(B)$. Hence, without loss of generality, we shall consider in the following only this last case.

If Φ is a real functional defined on $\mathbb{K}$, one says that Φ is *Gateaux differentiable* at a point $u_0 \in \mathbb{K}$ and that its Gateaux gradient at u_0 is $v \in H$ if

$$\Phi(u_0 + th) - \Phi(u_0) = (v, th) + \omega(u_0, th) \tag{1.3}$$

with $\lim_{t \to 0}[\omega(u_0, th)/t] = 0$, for all h such that $u_0 + h \in \mathbb{K}$ and all $t \in]0, 1]$. We then write $v = \nabla\Phi(u_0)$. If $\Phi: \mathbb{K} - \mathbb{K}_{\rho_0} \to \mathbb{R}$ and if $u_0 \in \mathbb{K} - \mathbb{K}_{\rho_0}$, we consider in (1.3) only the increments h such that $\|h\| < \|u_0\| - \rho_0$.

We shall use the following assumption in the Theorem 1.1.

ASSUMPTION 1.1. *For every $u \in \mathbb{K}$ such that $(\Gamma(u), u) > 0$ and $\Gamma(u) \neq \lambda u$, $\forall \lambda \in \mathbb{R}$, there exists a $v \in \mathbb{K}_\rho$ ($\rho = \|u\|$) such that $v \neq u$ and $(\Gamma(u), v) \geqslant (\Gamma(u), u)$.*

The following assumption will be used in Theorem 1.2.

ASSUMPTION 1.2. *For every $u \in \mathbb{K}$ with $\|u\| > \rho_0$ such that $(\Gamma(u), u) > 0$ and $\Gamma(u) \neq \lambda u$, $\forall \lambda \in \mathbb{R}$, there exists a $v \in \mathbb{K}_\rho$ ($\rho = \|u\|$) such that $v \neq u$ and $(\Gamma(u), v) \geqslant (\Gamma(u), u)$.*

It is easily verified that for points $u \in \operatorname{int} \mathbb{K}$ Assumptions 1.1 and 1.2 hold with $(\Gamma(u), v) > (\Gamma(u), u)$. Moreover we have the following proposition.

PROPOSITION 1.3. *Assumptions 1.1 and 1.2 hold in a point u (with $(\Gamma(u), v) > (\Gamma(u), u)$) if there exists $\epsilon = \epsilon(u) > 0$ such that $u + \epsilon\Gamma(u) \in \mathbb{K}$.*

In particular, the following corollaries hold.

COROLLARY 1.1. *Assumption 1.1 holds if $\Gamma\mathbb{K} \subset \mathbb{K}$ or if $\mathbb{K} = H$.*

COROLLARY 1.2. *Assumption 1.2 holds if $\Gamma(\mathbb{K} - \mathbb{K}_{\rho_0}) \subset \mathbb{K}$ or if $\mathbb{K} = H$.*

Note that Propositions 1.1 and 1.2 have a local character. Similarly, in the following theorem it is sufficient that Φ and Γ be defined only in a $\mathbb{K}_{\rho_0}$ with $\rho_0 > 0$ arbitrary small.

THEOREM 1.1. *Let $\Phi: \mathbb{K} \to \mathbb{R}$, $\Phi(0) = 0$, be a weakly upper semicontinuous and Gateaux differentiable functional. Let $\Gamma = \nabla\Phi$ verify Assumption 1.1 and $\Gamma = B + \omega$ with B positively homogeneous in $\mathbb{K}$ and ω satisfying (1.1). Then if $M_{\mathbb{K}}(B)$ is positive, $M_{\mathbb{K}}(B)$ is a bifurcation point for Γ in $\mathbb{K}$.*

In the next theorem condition (i) is obviously fulfilled if Φ is defined and weakly upper semicontinuous on all of $\mathbb{K}$.

THEOREM 1.2. *Let $\Phi: \mathbb{K} - \mathbb{K}_{\rho_0} \to \mathbb{R}$ be a Gateaux differentiable functional on $\mathbb{K} - \mathbb{K}_{\rho_0}$ and let $\Gamma = \nabla\Phi$ verify Assumption 1.2. Assume that $\Gamma = B + \omega$ with B positively homogeneous in $\mathbb{K}$ and ω satisfying (1.2). Finally let $\Phi(u) - \frac{1}{2}(B(u), u)$ be bounded on each $\mathbb{K} \cap S_\rho$ for $\rho > \rho_0$, and assume that one of the following conditions holds:*

(i) *Φ is weakly upper semicontinuous on $\overline{\mathbb{K} - \mathbb{K}_{\rho_0}}$, or*
(ii) *Φ is weakly upper semicontinuous on $\mathbb{K} - \mathbb{K}_{\rho_0}$ and $(B(u), u)$ is weakly upper semicontinuous on $\mathbb{K}$.*

Then if $M_{\mathbb{K}}(B)$ is positive, $M_{\mathbb{K}}(B)$ is an asymptotic bifurcation point for Γ on $\mathbb{K}$.

As a corollary to Theorem 1.2 one has

THEOREM 1.3. *Let $\Phi: \mathbb{K} \to \mathbb{R}$ be a weakly upper semicontinuous and Gateaux differentiable functional. Let $\Gamma = \nabla\Phi$ verify Assumption 1.2 and $\Gamma = B + \omega$ with B positively homogeneous in $\mathbb{K}$ and ω satisfying (1.2). Assume that $\Phi(u)$ and $(B(u), u)$ are bounded on bounded sets. Then if $M_{\mathbb{K}}(B)$ is positive, $M_{\mathbb{K}}(B)$ is an asymptotic bifurcation point for Γ on $\mathbb{K}$.*

Remark 1.2. From hypothesis (1.2) it follows that, for fixed $K > 0$, there exists $\rho_1 > \rho_0$ such that $|(\omega(u), u)| \leqslant K\|u\|^2$, $\forall u \in \mathbb{K}$ with $\|u\| \geqslant \rho_1$. Since

$$\Phi(\xi u) = \Phi(u) + \int_1^\xi (\Gamma(tu), u)\, dt, \qquad \xi > 1,$$

it easily follows that if $\Phi(u) - \frac{1}{2}(B(u), u)$ is bounded on $\mathbb{K} \cap S_{\rho_1}$, then it is bounded on each $\mathbb{K} \cap S_\rho$ for all $p \geqslant \rho_1$. Hence, if in Theorem 1.2 one chooses ρ_0 sufficiently large ($\rho_0 \geqslant \rho_1$), it is sufficient to assume that $\Phi(u) - \frac{1}{2}(B(u), u)$ is bounded only on $S_{\rho_0} \cap \mathbb{K}$.

2. PROOF OF THE MAIN RESULTS

In this section we prove the results described in the preceding one. In order to prove Theorem 1.1 we state some lemmas.

LEMMA 2.1. *Let $\Phi: \mathbb{K} \to \mathbb{R}$, $\Phi(0) = 0$, be radially continuous at the origin.*[2] *Assume that Φ is Gateaux differentiable on $\mathbb{K} - \{0\}$ and put $\Gamma = \nabla\phi$. Let*

[2] That is, for any $u \in \mathbb{K}$, the function $t \to \Phi(tu)$, $t \geqslant 0$, is continuous for $t = 0$.

$\Gamma = B + \omega$ with B positively homogeneous in $\mathbb{K}$ and ω verifying (1.1). *Then*

$$\lim_{\substack{u \to 0 \\ 0 \neq u \in \mathbb{K}}} |\Phi(u) - \tfrac{1}{2}(B(u), u)|/\|u\|^2 = 0. \tag{2.1}$$

In particular, Φ is Gateaux differentiable at the origin (Frechet differentiable if $m_{\mathbb{K}}(B)$ and $M_{\mathbb{K}}(B)$ are finite) with $\Gamma(0) = 0$.

Proof. Let $u \neq 0$, $u \in \mathbb{K}$, and consider the function $f(t) = \Phi(tu)$, $t \in [0, 1]$. Since $f(t)$ is differentiable on $]0, 1]$, with bounded derivative $f'(t) = (\Gamma(tu), u)$, and continuous at $t = 0$, it follows that

$$\Phi(u) = \tfrac{1}{2}(B(u), u) + \int_0^1 (\omega(tu), u)\, dt.$$

Denoting by $\delta(\epsilon)$ a positive number such that $|(\omega(u), u)| \leqslant \epsilon\|u\|^2$ if $u \in \mathbb{K}$, $\|u\| \leqslant \delta(\epsilon)$, it follows that

$$|\Phi(u) - \tfrac{1}{2}(B(u), u)| \leqslant \tfrac{1}{2}\epsilon\|u\|^2 \tag{2.2}$$

if $u \in \mathbb{K}$, $\|u\| \leqslant \delta(\epsilon)$.

LEMMA 2.2. *Suppose that $u \in \mathbb{K}$, $\|u\| = \rho$, and $(\Gamma(u), u) > 0$. If there exists a $v \in \mathbb{K}_\rho$, $v \neq u$, such that $(\Gamma(u), v) \geqslant (\Gamma(u), u)$, then there exists a $v' \in \mathbb{K}_\rho$ such that $(\Gamma(u), v') > (\Gamma(u), u)$.*

Proof. Suppose that $(\Gamma(u), v) = (\Gamma(u), u)$ and put $v_t = tu + (1 - t)v$, $t \in]0, 1[$. Since we use only values of t close to 1 we assume, without loss of generality, that $v_t \neq 0$. Define $v_t' = \rho v_t/\|v_t\|$. Then $v_t' \in \mathbb{K}_\rho$ and $(\Gamma(u), v_t' - u) = (\Gamma(u), v_t' - v_t)$ since $v_t - u = (1 - t)(v - u)$. Suppose now that the lemma is false. It follows that $(\Gamma(u), v_t' - v_t) \leqslant 0$ for all t in a neighborhood of 1. Since $v_t' - v_t = \alpha v_t$ with $\alpha > 0$ it follows that $(\Gamma(u), v_t) \leqslant 0$; letting $t \to 1$ we conclude that $(\Gamma(u), u) \leqslant 0$ which is in contradiction to our assumptions.

LEMMA 2.3. *Let $u \in \mathbb{K}$, $u \neq 0$, and $\Gamma(u) \neq \lambda u$ for all $\lambda \in \mathbb{R}$. Assume that there exists $\epsilon = \epsilon(u) > 0$ such that $u + \epsilon\Gamma(u) \in \mathbb{K}$. Then, putting $\rho = \|u\|$, there exists $v' \in \mathbb{K}_\rho$ such that $(\Gamma(u), v') > (\Gamma(u), u)$.*

Proof. Put

$$v' = \frac{\rho}{\|u + \epsilon\Gamma(u)\|}(u + \epsilon\Gamma(u)).$$

One has $v' \in \mathbb{K}_\rho$. Moreover $v' \neq u$ since $v' = u$ implies that $\Gamma(u) = \lambda u$. Put $w = u + \epsilon\Gamma(u)$. One has $\|w\|(\Gamma(u), v' - u) = \epsilon^{-1}(w - u, \|u\|w - \|w\|u)$ and consequently

$$(\Gamma(u), v' - u) = \frac{\|u\| + \|w\|}{\epsilon\|w\|}(\|u\|\,\|w\| - (u, w)) > 0$$

as desired.

Proof of Proposition 1.3. The proof follows directly from Lemma 2.2.

Remark 2.1. Note that in Proposition 1.3 and Lemmas 2.2 and 2.3, $\Gamma(u)$ is an arbitrary element of H.

Proof of Theorem 1.1. Given $\epsilon \in]0, M[, M = M_{\mathbb{K}}(B)$, let $\rho \in]0, \delta(\epsilon/3)]$ be fixed. If $u \in \mathbb{K}$, $\|u\| \leqslant \rho$, one has

$$|(\omega(u), u)| \leqslant \tfrac{1}{3}\epsilon\|u\|^2 \tag{2.3}$$

and

$$|\Phi(u) - \tfrac{1}{2}(B(u), u)| \leqslant \tfrac{1}{6}\epsilon\|u\|^2. \tag{2.4}$$

From the definition of M follows the existence of a sequence $u_n \in \mathbb{K}$ such that $\|u_n\| = \rho$ and $(B(u_n), u_n) > M\rho^2 - (1/n)$. Hence we get from (2.4) that

$$\begin{aligned}\Phi(u_n) &\geqslant \tfrac{1}{2}(B(u_n), u_n) - |\Phi(u_n) - \tfrac{1}{2}(B(u_n), u_n)| \\ &\geqslant \tfrac{1}{2}M\rho^2 - (1/2n) - \tfrac{1}{6}\epsilon\rho^2;\end{aligned}$$

consequently

$$\sup_{u \in \mathbb{K}_\rho} \Phi(u) \geqslant (\tfrac{1}{2}M - \tfrac{1}{6}\epsilon)\rho^2. \tag{2.5}$$

On the other hand, Φ attains its maximum in $\mathbb{K}_\rho$ at (at least) one point u_0 since it is weakly upper semicontinuous in $\mathbb{K}$, and (2.5) yields

$$\Phi(u_0) \geqslant (\tfrac{1}{2}M - \tfrac{1}{6}\epsilon)\rho^2.$$

For convenience put $\|u_0\| = \rho'$. One has $\tfrac{1}{2}(B(u_0), u_0) + \tfrac{1}{2}(\omega(u_0), u_0) \geqslant (\tfrac{1}{2}M - \tfrac{1}{6}\epsilon)\rho^2 - \tfrac{1}{6}\epsilon\rho'^2 - \tfrac{1}{6}\epsilon\rho'^2$, consequently

$$\tfrac{1}{2}(\Gamma(u_0), u_0) \geqslant (\tfrac{1}{2}M - \tfrac{1}{2}\epsilon)\rho^2 > 0; \tag{2.6}$$

and in particular $\Gamma(u_0) \neq 0$. We claim that

$$\Gamma(u_0) = \lambda u_0 \tag{2.7}$$

for a real number λ. Notice first that $\|u_0\| = \rho$. If not, then putting $h = [(\rho - \rho')/\rho']u_0$ in (1.3) we obtain

$$\Phi(u_0 + th) = \Phi(u_0) + t[(\rho - \rho')/\rho](\Gamma(u_0), u_0) + \omega(u_0, th)$$

and this yields $\Phi(u_0 + th) > \Phi(u_0)$ for $t > 0$ sufficiently small which is false since $\Phi(u_0)$ is a maximum on $\mathbb{K}_\rho$; hence $\|u_0\| = \rho$.

Suppose now, by contradiction, that (2.7) does not hold. Then from Assumption 1.1 and Lemma 2.2 the existence follows of a $v' \in \mathbb{K}_\rho$ such that $(\Gamma(u_0), v' - u_0) > 0$. Hence by (1.3) one has $\Phi(u_0 + t(v' - u_0)) > \Phi(u_0)$ for $t > 0$ sufficiently small, which is not possible since $\Phi(u_0)$ is a maximum on $\mathbb{K}_\rho$. Thus (2.7) holds.

Finally $(B(u_0), u_0) + (\omega(u_0), u_0) = \lambda\rho^2$ consequently $\lambda\rho^2 \leqslant M\rho^2 + \frac{1}{3}\epsilon\rho^2$, i.e.,

$$\lambda \leqslant M + \tfrac{1}{3}\epsilon. \tag{2.8}$$

On the other hand, from (2.6) and (2.7) it follows easily that

$$\lambda \geqslant M - \epsilon. \tag{2.9}$$

We remark that we have proved that for any $\epsilon \in]0, M[$ and any $\rho \in]0, \delta(\epsilon/3)]$ there exists $u_0 \in \mathbb{K}_\rho$ and $\lambda \in [M - \epsilon, M + \frac{1}{3}\epsilon]$ such that $\Gamma(u_0) = \lambda u_0$. Furthermore $\Phi(u_0) = \max \Phi(u)$ on $\mathbb{K}_\rho$.

Remark 2.2. If the condition described in Proposition 1.3 (or in Corollary 1.1) holds, then Lemma 2.2 is unnecessary since the existence of a point $v' \in \mathbb{K}_\rho$ such that $(\Gamma(u_0), v' - u_0) > 0$ follows directly from Lemma 2.3.

We now proceed to prove Theorem 1.2. We begin by stating the following lemma.

LEMMA 2.4. *Let $\Phi: \mathbb{K} - \mathbb{K}_{\rho_0} \to \mathbb{R}$ be a Gateaux differentiable functional and put $\Gamma = \nabla\Phi$. Let $\Gamma = B + \omega$ with B positively homogeneous in $\mathbb{K}$ and ω satisfying* (1.2) *and suppose that $|\Phi(u) - \frac{1}{2}(B(u), u)|$ is bounded in $\mathbb{K} \cap S_\rho$ for each $\rho > \rho_0$. Then*

$$\lim_{\|u\| \to +\infty} |\Phi(u) - \tfrac{1}{2}(B(u), u)|/\|u\|^2 = 0. \tag{2.10}$$

Proof. For every $\epsilon > 0$ there exists $R_\epsilon > \rho_0$ such that $|(\omega(u), u)| \leqslant \epsilon\|u\|^2$ if $u \in K$, $\|u\| \geqslant R_\epsilon$. Let $u \in K$, $\|u\| > R_\epsilon$. One has

$$\Phi(u) - \tfrac{1}{2}(B(u), u) = \Phi\left(\frac{R_\epsilon}{\|u\|}u\right) - \frac{1}{2}\left(B\left(\frac{R_\epsilon}{\|u\|}u\right), \frac{R_\epsilon}{\|u\|}u\right) + \int_{R_\epsilon/\|u\|}^1 (\omega(tu), u)\, dt. \tag{2.11}$$

On the other hand,

$$\left|\int_{R_\epsilon/\|u\|}^1 (\omega(tu), u)\, dt\right| < \tfrac{1}{2}\epsilon\|u\|^2. \tag{2.12}$$

On putting

$$S_\epsilon = \sup_{\substack{\|u\|=R_\epsilon \\ u\in\mathbb{K}}} \left|\Phi(u) - \tfrac{1}{2}(B(u), u)\right|,$$

it follows from (2.11) and (2.12) that

$$\left|\Phi(u) - \tfrac{1}{2}(B(u), u)\right| \leqslant \epsilon\|u\|^2, \qquad \forall u \in \mathbb{K}, \quad \|u\| \geqslant N_\epsilon, \tag{2.13}$$

where

$$N_\epsilon = \max(R_\epsilon, (2S_\epsilon/\epsilon)^{1/2}).$$

Proof of Theorem 1.2. First we verify that $M = M_{\mathbb{K}}(B) < +\infty$. Let $\rho > \rho_0$ and suppose, by contradiction, that $M_{\mathbb{K}}(B) = +\infty$. Then there exists a sequence $u_n \in \mathbb{K} \cap S_\rho$ such that $(B(u_n), u_n) \to +\infty$. If hypothesis (i) of Theorem 1.2 holds, one has $\Phi(u_n) \to +\infty$ and (at least for a subsequence) $u_n \rightharpoonup u_0 \in \overline{\mathbb{K} - \mathbb{K}_{\rho_0}}$. But this contradicts (i). If hypothesis (ii) holds, then (at least for a subsequence) $u_n \rightharpoonup u_0 \in \mathbb{K}$, which contradicts the weak upper semicontinuity of $(B(u), u)$ on $\mathbb{K}$; thus $M < +\infty$.

Given $\epsilon \in]0, M[$, let ρ_1, $\rho \in [N(\epsilon/6), +\infty[$, with $\rho_0 \leqslant \rho_1 < \rho$ fixed. If hypothesis (i) holds, we take ρ sufficiently large so that

$$\tfrac{1}{3}M\rho^2 \geqslant \sup_{u \in \mathbb{K}_{\rho_1} \cap \overline{\mathbb{K} - \mathbb{K}_{\rho_1}}} \bar{\Phi}(u). \tag{2.14i}$$

If (ii) holds, we take ρ such that

$$\rho^2 \geqslant 3\rho_1{}^2. \tag{2.14ii}$$

From the definition of M the existence follows of a sequence $u_n \in S_\rho \cap \mathbb{K}$ such that $(B(u_n), u_n) > M\rho^2 - (1/n)$. By using estimate (2.13) one obtains

$$\sup_{\mathbb{K}_\rho - \mathbb{K}_{\rho_1}} \Phi(u) \geqslant (\tfrac{1}{2}M - \tfrac{1}{6}\epsilon)\rho^2.$$

Now let $v_n \in \mathbb{K}_\rho - \mathbb{K}_{\rho_1}$ be a maximizing sequence for Φ such that $v_n \to u_0$. We claim that $u_0 \in \mathbb{K}_\rho - \mathbb{K}_{\rho_1}$. Suppose that condition (i) holds and suppose, by contradiction, that $u_0 \in \mathbb{K}_{\rho_1}$. Since $u_0 \in \overline{\mathbb{K}_\rho - \mathbb{K}_{\rho_1}}$ it follows from (2.14i) that $\bar{\Phi}(u_0) \leqslant \frac{1}{3}M\rho^2 < \lim \Phi(v_n)$, which contradicts (i). If condition (ii) holds, one has $\frac{1}{2}(B(v_n), v_n) \geqslant \Phi(v_n) - \frac{1}{6}\epsilon\rho^2$; passing to the limit and using (2.14ii), one has $\frac{1}{2}(B(u_0), u_0) > \frac{1}{2}M\rho_1{}^2$ and consequently $\|u_0\| > \rho_1$.

Hence u_0 is a maximum point for Φ on $\mathbb{K}_\rho - \mathbb{K}_{\rho_1}$; moreover,

$$\Phi(u_0) \geqslant (\tfrac{1}{2}M - \tfrac{1}{6}\epsilon)\rho^2.$$

Continuing the proof as for Theorem 1.1 we see that $\Gamma(u_0) = \lambda u_0$ with $\|u_0\| = \rho$ and $\lambda \in [M - \epsilon, M + \frac{1}{6}\epsilon]$. Theorem 1.2 is proved.

Remark 2.3. Notice that in Theorem 1.1 we have

$$M_{\mathbb{K}}(B) = \limsup_{\substack{u \in \mathbb{K} \\ u \to 0}} \Phi(u)/\tfrac{1}{2}\|u\|^2$$

as follows from (2.1). Similarly, from (2.10), it follows that in Theorem 1.2,

$$M_{\mathbb{K}}(B) = \limsup_{\substack{u \in \mathbb{K} \\ \|u\| \to +\infty}} \Phi(u)/\tfrac{1}{2}\|u\|^2.$$

Remark 2.4. If B is a weakly continuous potential operator (in particular a linear, continuous, compact, and self-adjoint operator) satisfying assumption 1.1, then $M_{\mathbb{K}}(B)$ is the largest positive λ for which $B(u) = \lambda u$ admits a nonvanishing solution $u \in \mathbb{K}$.

3. On the Basic Conditions (1.1) and (1.2)

In this section we make some observations on hypotheses (1.1) and (1.2).

Definition 3.1. Let B and $\bar{B}$ be two positively homogeneous operators on $\mathbb{K}$. Then B and $\bar{B}$ are said to be *equivalent* (and we write $B \sim \bar{B}$) if $(B(u), u) = (\bar{B}(u), u)$, $\forall u \in \mathbb{K}$.

Obviously if $B \sim \bar{B}$, one has $m_{\mathbb{K}}(B) = m_{\mathbb{K}}(\bar{B})$ and $M_{\mathbb{K}}(B) = M_{\mathbb{K}}(\bar{B})$.

Remark 3.1. If B is a linear continuous operator, then there exists a *unique* self-adjoint, linear, continuous operator B_p such that $B_p \sim B$. More precisely, $B_p = \tfrac{1}{2}(B + B^*)$.

Definition 3.2. Suppose that Γ is defined on $\mathbb{K}_\rho$, $\rho > 0$ [respectively, on $\mathbb{K} - \mathbb{K}_{\rho_0}$] and $\Gamma = B + \omega$ with B positively homogeneous on $\mathbb{K}$ and ω verifying (1.1) [respectively, (1.2)]. Under these conditions $B + \omega$ is said to be an a.d. (*admissible decomposition*) for Γ (at the origin) [respectively, at infinity].

Proposition 3.1. *If $\Gamma = B + \omega$ is an a.d., then $\bar{B} \sim B$ if and only if there exists an a.d. $\Gamma = \bar{B} + \bar{\omega}$.*

The proof is left to the reader.

Now let $\Gamma = A + \omega$ with A *positively homogeneous* on $\mathbb{K}$ and ω verifying

$$\lim_{t \to 0^+} \omega(tu)/t = 0, \qquad \forall x \in \mathbb{K}. \tag{3.1}$$

We then say that A is the *Gateaux derivative of Γ at the origin* and we write

$D\Gamma(0) = A$. Similarly if one has (3.1) when $t \to +\infty$, we say that A is the *Gateaux derivative of* Γ *at infinity* and we write $D\Gamma(\infty) = A$.

PROPOSITION 3.2. *If* $\Gamma = B + \omega$ *is an a.d. at the origin [respectively, at infinity] and if* Γ *is Gateaux differentiable at the origin [respectively, at infinity], then* $B \sim D\Gamma(0)$ *[respectively,* $B \sim D\Gamma(\infty)$*].*

Thus $D\Gamma(0)$ [respectively $D\Gamma(\infty)$] is not always the simplest operator in its equivalence class. For instance, the operator ω defined by (0.10) [respectively (0.17)] admits the a.d. $\omega = 0 + \omega$, but $D\omega(0) = A$ [respectively, $D\omega(\infty) = A$] exists in the G-sense with $A \neq 0$ since one has $A(x) = \phi(0)\eta e_1$ if $\xi = 0$.

If we replace (3.1) by the stronger condition

$$\lim_{\substack{u \to 0 \\ u \in \mathbb{K}}} \|\omega(u)\|/\|u\| = 0, \tag{3.2}$$

then we say that A is the *Fréchet derivative of* Γ *at the origin* (a similar definition holds for infinity by putting, in (3.2), $\|u\| \to +\infty$ instead of $u \to 0$). Obviously $\Gamma = D\Gamma(0) + (\Gamma - D\Gamma(0))$ is then an a.d. To be more precise, if one decomposes the vector field $\omega(u)$ in its radial and tangential components

$$\omega_r(u) = [(\omega(u), u)/\|u\|^2]u, \qquad \omega_t(u) = \omega(u) - [(\omega(u), u)/\|u\|^2]u$$

and if one considers the two conditions

$$\lim_{\substack{u \to 0 \\ u \in \mathbb{K}}} \|\omega_r(u)\|/\|u\| = 0 \tag{3.3i}$$

$$\lim_{\substack{u \to 0 \\ u \in \mathbb{K}}} \|\omega_t(u)\|/\|u\| = 0, \tag{3.3ii}$$

then (3.2) is equivalent to (3.3i) plus (3.3ii) while (1.1) is equivalent to (3.3i) only. These facts hold if we replace (1.1) by (1.2) and if we put $\|u\| \to +\infty$ instead of $u \to 0$ [in (3.2) and (3.3i or ii)].

Finally, for a positively homogeneous remainder, the following result holds.

PROPOSITION 3.3. *Suppose that* $\omega : \mathbb{K} \to H$ *verifies, for some* $\alpha > 0$,

$$(\omega(tu), u) = t^\alpha(\omega(u), u), \qquad \forall u \in \mathbb{K}, \quad \forall t > 0,$$

with $(\omega(u), u)$ *not identically zero in* $\mathbb{K}$. *Then* ω *verifies* (1.1) *[respectively,* (1.2)*] if and only if* $\alpha > 1$ *[respectively,* $\alpha < 1$*] and* $(\omega(u), u)$ *is bounded in* S_ρ *for one positive* ρ.

The proof is left an easy exercise.

4. Systems of the Hammerstein Type

Assumptions and Notation

In this section we apply the preceding results to systems of the Hammerstein type. To study this problem Theorems 1.1 and 1.3 together with some classical methods (see [6] and [10]) are used. For the study of Hammerstein operators we refer also to the papers by Amann [1, and references] and to a paper by H. Brézis and F. E. Browder [4].

In the following B is a subset of finite Lebesgue measure of the Euclidean space $\mathbb{R}^n$. Let $y = (y_1, \ldots, y_m)$ denote a generic $\mathbb{R}^m$-element. The norm and scalar product in this space are defined as $|y|^2 = |y_1|^2 + \cdots + |y_m|^2$ and $(y, \overline{y}) = y_1\overline{y}_1 + \cdots + y_m\overline{y}_m$. Also $p \geqslant 2$ is a real number and q is defined by the equation $(1/p) + (1/q) = 1$. We consider the reflexive Banach space $L^p(B)$ with the usual norm

$$\|u_1\|_p = \left(\int_B |u_1(x)|^p \, dx \right)^{1/p},$$

and also the Banach space $L^q(B)$ with norm $\| \; \|_q$ and the Hilbert space $L^2(B)$ with norm $\| \; \|_2$. We denote by $\langle \; , \; \rangle$ both the scalar product on $L^2(B)$ and the pairing between $L^q(B)$ and the dual space $L^p(B)$:

$$\langle v_1, u_1 \rangle = \int_B v_1(x) u_1(x) \, dx$$

for $v_1(x), u_1(x) \in L^2(B)$ or $v_1(x) \in L^q(B)$ and $u_1(x) \in L^p(B)$. We consider also the reflexive Banach space $L_m{}^p(B) = L^p(B) \times \cdots \times L^p(B)$ (m times), the norm of which is, as for $L^p(B)$, denoted by $\| \; \|_p$ and defined by

$$\|u\|_p = \left(\int_B |u(x)|^p \, dx \right)^{1/p},$$

where $u = (u_1, \ldots, u_m) \in L_m{}^p(B)$. Similar definitions hold for $\| \; \|_q$ and $\| \; \|_2$. The scalar product on $L_m{}^2(B)$ and the pairing between $L_m{}^q(B)$ and the dual space $L_m{}^p(B)$ are denoted by the same symbol $\langle \; , \; \rangle$, which is defined by

$$\langle v, u \rangle = \sum_{i=1}^{m} \langle v_i, u_i \rangle$$

with $v = (v_1, \ldots, v_m)$ and $u = (u_1, \ldots, u_m)$ belonging to $L_m{}^2(B)$ or with $v \in L_m{}^q(B)$ and $u \in L_m{}^p(B)$.

Let $g_i(x, y)$, $i = 1, \ldots, m$, be real functions defined on $B \times \mathbb{R}^m$ and satisfying the Carathéodory conditions, i.e., continuous in y for almost all $x \in B$ and measurable in x for all $y \in \mathbb{R}^m$. Moreover, assume that there exists a real function $G(x, y)$ defined on $B \times \mathbb{R}^m$ such that for almost all $x \in B$,

$$g_i(x, y) = \partial G(x, y)/\partial y_i, \qquad i = 1, \ldots, m,$$

and suppose that $G(x, 0) = 0$ and $g_i(x, 0) = 0$, $i = 1, \ldots, m$, for almost all $x \in B$. Finally defined a function $g: B \times \mathbb{R}^m \to \mathbb{R}^m$ by

$$g(x, y) = (g_1(x, y), \ldots, g_m(x, y)).$$

We denote also by the symbols g_i and g the *Nemytsky operators* defined by $[g_i(u)](x) = g_i(x, u(x))$ a.e. in B and $[g(u)](x) = g(x, u(x))$ a.e. in B, and we assume that each g_i acts from $L_m{}^p(B)$ into $L^q(B)$, i.e., that $g_i(x, u(x)) \in L^q(B)$, $\forall u(x) \in L_m{}^p(B)$. Under this condition the operator g_i is bounded and continuous (cf. Krasnoselskii [6, Chapter I, §2]) and $g: L_m{}^p(B) \to L_m{}^q(B)$ is bounded and continuous. Moreover g is a potential operator, i.e., $g = \nabla\Phi$ in the Fréchet sense where $\Phi: L_m{}^p(B) \to \mathbb{R}$ is defined by

$$\Phi(u) = \int_B G(x, u(x))\, dx$$

(cf. [10, Theorem 21.1]).

Consider now m real symmetric kernels $K_i(x, \xi)$, measurable on $B \times B$, and the associated linear integral operators

$$[A_i(f)](x) = \int_B K_i(x, \xi) f(\xi)\, d\xi \quad \text{a.e. on } B. \tag{4.1}$$

We assume that the operators A_i act and are *completely continuous from* $L^q(B)$ *into* $L^p(B)$. Hence they are completely continuous as operators on $L^2(B)$. Moreover we assume that each A_i, $i = 1, \ldots, m$ is *self-adjoint and positive on* $L^2(B)$. Obviously the operator $A(v) = (A_1(v_1), \ldots, A_m(v_m))$ verifies the corresponding properties as an operator from $L_m{}^q(B)$ into $L_m{}^p(B)$ and as an operator on $L_m{}^2(B)$. Moreover the positive square root of A on $L_m{}^2(B)$ is given by $A^{1/2} = (A_1^{1/2}, \ldots, A_m^{1/2})$, where $A_i^{1/2}$ denotes the positive root of A_i on $L^2(B)$. Recall that each $A_i^{1/2}$ (consequently $A^{1/2}$) is self-adjoint and positive (cf. [8, n. 104].

One can prove (see [10, Corollary 23.1] that the range of $A^{1/2}$ is contained in $L_m{}^p(B)$ and that *the operator* $A^{1/2} = L_m{}^2(B) \to L_m{}^p(B)$ *is completely continuous*. It follows that the adjoint operator is a completely continuous operator from $L_m{}^q(B)$ into $L_m{}^2(B)$. Since it coincides with $A^{1/2}$ on $L_m{}^2(B)$, it will be denoted by the same symbol $A^{1/2}$.

It is easy to see that the functional $\Phi A^{1/2}: L_m{}^2(B) \to \mathbb{R}$ is Fréchet differentiable with

$$\nabla(\Phi A^{1/2}) = A^{1/2}(\nabla\Phi)A^{1/2}. \tag{4.2}$$

We return now to the Nemytsky operator g. We assume that the functions g_i can be written in the form

$$g_i(x, y) = \sum_{j=1}^{m} g_{ij}(x)y_j + \omega_i(x, y), \tag{4.3}$$

where $g_{ij}(x)$, $i, j = 1, \ldots, m$, belongs to $L^s(B)$ with $s = p/(p-2)$, $g_{ij}(x) = g_{ji}(x)$[3] and the remainders $\omega_i(x, y)$ satisfy the Charathéodory conditions and act from $L_m{}^p(B)$ into $L^q(B)$. We define the Nemytsky operator $\omega(u)$ by

$$[\omega(u)](x) = (\omega_1(x, u(x)), \ldots, \omega_m(x, u(x))).$$

The Nemytsky operator g is then the sum of the linear operator g_0 defined by

$$[g_0(u)](x) = \left(\sum_{j=1}^{m} g_{1j}(x)u_j(x), \ldots, \sum_{j=1}^{m} g_{mj}(x)u_j(x) \right)$$

with the remainder ω.

The Main Results

The following results hold.

THEOREM 4.1. *Assume that the conditions described in this section hold and that*

$$\lim_{\|u\|_p \to 0} \langle \omega(u), u \rangle / \|u\|_p^2 = 0. \tag{4.4}$$

Then [*in the space* $L_m{}^p(B)$] *the largest positive eigenvalue for the linear problem*

$$Ag_0(u) = \lambda u \tag{4.5}$$

is the largest positive bifurcation point for the problem

$$Ag(u) = \lambda u. \tag{4.6}$$

The result holds again if we replace "largest positive" with "smallest negative."

THEOREM 4.2. *Assume that the conditions described in this section hold and that*

$$\lim_{\|u\|_p \to +\infty} \langle \omega(u), u \rangle / \|u\|_p^2 = 0. \tag{4.7}$$

[3] If the functions $g_i(x, y)$ are regular, then $g_{ij}(x) = [\partial g_i(x, y)/\partial y_j]_{y=0}$; an analogous remark holds for the asymptotic case.

Then [*in the space* $L_m{}^p(B)$] *the largest positive eigenvalue for the linear problem* (4.5) *is the largest positive asymptotic bifurcation point for problem* (4.6). *The result holds again if one replace "largest positive" with "smallest negative."*

Recall that (4.5) and (4.6) can be written more explicitly as

$$\int_B K_i(x, \xi)\left[\sum_{j=1}^{m} g_{ij}(\xi)u_j(\xi)\right]d\xi = \lambda u_i(x), \qquad i = 1, \ldots, m,$$

and

$$\int_B K_i(x, \xi)g_i(\xi, u_1(\xi), \ldots, u_m(\xi))\, d\xi = \lambda u_i(x), \qquad i = 1, \ldots, m,$$

respectively.

Remark 4.1. In Theorems 4.1 and 4.2 it is not necessary that the operators A_i, $1 \leqslant i \leqslant m$, are integral operators [see (4.1)]. In fact the results and proofs hold if these operators verify the following properties:

(a) A_i is completely continuous as an operator from $L^q(B)$ into $L^p(B)$;
(b) A_i is self-adjoint and positive as an operator on $L^2(B)$;
(c) $A_i^{1/2}(L^2(B)) \subset L^p(B)$, and $A_i^{1/2}$ is completely continuous as an operator from $L^2(B)$ into $L^p(B)$.

Proof of Theorem 4.1. The functional $\Phi A^{1/2}$ is weakly continuous on $L_m{}^2(B)$ since $A^{1/2}$ is completely continuous from $L_m{}^2(B)$ into $L_m{}^p(B)$. On the other hand, Φ is Fréchet differentiable and (4.2) holds. It follows that

$$\nabla(\Phi A^{1/2}) = A^{1/2}g_0A^{1/2} + A^{1/2}\omega A^{1/2},$$

where $A^{1/2}g_0A^{1/2}$ is linear and completely continuous on $L_m{}^2(B)$. Moreover from (4.4),

$$\lim_{\|u\|_2 \to 0} \langle (A^{1/2}\omega A^{1/2})(u), u\rangle / \|u\|_2^2 = 0$$

since $\|A^{1/2}(u)\|_p \geqslant c\|u\|_2$. Thus (1.1) holds. Then, by Theorem 1.1, the largest positive bifurcation point for

$$(A^{1/2}gA^{1/2})(u) = \lambda u \qquad \text{in} \quad L_m{}^2(B) \tag{4.8}$$

is given by

$$M(A^{1/2}g_0A^{1/2}) \equiv \sup_{\|u\|_2 = 1} \langle (A^{1/2}g_0A^{1/2})(u), u\rangle,$$

if this quantity is positive. Since $A^{1/2}g_0A^{1/2}$ is a self-adjoint, completely continuous operator, M is the largest positive eigenvalue for

$$(A^{1/2}g_0A^{1/2})(u) = \lambda u \qquad \text{in} \quad L_m{}^2(B). \tag{4.9}$$

A corresponding result holds for the smallest negative bifurcation point in (4.8).

To conclude the proof it suffices to verify that $\lambda \neq 0$ is a bifurcation point for (4.8) in $L_m{}^2(B)$ if and only if it is a bifurcation point for (4.6) in $L_m{}^p(B)$.[4] This can be done with a standard method: If λ is a bifurcation point for (4.8), one has $(A^{1/2}gA^{1/2})(u_\epsilon) = \lambda_\epsilon u_\epsilon$ with $0 \neq \|u_\epsilon\|_2 \to 0$ and $\lambda_\epsilon \to \lambda$. By putting $v_\epsilon = A^{1/2}(u_\epsilon)$, one has $(Ag)(v_\epsilon) = \lambda_\epsilon v_\epsilon$ with $\|v_\epsilon\|_p \to 0$ and $v_\epsilon \neq 0$ (since $(A^{1/2}g)(v_\epsilon) = \lambda_\epsilon u_\epsilon \neq 0$). Reciprocally, if λ is a bifurcation point for (4.6), one has $(Ag)(v_\epsilon) = \lambda_\epsilon v_\epsilon, \lambda_\epsilon \to \lambda$, $0 \neq \|v_\epsilon\|_p \to 0$. By putting $u_\epsilon = (1/\lambda_\epsilon)(A^{1/2}g)(v_\epsilon)$, it follows that $\|u_\epsilon\|_2 \leqslant c|\lambda_\epsilon|^{-1}\|g(v_\epsilon)\|_q \to 0$. Moreover $A^{1/2}(u_\epsilon) = v_\epsilon$, consequently $u_\epsilon \neq 0$ and

$$(A^{1/2}gA^{1/2})(u_\epsilon) = (A^{1/2}g)(v_\epsilon) = \lambda_\epsilon u_\epsilon .$$

Proof of Theorem 4.2. If $A^{1/2}\omega A^{1/2}$ does not verify (1.2), there exists a real $\epsilon_0 > 0$ and a sequence $u_k \in L_m{}^2(B)$ such that $\|u_k\|_2 \to +\infty$ and

$$\langle(\omega A^{1/2})(u_k), A^{1/2}(u_k)\rangle/\|u_k\|_2^2 \geqslant \epsilon_0, \qquad \forall k. \tag{4.10}$$

Since ω is a bounded operator from $L_m{}^p(B)$ into $L_m{}^q(B)$, it follows from (4.10) that the sequence $A^{1/2}(u_k)$ is unbounded in $L_m{}^p(B)$. Moreover (4.10) yields

$$|\langle(\omega A^{1/2})(u_k), A^{1/2}(u_k)\rangle|/\|A^{1/2}(u_k)\|_p^2 \geqslant c\epsilon_0, \qquad \forall k,$$

and with the help of (4.7) it follows that $A^{1/2}(u_k)$ is bounded in $L_m{}^p(B)$, which is in contradiction with the preceding conclusion. The remainder of the proof is similar to the preceding one.

Sufficient Conditions for (4.4) *and* (4.7)

Now we state some sufficient conditions for (4.4) and (4.7).

PROPOSITION 4.1. *Assume that the vector function ω verifies the condition*

$$|(\omega(x, y), y)| \leqslant b(x)|y|^2 + c_1|y|^p, \tag{4.11}$$

[4] In the same way one proves a similar result for the eigenvalues of (4.9) and (4.5).

where $b(x) \in L^s(B)$ *and* c_1 *is a constant, and also the condition*

$$\lim_{|y| \to 0} (\omega(x, y), y)/|y|^2 = 0 \tag{4.12}$$

for almost all $x \in B$. *Then* (4.4) *holds.*

Remark 4.2. A sufficient condition for (4.11) is

$$|\omega(x, y)| \leqslant b(x)|y| + c_1|y|^{p-1}, \tag{4.13}$$

where $b(x)$ and c_1 are defined as before (note that $p/q = p - 1$).

Moreover terms of the form $d(x)|y^r$, with $1 \leqslant r \leqslant p - 1$ and $d(x) \in L^{p/(p-r-1)}(B)$, are implicitly included in the second term of (4.13) since they can be interpolated by $b(x)|y|$ and $c_1|y|^{p/q}$.

In connection with (4.3) recall that ω acts from $L_m{}^p(B)$ into $L_m{}^q(B)$ if and only if $|\omega(x, y)| \leqslant a(x) + b(x)|y| + c_1|y|^{p-1}$ with $a(x) \in L^q(B)$.

PROPOSITION 4.2. *Assume that the vector function* ω *verifies the condition*

$$|(\omega(x, y), y)| \leqslant a_0(x) + b(x)|y|^2, \tag{4.14}$$

where $a_0(x) \in L^1(B)$ *and* $b(x) \in L^s(B)$, *and also the condition*

$$\lim_{|y| \to +\infty} (\omega(x, y), y)/|y|^2 = 0 \tag{4.15}$$

for almost all $x \in B$. *Then* (4.7) *holds.*

PROPOSITION 4.3. *Assume that* (4.15) *holds unforrnly for almost all* $x \in B$. *Then* (4.7) *holds.*

Proof of Proposition 4.1. Assume that (4.4) fails. Then there exists and $\epsilon_0 > 0$ and a sequence $u_k(x) \in L_m{}^p(B)$ such that $\|u_k\|_p \to 0$ and

$$|\langle \omega(u_k), u_k \rangle| > (1 + c_2 - c_3)\epsilon_0\|u_k\|_p^2, \qquad \forall k, \tag{4.16}$$

where the positive constants c_2 and c_3 are defined in the following. Consider now the real measurable functions defined on B by

$$f_k(x) = \begin{cases} |(\omega(x, u_k(x)), u_k(x))|/ u_k(x)|^2 & \text{if } |u_k(x)| \neq 0, \\ 0 & \text{if } |u_k(x)| = 0. \end{cases} \tag{4.17}$$

Since the integral is absolutely continuous with respect to the measure, there exists a $\delta_0 > 0$ such that

$$|E| \leqslant \delta_0 \Rightarrow \int_E |b(x)|^s \, dx \leqslant \epsilon_0, \tag{4.18}$$

where E is a measurable subset of B and $|E|$ denotes the measures of E.

On the other hand by assuming (without loss of generality) that $u_k(x) \to 0$ a.e. on B, it follows by (4.12) that $f_k(x) \to 0$ a.e. on B. Thus $f_k(x)$ converges in measure to 0. Hence there exists an integer k_0 such that

$$k > k_0 \Rightarrow |B_0^{(k)}| \leqslant \delta_0, \tag{4.19}$$

where

$$B_0^{(k)} = \{x \in B : f_k(x) \geqslant \epsilon_0\}.$$

Since

$$|\langle \omega(u_k), u_k \rangle| \leqslant \int_{B - B_0^{(k)}} |(\omega(x, u_k(x)), u_k(x))|\, dx + \int_{B_0^{(k)}} |(\omega(x, u_k(x)), u_k(x))|\, dx,$$

it follows from the definition of $B_0^{(k)}$ and (4.11) that

$$|\langle \omega(u_k), u_k \rangle| \leqslant \epsilon_0 \|u_k\|_2^2 + \int_{B_0^{(k)}} b(x)|u_k(x)|^2\, dx + c_1 \|u_k\|_p^p,$$

and by Hölder's inequality,

$$|\langle \omega(u_k), u_k \rangle| \leqslant c_2 \epsilon_0 \|u_k\|_p^2 + c_1 \|u_k\|_p^p + \left(\int_{B_0^{(k)}} b(x)^s\, dx \right)^{1/s} \|u_k\|_p^2,$$

where c_2 is a positive constant such that $\| \ \|_2^2 \leqslant c_2 \| \ \|_p^2$. Assuming k sufficiently large in order that $k > k_0$ and $\|u_k\|_p^{p-2} \leqslant \epsilon_0$, it follows, with the help of (4.19) and (4.18), that

$$|\langle \omega(u_k), u_k \rangle| \leqslant (1 + c_1 + c_2)\epsilon_0 \|u_k\|_p^2,$$

which contradicts (4.16).

Proof of Proposition 4.2. Assume that (4.7) fails. Then there exists a positive ϵ_0 and a sequence $u_k \in L_m{}^p(B)$ such that $\|u_k\|_p \to +\infty$ and

$$|\langle \omega(u_k), u_k \rangle| > (2 + c_2)\epsilon_0 \|u_k\|_p^2, \qquad \forall k. \tag{4.20}$$

Let $\delta_0 > 0$ be such that (4.18) holds and define a function $f : B \times \mathbb{R}^m \to \mathbb{R}$ by

$$f(x, y) = \begin{cases} |(\omega(x, y), y)|/|y|^2 & \text{if } \ y \neq 0, \\ 0 & \text{if } \ y = 0. \end{cases} \tag{4.21}$$

Then there exists a constant $M_0 > 0$ such that

$$|\{x \in B : f(x, u(x)) \geqslant \epsilon_0\}| \leqslant \delta_0 \tag{4.22}$$

holds for any measurable $u : B \to \mathbb{R}^m$ such that $|u(x)| \geqslant M_0$ a.e. in B. If not, to any positive integer M there corresponds a measurable function $u_M(x)$ such that $|u_M(x)| \geqslant M$ a.e. in B and

$$|\{x \in B : f(x, u_M(x)) \geqslant \epsilon_0\}| > \delta_0. \tag{4.23}$$

Since $|u_M(x)| \to +\infty$ as $M \to +\infty$, it follows from (4.15) that $f(x, u_M(x)) \to 0$ a.e. in B and in particular $f(x, u_M(x)) \to 0$ in measure. But this contradicts (4.23).

Now define $B_0^{(k)}$ and $C^{(k)}$ by

$$B_0^{(k)} = \{x \in B : f(x, u_k(x)) \geqslant \epsilon_0\} \qquad \text{and} \qquad C^{(k)} = \{x \in B : |u_k(x)| \geqslant M_0\},$$

respectively. We prove that

$$|C^{(k)} \cap B_0^{(k)}| \leqslant \delta_0, \qquad \forall k. \tag{4.24}$$

Put

$$\bar{u}_k(x) = \begin{cases} u_k(x) & \text{if } x \in C^{(k)}, \\ y_0 & \text{if } x \notin C^{(k)}, \end{cases}$$

where $y_0 \in \mathbb{R}^m$ verifies $|y_0| = M_0$. Since $|\bar{u}_k(x)| \geqslant M_0$ on B it follows, from (4.22), that

$$|\{x \in B : f(x, \bar{u}_k(x)) \geqslant \epsilon_0\}| \leqslant \delta_0.$$

Moreover one easily sees that $C^{(k)} \cap B_0^{(k)} \subset \{x \in B : f(x, \bar{u}_k(x)) \geqslant \varepsilon_0\}$ and consequently (4.24) holds.

Finally one has, for any k,

$$\begin{aligned} |\langle \omega(u_k), u_k \rangle| \leqslant & \int_{B - C^{(k)}} |(\omega(x, u_k(x)), u_k(x))| \, dx \\ & + \int_{C^{(k)} \cap (B - B_0^{(k)})} |(\omega(x, u_k(x)), u_k(x))| \, dx \\ & + \int_{C^{(k)} \cap B_0^{(k)}} |(\omega(x, u_k(x)), u_k(x))| \, dx \end{aligned} \tag{4.25}$$

The first integral on the right is bounded by a constant (dependent on M_0) as follows from (4.14). The second is bounded by $\epsilon_0 \int_B |u_k(x)|^2 \, dx$ as follows from the definitions of $B_0^{(k)}$ and f. The third integral on the right of (4.25) is less than or equal to

$$c + \left(\int_{C^{(k)} \cap B_0^{(k)}} b(x)^s \, dx \right)^{1/s} \|u_k\|_p^2$$

as follows from (4.14) and Hölder's inequality. Thus, by using (4.24) and (4.18), the referred integral is bounded by

$$c + \epsilon_0 \|u_k\|_p^2.$$

Hence (4.25) yields

$$|\langle \omega(u_k), u_k \rangle| \leqslant c_3 + (1 + c_2)\epsilon_0 \|u_k\|_p^2, \qquad \forall k,$$

which contradicts (4.20) for any index k such that $\|u_k\|_p \geqslant c_3 \varepsilon_0^{-1}$.

Proof of Proposition 4.3. If (4.7) fails, there exists an $\epsilon_0 > 0$ and a sequence $u_k \in L_m{}^p(B)$ such that $\|u_k\|_p \to +\infty$ and (4.20) holds. On the other hand there exists $M_0 > 0$ such that

$$|y| > M_0 \Rightarrow |(\omega(x, y), y)| \leqslant \epsilon_0 |y|^2$$

for almost all $x \in B$. Hence

$$\begin{aligned} |\langle \omega(u_k), u_k \rangle| &\leqslant \int_{\{|u_k(x)| > M_0\}} |(\omega(x, u_k(x)), u_k(x))| \, dx \\ &\quad + \int_{\{|u_k(x)| \leqslant M_0\}} |(\omega(x, u_k(x)), u_k(x))| \, dx \\ &\leqslant c_2 \epsilon_0 \|u_k\|_p^2 + c(M_0). \end{aligned}$$

By choosing k sufficiently large in order that $\|u_k\|_p^2 \geqslant c(M_0)/2\epsilon_0$ one contradicts (4.20).

Examples with a Nondifferentiable Operator g

There exist operators g for which the sufficient conditions stated in Proposition 4.1 hold (hence Theorem 4.1 applies) but which are not F-differentiable at the origin. Similarly there exist operators g for which the sufficient conditions stated in Propositions 4.2 or 4.3 hold (hence Theorem 4.2 applies) but which are not F-differentiable at infinity. This holds if $m \geqslant 2$, even for functions g independent of the variable x. To show this we give two examples of remainders ω, one verifying the conditions of Proposition 4.1 but not

$$\lim_{\|u\|_p \to 0} \|\omega(u)\|_q / \|u\|_p = 0, \tag{4.26}$$

and the other verifying the conditions of Proposition 4.2 but not

$$\lim_{\|u\|_p \to +\infty} \|\omega(u)\|_q / \|u\|_p = 0. \tag{4.27}$$

Note that (4.26) means that $Dg(0) = g_0$ in the F-sense and consequently it implies that $D(A^{1/2} g A^{1/2})(0) = A^{1/2} g_0 A^{1/2}$ in the F-sense. A similar remark, concerning (4.27), holds at infinity.

EXAMPLE 1. Consider the vector function $\omega(y)$ whose components $\omega_1(y)$ and $\omega_2(y)$ are defined by (0.10) with $\xi = y_1$ and $\eta = y_2$; recall that $\phi, \phi(0) \neq 0$, is an arbitrary, real, continuously differentiable function defined and with compact support on $\mathbb{R}$. This function $\omega(y)$ verifies the hypothesis of Proposition 4.1. In fact (4.12) and (4.11) hold (with $b(x)$ constant and $c_1 = 0$) as follows from (0.13) and (0.15), respectively. Defining $g_i(x, y)$, $i = 1, 2$, by (4.3) with arbitrary $g_{ij}(x) \in L^{p/(p-2)}(B)$, $g_{ij}(x) = g_{ji}(x)$ [note that $g_i(x, y) =$

$\partial G(x, y)/\partial y_i$ with G given by $G(x, y_1, y_2) = \frac{1}{2}\sum_{i,j=1}^{m} g_{ij}(x) y_i y_j + \phi(y_1/y_2^2) y_1 y_2$; see (0.9)], it follows from Theorem 4.1 that the largest positive and the smallest negative eigenvalues for the linear system

$$\int_B K_i(x, \xi)[g_{i1}(\xi)u_1(\xi) + g_{i2}(\xi)u_2(\xi)]\, d\xi = \lambda u_i(x), \qquad i = 1, 2, \quad (4.28)$$

are, respectively, the largest positive and the smallest negative bifurcation points for the nonlinear system

$$\int_B K_i(x, \xi) g_i(\xi, u_1(\xi), u_2(\xi))\, d\xi = \lambda u_i(x), \qquad i = 1, 2. \quad (4.29)$$

However, (4.26) does not hold. This is easily verified by considering in (4.26) the constant (vector) functions $u_t(x)$ whose two components are, respectively, the constant functions 0 and t ($t \neq 0$). Thus $\omega(u_t)$ has as components the constant functions $t\phi(0)$, and 0 and consequently

$$\|\omega(u_t)\|_q / \|u_t\|_p = |\phi(0)| (\text{mes } B)^{(p-2)/p} \quad (4.30)$$

does not verify (4.26) when $t \to 0$.

EXAMPLE 2. (Asymptotic case). In Example 1 replace (0.10) by (0.18) [and $\phi(y_1/y_2^2)$ by $\phi(y_1^2/y_2)$]. From (0.19) and (0.15) it follows that Proposition 4.2 (as well as Proposition 4.3) applies to our remainder ω. It follows from Theorem 4.2 that the largest positive and the smallest negative eigenvalues for the linear system (4.28) are, respectively, the largest positive and the smallest negative bifurcation points for the nonlinear system (4.29). However, condition (4.27) fails since (4.30) holds for the vector functions $u_t(x)$ used in Example 1.

Remark 4.3. Since Theorems 1.1 and 1.2 hold for nonlinear positively homogeneous operators B it follows that the results of this section apply also when g_0 is an operator of this type. For instance, let $m = 1$ and put $g(x, y) = g_0(x, y) + \omega(x, y)$ with $g_0(x, y) = g_0^{+}(x) y$ if $y \geqslant 0$, $g_0(x, y) = g_0^{-}(x) y$ if $y \leqslant 0$, where $g_0^{+}(x)$ and $g_0^{-}(x)$ are given functions and ω is a suitable remainder. Then the Nemytsky operator g_0 is positively homogeneous and consequently the results stated in Theorem 4.1 or in Theorem 4.2 hold with obvious changes. Roughly speaking, g_0 corresponds to the case in which $y \to g(x, y)$ is right and left differentiable at $y = 0$, for almost all $x \in B$, but is not necessarily differentiable. A similar example holds for the study of the largest negative bifurcation point λ for Problem (2.1) of [2][5]) if one assumes the graphs β and γ only right and left differentiable at the origin

[5] In this case λ is the principal bifurcation point for (2.1) and corresponds to the largest positive bifurcation point $1/\mu$ for Equation (2.4) of [2].

(differentiability is understood in the sense introduced in Beirão da Veiga [2]). This holds because the proof of Theorem I of Beirão da Veiga [2] holds (with obvious changes) if the graphs β and γ are right and left differentiable instead of differentiable (this was remarked to the author by J. Hernandez).

Remark 4.4. We use in the following the usual notations for Sobolev spaces. Consider the bifurcation problem

$$\begin{aligned} Lu(x) &= \mu g(x, u(x)) \quad &\text{in} \quad B,\\ u(x) &= 0 \quad &\text{on} \quad \partial B, \end{aligned} \tag{4.31}$$

where B is an open bounded set with boundary ∂B, and

$$Lu \equiv - \sum_{i,\, j=1}^{m} \frac{\partial}{\partial x_i}\left(a_{ij}(x)\frac{\partial u}{\partial x_j}\right), \qquad a_{ij}(x) = a_{ji}(x), \tag{4.32}$$

is an uniformly elliptic operator on B and g acts from $L^p(B)$ into $L^q(B)$ with $p < 2^* = 2n/(n-2)$ (assume for the sake of convenience that $n > 2$). It is well known that the Green's operator $A(f) = u$ related to the problem $Lu = f$ in B, $u = 0$ on ∂B, acts from $H^{-1,2}(B)$ onto $H_0^{1,2}(B)$ and, in particular, it is completely continuous from $L^q(B)$ into $L^p(B)$. Moreover (4.32) is equivalent to the problem (in $L^p(B)$)

$$(Ag)(u) = \lambda u, \qquad \lambda = 1/\mu,^{6} \tag{4.31'}$$

which can be reduced, by splitting the linear operator A, to

$$(A^{1/2} g A^{1/2})(u) = \lambda u. \tag{4.31''}$$

This last problem concerns a potential operator in the Hilbert space $L^2(B)$.

Assume now that instead of the linear operator L one has a nonlinear operator, as *for instance*,

$$Lu \equiv - \sum_{i=1}^{n} \frac{\partial}{\partial x_i}\left(|\nabla u|^{r-2}\frac{\partial u}{\partial x_i}\right), \tag{4.33}$$

and assume that g acts from $L^p(B)$ into $L^q(B)$ with $p < r^* = rn/(r-n)$ (assume for convenience that $n > r$). The inverse operator $A(f) = u$ related to the problem $Lu = f$ in B, $u = 0$ on ∂B, acts from $H^{-1,r'}(B)$ onto $H_0^{1,r}(B)$ and is weakly continuous from $L^q(B)$ into $L^p(B)$. However, the problem does not reduce to (4.31″) since A is nonlinear. This leads to the study of problem (4.31) directly. Since L is defined on $H_0^{1,r}(B)$ but not on all of $L^p(B)$ (and g is defined in both of these spaces) one is lead to consider L and g as

[6] The largest positive λ for (4.31′) corresponds to the smallest positive μ for (4.31) which is the principal bifurcation point for a wide class of problems.

operators from $H_0^{1,r}(B)$ into $H^{-1,r'}(B)$. In this context the operator g is weakly continuous. Furthermore g and L are potential operators with $L = \nabla\Psi$ where

$$\Psi(u) = \int_B |\nabla u|^r \, dx, \qquad u \in H_0^{1,r}(B).$$

is weakly lower semicontinuous.

Thus the bifurcation problem (4.31), (4.33) can be studied as a particular case ($\Lambda = L, \Gamma = g$) of the general problem

$$\Gamma(u) = \lambda\Lambda(u), \qquad \lambda = 1/\mu, \tag{4.34}$$

where $\Gamma = \nabla\Phi$ and $\Lambda = \nabla\Psi$ are potential operators acting from a Banach space V into the dual space V', Φ being weakly continuous (more generally weakly upper semicontinuous) and Ψ weakly lower semicontinuous. The results of the first part of this chapter can be extended to this more general problem and applied, in particular, to the referred example or to more general problems where $L = (L_1, \ldots, L_m)$ is a system of nonlinear operators and the boundary conditions are not of the Dirichlet type.

These results will be proved in a forthcoming paper and generalize in particular a result of Naumann which is applied by this author to the study of bifurcation buckling of thin elastic shells (cf. Naumann [7], Theorem 5.1'.[7]

Appendix

The proof of Theorem 1.1 is easily adapted to prove an analogous result concerning the case $M_{\mathbb{K}}(B) = 0$. Consider the following condition.

ASSUMPTION A.1. *For any $u \in \mathbb{K}$, $u \neq 0$, such that $(\Gamma(u), u) \geqslant 0$ and $\Gamma(u) \neq \lambda u$, $\forall \lambda \in \mathbb{R}$, there exists a $v \in \mathbb{K}_\rho$ ($\rho = \|u\|$), $v \neq u$, such that $(\Gamma(u), v) > (\Gamma(u), u)$.*

One has, in particular, from Lemma 2.3:

Remark A.1. Proposition 1.3 and Corollary 1.1 hold if one replaces in these statements Assumption 1.1 by Assumption 5.1.

The following result holds (a similar result holds for the asymptotic bifurcation).

[7] Similar problems under unilateral conditions were studied by Do [5, and references]. It seems that our results regarding Equation (4.34) can be adapted in order to be applicable in this case.

THEOREM A.1. *Assume that the conditions of Theorem* 1.1 *hold with* $M_{\mathbb{K}}(B) = 0$ *and with Assumption* 1.1 *replaced by Assumption* 5.1. *Furthermore assume that*:

(i) *for any* $\rho > 0$ *there exists* $u \in \mathbb{K}_\rho$ *such that* $\Phi(u) > 0$;
(ii) *if* $\Phi(u) > 0$ *and* $u \in \mathbb{K}$, *then* $\Gamma(u) \neq 0$.

Then 0 *is the largest bifurcation point for* Γ *in* $\mathbb{K}$.

To finish this chapter we remark that the method used here is applicable if B is positively homogeneous of degree α (i.e., if $B(tu) = t^\alpha B(u), \forall u \in \mathbb{K}, \forall t > 0$) and if the remainder ω verifies (1.1) [respectively, (1.2)] with $\|u\|^\beta$ instead of $\|u\|^2$. In particular, one has the following results, the proofs of which are similar to those of Theorems 1.1 and 1.2.

THEOREM A.2. *Assume that the conditions of Theorem* 1.1 *hold with the following modifications: B is positively homogeneous of degree* $\alpha > 0$ (*instead of* $\alpha = 1$) *and*

$$\lim_{\substack{\|u\| \to 0 \\ u \in \mathbb{K}}} (\omega(u), u)/\|u\|^{1+\alpha} = 0$$

instead of (1.1).[8] *Then if* $\alpha > 1$, *they are not positive bifurcation points and* 0 *is a bifurcation point for* Γ (*unique if* $m_{\mathbb{K}}(B) > -\infty$). *If* $\alpha < 1$, *then* $+\infty$ *is a bifurcation point for* Γ (*unique if* $m_{\mathbb{K}}(B) > 0$).

THEOREM A.3. *Assume that the conditions of Theorem* 1.2 *hold with the following modifications: B is positively homogeneous of degree* $\alpha > -1$ (*instead of* $\alpha = 1$) *and*

$$\lim_{\substack{\|u\| \to +\infty \\ u \in \mathbb{K}}} (\omega(u), u)/\|u\|^{1+\alpha} = 0$$

instead of (1.2) *and* $\Phi(u) - [1/(\alpha + 1)](B(u), u)$ *replaces* $\Phi(u) - \frac{1}{2}(B(u), u)$.[8] *Then if* $\alpha < 1$, *they are not positive asymptotic bifurcation points and* 0 *is an asymptotic bifurcation point for* Γ (*unique if* $m_{\mathbb{K}}(B) > -\infty$). *If* $\alpha > 1$, *then* $+\infty$ *is an asymptotic bifurcation point* (*unique if* $m_{\mathbb{K}}(B) > 0$).

These results can be adapted to the case $M_{\mathbb{K}}(B) = 0$ as in Theorem 5.1.

REFERENCES

1. H. AMANN, Existence theorems for equations of Hammerstein type, *Applicable Anal.* **1** (1972), 385–397.

[8] We assume that $M_{\mathbb{K}}(B) > 0$.

2. H. BEIRÃO DA VEIGA, Differentiability for Green's operators of variational inequalities and applications to the calculus of bifurcation points, *Journal of Functional Analysis*, to appear; preprint from the Instituto de Física e Matemática, Lisbon, November 1974.
3. H. BEIRÃO DA VEIGA, Sul più grande punto di biforcazione positivo per una classe di operatori potenziali non differenziabili, *Atti Accad. Naz. Lincei* **59**, S. 8 (1975), 615–619.
4. H. BREZIS AND F. E. BROWDER, Nonlinear integral equations and systems of Hammerstein type, *Advances in Math.* **18** (1975), 113–147.
5. C. DO, Bifurcation theory for elastic plates subjected to unilateral conditions, *J. Math. Anal. Appl.* **60** (1977), 435–448.
6. M. A. KRASNOSEL'SKII, "Topological Methods in the Theory of Nonlinear Integral Equations," Pergamon, Oxford, 1963.
7. J. NAUMANN, On bifurcation buckling of thin elastic shells, *J. Mécanique* **13** (1974), 715–741.
8. F. RIESZ AND B.SZ.-NAGY, "Leçons d'analyse fonctionelle," Akadémiai Kiadó, Budapest, 1953.
9. R. E. L. TURNER, Transversality in nonlinear eigenvalue problems, *in* "Contributions to nonlinear functional analysis," Academic Press, New York, 1971.
10. M. M. VAINBERG, "Variational Methods for the Study of Nonlinear Operators" English edition), Holden–Day, San Francisco, California, 1964.

STUDIES IN ANALYSIS
ADVANCES IN MATHEMATICS SUPPLEMENTARY STUDIES, VOL. 4

Manifolds of C^r Mappings and Applications to Differentiable Dynamical Systems

JOHN FRANKS

Department of Mathematics
Northwestern University
Evanston, Illinois

Contents

INTRODUCTION

This chapter is largely expository, dealing with the properties of the space of r times continuously differentiable functions from one manifold to another and with applications to differentiable dynamical systems.

Our first object is to give an elementary proof that if M and N are finite-dimensional C^∞ manifolds and M is compact, then $C^r(M, N)$, the set of C^r mappings from M to N, has the structure of a C^∞ Banach manifold. This has been done in greater generality by Palais [12]. We also prove that if P is another compact manifold, then the map comp: $C^{r+s}(M, N) \times C^r(P, M) \to C^r(P, N)$ given by $\text{comp}(f, g) = f \circ g$ is a C^s map. This result seems to be well known (cf. Eells [3, p. 780]), but to the best of the author's knowledge an elementary proof is not readily available.

Two final sections give applications of these results to dynamical systems. In Section 5 we give a very simple proof that Anosov diffeomorphisms are structurally stable, which also shows that the function that assigns to each perturbation the appropriate conjugacy is a C^1 function. Section 6 is devoted to an elegant new proof by Irwin [6] of the stable manifold theorem for a fixed point.

ISBN 0-12-599150-9

1. Preliminaries

In this section we give a short summary of some facts about calculus in Banach spaces and differential topology which will be needed in the following sections.

Some familiarity with the concept of the derivative (and higher-order derivatives) of the map from one Banach space to another is assumed. A good general reference for this is by Lang [8].

Let E and F be Banach spaces and let $L_s^k(E, F)$ denote the space of symmetric k-linear maps from the k-fold product $E \times \cdots \times E$ into F.

THEOREM 1.1. *If U is open in E and $f: U \to F$ is C^k, then the kth derivative of f, $D^k f$, is a continuous map from U to $L_s^k(E, F)$.*

Proof. See [8].

THEOREM 1.2 (Taylor's formula). *Let U be open in E and let $f: U \to F$ be C^k. Let $x \in U$ and let $y \in E$ such that the segment $x + ty$, $0 \leqslant t \leqslant 1$, is contained in U. Denote by y^k the k-tuple $(y, y, \ldots, y)$. Then*

$$f(x + y) = f(x) + Df(x)(y) + \frac{1}{2!} D^2 f(x)(y^2) + \cdots$$

$$+ \frac{1}{(k-1)!} D^{k-1} f(x)(y^{k-1}) + R_k,$$

where

$$R_k = \int_0^1 \frac{(1-t)^{k-1}}{(k-1)!} D^k f(x + ty)(y^k)\, dt.$$

Proof. See [8].

As an immediate consequence we have

THEOREM 1.3 (Mean value theorem). *Let U be open in E and let x, $z \in U$ be such that the line segment between x and z is contained in U. Let $f: U \to F$ be C^1. Then*

$$\|f(z) - f(x)\| \leqslant \|z - x\| \sup \|Df(v)\|,$$

the sup *being taken over all v in the line segment joining x and z.*

THEOREM 1.4 (Converse to Taylor's theorem) *Let E and F be Banach spaces, let U be open in E, and suppose $f: U \to F$ satisfies*

$$f(x+y) = \varphi_0(x) + \varphi_1(x)(y) + \frac{1}{2!}\varphi_2(x)(y^2) + \cdots + \frac{1}{k!}\varphi_k(x)(y^k) + o(y^k),$$

where $\varphi_i: U \to L_s{}^i(E, F)$ is continuous for $0 \leqslant i \leqslant k$. Then f is C^k and $D^i f = \varphi_i$.

A very good proof of this is given in [11].

THEOREM 1.5 (Inverse function theorem). *Let U be open in E and let $f: U \to F$ be a C^k map. Suppose $x_0 \in U$ and $Df(x_0): E \to F$ is a continuous linear isomorphism.*

Then there is an open neighborhood V of x_0 in U such that f is a C^k diffeomorphism from V onto an open set in F.

Proof. See [8].

THEOREM 1.6 (Implicit function theorem). *Let E, F, and G be Banach spaces, let U, V be open in E and F, respectively, and let $f: U \times V \to G$ be a C^k mapping. Let $(a, b) \in U \times V$ and assume that $D_2 f(a, b): F \to G$ is a continuous linear isomorphism. Suppose $f(a, b) = 0$. Then there exists a C^k map $g: U_0 \to V$ defined on an open neighborhood U_0 of a such that $g(a) = b$ and*

$$f(x, g(x)) = 0 \qquad \text{for} \quad \text{all } x \in U_0.$$

If U_0 is taken to be a sufficiently small ball, then g is uniquely determined.

Proof. See [8].

THEOREM 1.7. *Let U be open in E and let $\{f_n\}$ be a sequence of C^1 maps of U into F. Assume that $\{f_n\}$ converges pointwise to a map f and also that the sequence of derivatives $\{Df_n\}$ converges uniformly to a mapping $g: U \to L(E, F)$. Then f is C^1 and $Df = g$.*

Proof. See [8].

Familiarity with the basic definitions concerning a Banach manifold and its tangent bundle will be assumed.

A good reference for this is [9] or [2]. We shall also need the following two well-known results of differential topology.

THEOREM 1.8 (Whitney embedding theorem). *Let M be a finite-dimensional C^∞ manifold. Then there exists a Euclidean space R^n and a C^∞ submanifold $M' \subset R^n$ such that M and M' are C^∞ diffeomorphic.*

THEOREM 1.9 (Tubular neighborhood theorem). *Suppose M is an m-dimensional C^∞ submanifold of R^n. Then there exists an open neighborhood U of M in R^n and a C^∞ map $\pi: U \to M$ such that*

(a) *$\pi(x) = x$ for all $x \in M$.*

(b) *If $x \in M$, $\pi^{-1}(x)$ is an open $(n - m)$-disk in R^n centered at x and normal to M.*

2. CONTINUITY OF COMPOSITION

Suppose that E and F are Banach spaces and U is an open set in E.

DEFINITION 2.1. Let $C_b{}^k(U, F)$ denote the set of C^k maps from U to F with the property that $f \in C_b{}^k(U, F)$ implies there is a constant K depending on f such that $\sup_{x \in U} \|D^i f(x)\| < K$ for all $0 \leqslant i \leqslant k$.

That is, $C_b{}^k(U, F)$ is the set of C^k maps from U to F whose first k derivatives are bounded on U.

If $f \in C_b{}^k(U, F)$, we define $\|f\|_k$ to be $\sum_{i=0}^k \sup_{x \in U} \|D^i f(x)\|$.

THEOREM 2.1. *The linear space $C_b{}^k(U, F)$ is a Banach space with $\| \; \|_k$ as norm.*

The proof, which is an easy consequence of Theorem 1.7, is left to the reader.

DEFINITION 2.2. Suppose M is a compact C^∞ manifold. Then a special atlas A on M is a finite set of charts $\{\alpha: U_\alpha \to R^n\}$ such that each α extends to a chart $\alpha: V_\alpha \to R^n$ with $\mathrm{cl}(U_\alpha) \subset V_\alpha$ and such that $\bigcup_{\alpha \in A} U_\alpha = M$.

If M is a compact manifold, $f: M \to F$ is C^k and if $\alpha: U \to R^n$ is a chart in a special atlas on M, then, since $\mathrm{cl}(U)$ is compact, it is easy to see that $f \circ \alpha^{-1} \in C_b{}^k(\alpha(U), F)$.

DEFINITION 2.3. Let $C^k(M, F)$ denote the set of C^k maps from M to F and if A is a special atlas on M, define $\|f\|_k^A = \max_{\alpha \in A} \|f \circ \alpha^{-1}\|_k$. If U is open in F, let $C^k(M, U) = \{f \in C^k(M, F) \mid f(M) \subset U\}$.

THEOREM 2.2. *Suppose M is a compact C^∞ manifold, A a special atlas on M, and F a Banach space. Then $C^k(M, F)$ is a Banach space with $\| \; \|_k^A$*

as norm. If B is another special atlas on M, the norms $\| \|_k^A$ and $\| \|_k^B$ are equivalent.

The proof is not difficult (cf. Theorem 2.3) and is left to the reader. Since, for our purposes, the particular norm (in an equivalence class) is unimportant, we will write $\| \|_k$ for the norm on $C_k(M, F)$ with the special atlas used to construct it being understood.

LEMMA 2.1. *Let U be an open set in a Banach space G and let $E_1, \ldots, E_n, F$ be Banach spaces. Suppose $f_i \in C_b{}^r(U, E_i)$ and $\theta: E_1 \times \cdots \times E_n \to F$ a continuous n-linear map. Then*

$$\|\theta \circ (f_1, f_2, \ldots, f_n)\|_r \leqslant n^r \|\theta\| \, \|f_1\|_r \cdots \|f_n\|_r.$$

The result also holds if M is a compact manifold and $f_i \in C^r(M, E_i)$.

Proof. We first consider the case $f_i \in C_b{}^r(U, E_i)$ and induct on r. For $r = 0$, the result is trivial. From the chain rule and the formula for the derivative of an n-linear map (see Lang [8]) we have

$$D(\theta \circ (f_1, \ldots, f_n)) = \sum_{i=1}^{n} (\theta_i \circ (f_1, \ldots, Df_i, \ldots, f_n)),$$

where $\theta_i: E_1 \times \cdots \times E_{i-1} \times L(G, E_i) \times E_{i+1} \times \cdots \times E_n \to F$ is the n-linear map induced by θ. Note $\|\theta_i\| = \|\theta\|$. Thus

$$\begin{aligned}
\|\theta \circ (f_1, \ldots, f_n)\|_r &= \|\theta \circ (f_1, \ldots, f_n)\|_0 + \|D(\theta \circ (f_1, \ldots, f_n))\|_{r-1} \\
&\leqslant \|\theta\| \, \|f_1\|_0 \cdots \|f_n\|_0 + \sum_{i=1}^{n} \|\theta_i \circ (f_1, \ldots, Df_i, \ldots, f_n)\|_{r-1} \\
&\leqslant \|\theta\| \, \|f_1\|_0 \cdots \|f_n\|_0 \\
&\quad + \sum_{i=1}^{n} n^{r-1} \|\theta\| \, \|f_1\|_{r-1} \cdots \|Df_i\|_{r-1} \cdots \|f_n\|_{r-1} \\
&\qquad \text{(by the induction hypothesis)} \\
&\leqslant n^{r-1} \|\theta\| (\|f_1\|_0 + \|Df_1\|_{r-1}) \|f_2\|_{r-1} \cdots \|f_n\|_{r-1} \\
&\quad + \sum_{i=2}^{n} n^{r-1} \|\theta\| \, \|f_1\|_{r-1} \cdots \|Df_i\|_{r-1} \cdots \|f_n\|_{r-1} \\
&\leqslant n^r \|\theta\| \, \|f_1\|_r \cdots \|f_n\|_r.
\end{aligned}$$

Thus the lemma holds when $f_i \in C_b{}^r(U, E_i)$.

Suppose now that $f_i \in C^r(M, E_i)$. Let A be the special atlas used to define $\| \|_r$ and let $\alpha \in A$. Then

$$\|\theta \circ (f_1 \circ \alpha^{-1}, \ldots, f_n \circ \alpha^{-1})\|_r \leqslant n^r \|\theta\| \, \|f_1 \circ \alpha^{-1}\|_r \cdots \|f_n \circ \alpha^{-1}\|_r .$$

But $\|f_i \circ \alpha^{-1}\|_r \leqslant \|f_i\|_r$ by the definition of $\| \|_r$ on $C^r(M, E_i)$. Hence

$$\|\theta \circ (f_1 \circ \alpha^{-1}, \ldots, f_n \circ \alpha^{-1})\|_r \leqslant n^r \|\theta\| \, \|f_1\|_r \cdots \|f_n\|_r .$$

Since this holds for each $\alpha \in A$,

$$\begin{aligned} \|\theta \circ (f_1, \ldots, f_n)\|_r &= \max_{\alpha \in A} \|\theta \circ (f_1 \circ \alpha^{-1}, \ldots, f_n \circ \alpha^{-1})\|_r \\ &\leqslant n^r \|\theta\| \, \|f_1\|_r \cdots \|f_n\|_r . \end{aligned}$$

Q.E.D.

THEOREM 2.3. *Let M be a compact C^∞ manifold and let E and F be Banach spaces. Suppose U is open in E. Then $C^k(M, U)$ is open in $C^k(M, E)$ and the map $C_b^{\,k}(U, F) \times C^k(M, U) \to C^k(M, F)$ given by composition is continuous. If $k = 0$, then M can be any compact space.*

Proof. Since M is compact and U is open, if $g \in C^k(M, U)$, there exists $\rho > 0$ such that $\|g - g'\|_k < \rho$ implies $g' \in C^k(M, U)$. Hence $C^k(M, U)$ is open.

We prove the rest of the theorem by induction on k. First consider $k = 0$ with M any compact space. We shall show continuity at (f, g). Suppose $\epsilon > 0$ is given. If $y \in g(M)$, there exists δ_y such that if $z, z' \in B(y, \delta_y)$, the ball of radius δ_y about y in E, then $|f(z) - f(z')| < \frac{1}{4}\epsilon$. The balls $\{B(y, \frac{1}{2}\delta_y) \mid y \in g(M)\}$ cover $g(M)$ and, since $g(M)$ is compact, there is a finite subcover, say the balls centered at $y_1, y_2, \ldots, y_n$. Let $\delta = \min\{\delta_{y_i}\}$.

Now suppose $\|f - f'\|_0 < \frac{1}{2}\epsilon$ and $\|g - g'\|_0 < \frac{1}{2}\delta$. Then

$$|f(g(x)) - f'(g'(x))| \leqslant |f(g(x)) - f(g'(x))| + |f(g'(x)) - f'(g'(x))|,$$

and $|f(g'(x)) - f'(g'(x))| < \frac{1}{2}\epsilon$. Also, since $|g(x) - g(x')| < \frac{1}{2}\delta$, there is a y_i with $|g(x) - y_i| < \frac{1}{2}\delta_{y_i}$ and $|g'(x) - y_i| < \frac{1}{2}\delta_{y_i} + \frac{1}{2}\delta \leqslant \delta_{y_i}$, so $|f(g(x)) - f(g'(x))| < \frac{1}{4}\epsilon$. Hence $|f(g(x)) - f'(g'(x))| < \epsilon$. Since this hold for all $x \in M$,

$$\|f \circ g - f' \circ g'\|_0 < \epsilon$$

when $\|f - f'\|_0 < \frac{1}{2}\epsilon$ and $\|g - g'\|_0 < \delta$.

We have thus proved the theorem in the case $k = 0$; we continue by inducting on k. Let A be the special atlas used to define $\| \|_k$ on $C^k(M, E)$, and let $\alpha: U \to V \subset R^n$ be a chart in A. If g, $g' \in C^k(M, E)$, let $h = g \circ \alpha^{-1}$ and $h' = g' \circ \alpha^{-1}$. We shall show continuity at (f, g).

Let $B: L(E, F) \times L(R^n, E) \to L(R^n, F)$ be the bilinear map given by composition, so $D(f \circ h) = B \circ (Df \circ h, Dh)$ and $\|B\| = 1$. Then

$$\begin{aligned}
\|f \circ h - f' \circ h'\|_k &= \|f \circ h - f' \circ h'\|_0 + \|D(f \circ h) - D(f' \circ h')\|_{k-1} \\
&= \|f \circ h - f' \circ h'\|_0 \\
&\quad + \|B \circ (Df \circ h, Dh) - B \circ (Df' \circ h', Dh')\|_{k-1} \\
&\leqslant \|f \circ h - f' \circ h'\|_0 \\
&\quad + \|B \circ (Df \circ h, Dh) - B \circ (Df \circ h, Dh')\|_{k-1} \\
&\quad + \|B \circ (Df \circ h, Dh') - B \circ (Df' \circ h', Dh')\|_{k-1} \\
&\leqslant \|f \circ h - f' \circ h'\|_0 + \|B \circ (Df \circ h - Df' \circ h', Dh')\|_{k-1} \\
&\quad + \|B \circ (Df \circ h, Dh - Dh')\|_{k-1} \\
&\leqslant \|f \circ h - f' \circ h'\|_0 + 2^{k-1}\|Df \circ h - Df' \circ h'\|_{k-1}\|Dh'\|_{k-1} \\
&\quad + 2^{k-1}\|Df \circ h\|_{k-1}\|Dh - Dh'\|_{k-1} \qquad \text{(by Lemma 2.6)} \\
&\leqslant \|f \circ g - f' \circ g'\|_0 + 2^{k-1}\|Df \circ g - Df' \circ g'\|_{k-1}\|Dg'\|_{k-1} \\
&\quad + 2^{k-1}\|Df \circ g\|_{k-1}\|Dg - Dg'\|_{k-1}.
\end{aligned}$$

Let $P = 2^{k-1}\|Df \circ g\|_{k-1}$ and let $\delta_1 = \min(\epsilon/P, 1)$. Let

$$Q = 2^{k-1}(\|Dg\|_{k-1} + 1).$$

By the induction hypothesis there exists $\delta_2 > 0$ such that $\|Df - Df'\|_{k-1}$, $\|g - g'\|_{k-1} < \delta_2$ implies $\|Df \circ g - Df' \circ g'\|_{k-1} < \epsilon/Q$ and $\|f \circ g - f' \circ g'\|_0 < \epsilon$.

Hence if $\delta = \min(\delta_1, \delta_2)$ and $\|f - f'\|_k < \delta$ and $\|g - g'\|_k < \delta$, then

$$\begin{aligned}
\|f \circ g \circ \alpha^{-1} - f' \circ g' \circ \alpha^{-1}\|_k &= \|f \circ h - f' \circ h'\|_k \\
&\leqslant \epsilon + 2^{k-1}(\epsilon/P)\|Df \circ g\|_{k-1} \\
&\quad + 2^{k-1}(\epsilon/Q)\|Dg'\|_{k-1} \\
&\leqslant 3\epsilon.
\end{aligned}$$

Since this holds for any chart α we have $\|f \circ g - f' \circ g'\|_k \leqslant 3\epsilon$. Q.E.D.

3. Differentiability of Composition

Lemma 3.1. *If $F: [a, b] \to C^r(M, F)$ is continuous, then*

$$\left\|\int_a^b F(t)\,dt\right\| \leqslant \int_a^b \|F(t)\|_r\,dt \qquad \text{and} \qquad \left(\int_a^b F(t)\,dt\right)(x) = \int_a^b F(t)(x)\,dt.$$

Proof. Both results hold trivially for step functions from $[a,b]$ to $C^r(M,F)$. Since every continuous function is the uniform limit of step functions, the desired result follows from taking limits.

LEMMA 3.2. *There exist continuous functions* $C_r: C^r(M,U) \to R$ *such that if* $f \in C_b{}^r(U,F)$, *then* $\|f \circ g\|_r \leqslant C_r(g)\|f\|_r$ *and if* $f \in C_b^{r+1}(U,F)$, *then* $\|f \circ g - f \circ g'\|_r \leqslant B_r(g)\|f\|_{r+1}\|g - g'\|_r$ *if* $\|g - g'\|_r$ *is sufficiently small and*

$$B_r(g) = 2^r(C_r(g) + 1).$$

Proof. We prove the first assertion by induction on r. For $r = 0$ we have $\|f \circ g\|_0$ so $C_0(g) \equiv 1$. Suppose we have constructed $C_0, C_1, \ldots, C_{r-1}$. Suppose $\alpha: V \to R^n$ is a chart in the special atlas used to define $\| \;\|_r$. Let $\hat{g} = g \circ \alpha^{-1}$ and let $\theta: L(E,F) \times L(R^n,E) \to L(R^n,F)$ be the bilinear map given by composition. Then

$$\begin{aligned}
\|f \circ \hat{g}\|_r &= \|f \circ \hat{g}\|_0 + \|D(f \circ \hat{g})\|_{r-1} \\
&\leqslant \|f\|_0 + \|\theta \circ (Df \circ \hat{g}, D\hat{g})\|_{r-1} \\
&\leqslant \|f\|_0 + 2^{r-1}\|Df \circ \hat{g}\|_{r-1}\|D\hat{g}\|_{r-1} \qquad \text{(by Lemma 2.6)}, \\
&\leqslant \|f\|_0 + 2^{r-1}\|Df \circ g\|_{r-1}\|Dg\|_{r-1} \\
&\leqslant \|f\|_0 + 2^{r-1}C_{r-1}(g)\|Df\|_{r-1}\|Dg\|_{r-1} \\
&\leqslant C_r(g)\|f\|_r \qquad \text{if} \quad C_r(g) = 2^{r-1}C_{r-1}(g)\|Dg\|_{r-1} + 1.
\end{aligned}$$

Since this holds for any special chart α, then $\|f \circ g\|_r \leqslant C_r(g)\|f\|_r$.

To prove the remainder of the lemma we let $g' \in C^r(M,U)$, $\hat{g}' = g' \circ \alpha^{-1}$, $h = g' - g$, and $\hat{h} = \hat{g}' - \hat{g}$. By Taylor's Theorem 1.2 with $k = 1$,

$$f(\hat{g}'(x)) - f(\hat{g}(x)) = \int_0^1 Df(\hat{g}(x) + t\hat{h}(x))\, dt(\hat{h}(x)).$$

So by Lemma 3.1,

$$f \circ \hat{g}' - f \circ g = \theta \circ \left(\int_0^1 Df \circ (\hat{g} + t\hat{h})\, dt, \hat{h} \right).$$

Hence

$$\begin{aligned}
\|f \circ \hat{g}' - f \circ \hat{g}\|_r &\leqslant 2^r \left\| \int_0^1 Df \circ (\hat{g} + t\hat{h})\, dt \right\|_r \|\hat{h}\|_r \\
&\leqslant 2^r \left(\int_0^1 \|Df \circ (g + th)\|_r\, dt \right) \|h\|_r \\
&\leqslant 2^r(C_r(g) + 1)\|f\|_{r+1}\|h\|_r
\end{aligned}$$

if $\|h\|_r$ is sufficiently small that $C_r(g + th) \leqslant C_r(g) + 1$ for all $t \in [0,1]$.

Since this holds for any chart α, then $\|f \circ g' - f \circ g\|_r \leqslant B_r(g)\|f\|_{r+1}\|g - g'\|_r$ whenever $\|g - g'\|_r$ is sufficiently small. Q.E.D.

THEOREM 3.1. *Let M be a compact C^∞ manifold, let E, F be Banach spaces, and let U be open in E. Then the map*

$$\mathrm{comp}: C_b^{r+s}(U, F) \times C^r(M, U) \to C^r(M, F)$$

given by $\mathrm{comp}(f, g) = f \circ g$ *is a C^s map. If $r = 0$, then M can be any compact topological space.*

Proof. We will use the converse to Taylor's Theorem 1.4. By Taylor's Theorem 1.3,

$$\begin{aligned}(f + \gamma) \circ (g + h)(x) &= f(g(x) + h(x)) + \gamma(g(x) + h(x)) \\ &= \sum_{k=0}^{s} \frac{1}{k!} D^k f(g(x))(h(x))^k + R_1(f, g, h)(x)(h(x))^s \\ &\quad + \sum_{k=0}^{s} \frac{1}{k!} D^k \gamma(g(x))(h(x))^k + R_2(\gamma, g, h)(x)(h(x))^s,\end{aligned}$$

where

$$R_1(f, g, h) = \int_0^1 \frac{(1 - t)^{s-1}}{(s - 1)!} (D^s f \circ (g + th) - D^s f \circ g)\, dt$$

and

$$R_2(\gamma, g, h) = \int_0^1 \frac{(1 - t)^{s-1}}{(s - 1)!} (D^s \gamma \circ (g + th) - D^s \gamma \circ g)\, dt.$$

Note by Theorem (2.3) and Lemma 3.1 that $\|R_1(f, g, h)\|_r$ and $\|R_2(\gamma, g, h)\|_r$ tend to 0 as $\|h\|_r$ tends to 0.

Now let $\theta_k: L_s^k(E; F) \times E \times \cdots \times E \to F$ be the $k + 1$ linear map defined by $\theta_k(w, v_1, \ldots, v_k) = w(v_1, \ldots, v_k)$, so $\|\theta_k\| = 1$. Then

$$\begin{aligned}(f + \gamma) \circ (g + h) &= \sum_0^s \frac{1}{k!} \theta_k \circ (Df^k \circ g, h^k) + \sum_0^s \frac{1}{k!} \theta_k \circ (D\gamma^k \circ g, h^k) \\ &\quad + \theta_s \circ (R_1(f, g, h), h^s) + \theta_s \circ (R_2(\gamma, g, h), h^s).\end{aligned}$$

Let $G = C_b^{r+s}(U, F) \times C^r(M, E)$ with norm given by

$$|(\gamma, h)| = \max(\|\gamma\|_{r+s}, \|h\|_r).$$

Then

$$(f + \gamma) \circ (g + h) = \sum_{k=0}^{s} \frac{1}{k!} \varphi_k(f, g)(\gamma, h)^k + \rho(f, g, \gamma, h),$$

where $\varphi_k: G \to L_s{}^k(G, C^r(M, F))$ is given by

$$\begin{aligned}\varphi_k(f, g)(\gamma_1, h_1)(\gamma_2, h_2) \cdots (\gamma_k, h_k) \\ = \theta_k \circ (D^k f \circ g, h_1, h_2, \ldots, h_k) \\ + \sum_{j=1}^{k} \theta_{k-1} \circ (D^{k-1}\gamma_j \circ g, h_1, \ldots, h_{j-1}, h_{j+1}, \ldots, h_k)\end{aligned}$$

if $k \geqslant 1$ and $\varphi_0(f, g) = f \circ g$; and

$$\rho(f, g, \gamma, h) = \theta_s \circ (R_1(f, g, h), h^s) + \theta_s \circ (R_2(\gamma, g, h), h^s) + \frac{1}{s!} \theta_s \circ (D^s\gamma, h^s).$$

We must show that φ_k is continuous and $\|\rho(f, g, \gamma, h)\|_r = o(|(\gamma, h)|^s)$. However,

$$\|\rho(f, g, \gamma, h)\|_r \leqslant (s + 1)^r |\theta_s| \, \|h\|_r^s (\|R_1\|_r + \|R_2\|_r + \|D^s\gamma\|_r)$$

by Lemma 2.1. Since $\|R_1\|_r$ and $\|R_2\|_r$ tend to zero as $\|h\|_r$ tends to zero and $\|D^s\gamma\|_r$ tends to zero as $\|\gamma\|_{r+s}$ tends to zero, it is in fact the case that $\|\rho(f, g, \gamma, h)\|_r$ is $o(|(\gamma, h)|^s)$.

Now to show φ_k is continuous we note

$$\begin{aligned}(\varphi_k(f, g) - \varphi_k(f', g'))(\gamma_1, h_1), \ldots, (\gamma_k, h_k) \\ = \theta_k \circ (D^k f \circ g - D^k f' \circ g', h_1, \ldots, h_k) \\ + \sum_{j=1}^{k} \theta_{k-1} \circ (D^{k-1}\gamma_j \circ g - D^{k-1}\gamma_j \circ g', h_1, \ldots, h_{j-1}, h_{j+1}, \ldots, h_k) \\ \leqslant (k + 1)^r \|D^k f \circ g - D^k f' \circ g'\|_r \|h_1\|_r \cdots \|h_k\|_r \\ + \sum_{j=1}^{k} k^r \|D^{k-1}\gamma_j \circ g - D^{k-1}\gamma_j \circ g'\|_r \|h_1\|_r \cdots \|h_{j-1}\|_r \|h_{j+1}\|_r \cdots \|h_k\|_r\end{aligned}$$

(by Lemma 2.1),

$$\begin{aligned}\leqslant (k + 1)^r \|D^k f \circ g - D^k f' \circ g'\|_r \|h_1\|_r \cdots \|h_k\|_r \\ + \sum_{j=1}^{k} k^r B_k(g) \|g - g'\|_r (\|h_1\|_r \cdots \|h_{j-1}\|_r \|D^k\gamma_j\|_r \|h_{j+1}\|_r \cdots \|h_k\|_r)\end{aligned}$$

(by Lemma 3.2, where

$B_k(g) = 2^k(C_k(g) + 1)$ and $\|g - g'\|_r$ is small)

$$\leqslant C(\|D^k f \circ g - D^k f' \circ g'\|_r + \|g - g'\|_r)|(\gamma_1, h_1)|\,|(\gamma_2, h_2)| \cdots |(\gamma_k, h_k)|$$

for some large constant C depending on g. However, $\|D^k f \circ g - D^k f' \circ g'\|_r$ tends to zero when $\|f - f'\|_{r+s}$ and $\|g - g'\|_r$ tends to zero by Theorem 2.3. So φ_k is continuous. Q.E.D.

COROLLARY 3.1. $((D\,\mathrm{comp}(f,g))(\gamma,h))(x) = Df(g(x))(h(x)) + \gamma(g(x))$.

Proof.

$$(D\,\mathrm{comp}(f,g))(\gamma,h) = (\varphi_1(f,g))(\gamma,h) = \theta_1 \circ (Df \circ g, h) + \gamma \circ g$$

Evaluating at x gives

$$((D\,\mathrm{comp}(f,g))(\gamma,h))(x) = Df(g(x))(h(x)) + \gamma(g(x)). \qquad \text{Q.E.D.}$$

Suppose $f \in C_b^{r+s}(U,F)$; we define $\omega_f : C^r(M,U) \to C^r(M,F)$ by $\omega_f(g) = f \circ g$.

COROLLARY 3.2. *ω_f is C^s and*

$$((D\omega_f(g))(h))(x) = Df(g(x))(h(x)).$$

This is an immediate consequence of Theorem 3.1 and Corollary 3.1.

DEFINITION 3.1. Let $UC_b^r(E,F)$ denote the set of $f \in C_b^r(E,F)$ such that $D^i f : E \to L_s^i(E,F)$ is uniformly continuous for all $0 \leqslant i \leqslant r$.

Note that $UC_b^r(E,F)$ is a closed subspace of $C_b^r(E,F)$. Irwin [7] proves the following theorem which is quite similar to Theorem 3 1 and could be proven in a very similar way.

THEOREM 3.2. *If E, F, and G are Banach spaces, the map*

$$\mathrm{comp} : UC_b^{r+s}(F,G) \times C_b^r(E,F) \to C_b^r(E,G)$$

is a C^s map.

4. MANIFOLDS OF C^r MAPPINGS

In this section we show that the set of C^∞ mappings from a compact C^∞ manifold M to a finite-dimensional C^∞ manifold N has the structure of a Banach manifold. This result was originally proved by Palais (see Abraham [1] or Palais [12]). Our proof differs somewhat from his.

Definition 4.1. If M is a compact C^∞ manifold and N a C^∞ submanifold of R^m, let $C^r(M, N) = \{f \in C^r(M, R^m) | f(M) \subset N\}$.

Theorem 4.1. *If M is a compact C^∞ manifold and N a C^∞ submanifold of R^m, then $C^r(M, N)$ is a C^∞ submanifold of $C^r(M, R^m)$. If $r = 0$, M can be any compact topological space.*

Proof. Let $\pi: U \to N$ be a tubular neighborhood of N in R^m. Let $TN = \{(y, v) \in R^{2m} | y \in N, v \in TN_y \subset R^m\}$ be the tangent bundle of N and let $N^\perp = \{(y, v) \in R^{2m} | y \in N, v \in TN_y{}^\perp \subset R^m\}$ be the normal bundle of N. Both TN and $N^\perp$ are C^∞ subbundles of the trivial bundle $N \times R^m$. Let $P: N \times R^m \to R^m$ be given by $P(y, v)$, the orthogonal projection of v on TN_y, and let $P^\perp: N \times R^m \to R^m$ be given by $P^\perp(y, v) = v - P(y, v)$. Both P and $P^\perp$ are C^∞ mappings.

Now choose a fixed $g \in C^r(M, N)$. Let $E = C^r(M, R^m)$, let $E^t = \{h \in E | (g(x), h(x)) \in TN \text{ for all } x \in M\}$, and let $E^n = \{h \in E | (g(x), h(x)) \in N^\perp \text{ for all } x \in M\}$. Clearly E^t and E^n are closed subspaces of E and $E^t \cap E^n = 0$. However, if $h \in E$, then $h = h^t + h^n$ where $h^t \in E^t$ is given by $h^t(x) = P(g(x), h(x))$ and $h^n \in E^n$ is given by $h^n(x) = P^\perp(g(x), h(x))$. Hence $E = E^t \oplus E^n$.

Let B^t and B^n be balls centered at 0 in E^t and E^n, respectively, which are sufficiently small that if $h_1 \in B^t$ and $h_2 \in B^n$, then $g + h_1 + h_2 \in C^r(M, U)$. Let $B = B^t \times B^n$ and define $\alpha: B \to E$ by $\alpha(h_1, h_2) = \omega_\pi(g + h_1) + h_2$. Clearly $\alpha(0, 0) = g$ and $\alpha^{-1}(C^r(M, N)) = B^t \times 0$.

Then $\pi: U \to N$ is C^∞, so by Corollary 3.2, ω_π is C^∞ and hence α is C^∞. Also $(D\alpha(0, 0))(\gamma_1, \gamma_2) = D\omega_\pi(g))(\gamma_1) + \gamma_2$, so $((D\alpha(0, 0))(\gamma_1, \gamma_2))(x) = (D\pi(g(x)))(\gamma_1(x)) + \gamma_2(x)$. However, $\pi | N$ is the identity, so $D\pi(y) | TN_y$ is the identity for all y. Hence $(D\alpha(0, 0))(\gamma_1, \gamma_2) = \gamma_1 + \gamma_2$ and $D\alpha(0, 0): E^t \times E^n \to E$ is the obvious isomorphism.

Thus by the Inverse function theorem 1.5 there are smaller balls centered at 0, $B_0{}^t \subset B^t$ and $B_0{}^n \subset B^n$, such that if $B_0 = B_0{}^t \times B_0{}^n$, then $\alpha | B_0: B_0 \to \alpha(B_0)$ is a C^∞ diffeomorphism. Let $V = \alpha(B_0)$ and $\varphi = (\alpha | B_0)^{-1}$. Then (φ, V) is a C^∞ chart on $C^r(M, U)$ and $\varphi(V \cap C^r(M, N)) = B_0{}^t \times 0$. Since we can construct such a chart for any $g \in C^r(M, N)$, it follows that $C^r(M, N)$ is a C^∞ submanifold of $C^r(M, R^m)$. Q.E.D.

Remark. Since $\alpha | B_0{}^t \times 0: B_0{}^t \to C^r(M, N) \cap V$ is a diffeomorphism, we can identify $T(C^r(M, N))_g$ with $D\alpha(0, 0)(T(B_0{}^t)_0)$. However, $T(B_0{}^t)_0 = E^t$ and $D\alpha(0, 0)(E^t) = E^t$.

Definition 4.2. Let $\Gamma_g{}^r = \{\gamma \in C^r(M, TN) | \gamma(x) \in TN_{g(x)}\}$. Then $\Gamma_g{}^r$ is called the space of sections over g. It is a Banachable space which is naturally isomorphic to $E^t = \{\gamma \in C^r(M, R^{2m}) | (g(x), \gamma(x)) \in TN\}$. Henceforth we shall generally identify $T(C^r(M, N))_g$ with $\Gamma_g{}^r$.

PROPOSITION 4.1. *If M is a compact C^∞ manifold and N_1, N_2 are C^∞ diffeomorphic submanifolds of R^m and R^p, respectively, then $C^r(M, N_1)$ and $C^r(M, N_2)$ are C^∞ diffeomorphic. If $r = 0$, then M can be any compact space.*

Proof. Let $\varphi: N_1 \to N_2$ be a C^∞ diffeomorphism and let $\pi: U \to N_1$ be a tubular neighborhood of N_1 in R^m. Let $\psi: U \to N_2$ be $\varphi \circ \pi$. Then $\omega_\psi: C^r(M, U) \to C^r(M, N_2)$ is C^∞ by Theorem 3.2.

Now $\omega_\psi | C^r(M, N_1) = \omega_\varphi: C^r(M, N_1) \to C^r(M, N_2)$, so ω_φ is C^∞. Similarly $\omega_{\varphi^{-1}} = (\omega_\varphi)^{-1}$ is C^∞ so ω_φ is a C^∞ diffeomorphism from $C^r(M, N_1)$ to $C^r(M, N_2)$. Q.E.D.

A consequence of this proposition is the following.

COROLLARY 4.1. *If M is a compact C^∞ manifold and N a finite-dimensional C^∞ manifold, then there is a unique C^∞ manifold structure on $C^r(M, N)$ such that $\omega_h: C^r(M, N) \to C^r(M, R^m)$ is a C^∞ embedding when $h: N \to R^m$ is a C^∞ embedding. If $r = 0$, M can be any compact space.*

Proof. By the Whitney embedding theorem there is a C^∞ embedding $\varphi: N \to R^n$ for some R^n. We define a C^∞ manifold structure on $C^r(M, N)$ by asserting $\omega_\varphi: C^r(M, N) \to C^r(M, \varphi(N))$ is a C^∞ diffeomorphism. Uniqueness follows from Proposition 4.1.

Remark. We generally identify $T(C^r(M, N))_g$ with $\Gamma_g{}^r$ (see Definition 4.2). Note that the definition of $\Gamma_g{}^r$ is independent of any embedding of N.

LEMMA 4.1. *Let N be a compact C^∞ submanifold of R^m, $\pi: U \to N$ a tubular neighborhood, and E a Banach space. Then $\alpha_\pi: C^r(N, E) \to C_b{}^r(U, E)$ defined by $\alpha_\pi(f) = f \circ \pi$ is a continuous linear map.*

Proof. Clearly α_π is linear so it suffices to check continuity at $0 \in C^r(N, E)$. This can be done by a simple computation similar to Lemma 3.2.

THEOREM 4.2. *Suppose M, N, and P are finite-dimensional C^∞ manifolds with M and N compact. Then the map* comp$: C^{r+s}(N, P) \times C^r(M, N) \to C^r(M, P)$ *given by* comp$(f, g) = f \circ g$ *is C^s. If $r = 0$, then M can be any compact space.*

Proof. We may assume that N is a submanifold of R_n and P a submanifold of R^p. Let $\pi: U \to N$ be a tubular neighborhood. The map that the theorem asserts is C^s can be written as the composition

$$C^{r+s}(N, P) \times C^r(M, N) \xrightarrow{(\alpha_\pi, i)} C^{r+s}(U, R^p) \times C^r(M, U) \xrightarrow{\text{comp}} C^r(M, R^p),$$

where $i: C^r(M,N) \to C^r(M,U)$ is the inclusion and hence C^∞, $\alpha_\pi: C^{r+s}(N,P) \to C^{r+s}(U,R^p)$ is C^∞ by Lemma 4.1, and $\text{comp}: C^{r+s}(U,R^p) \times C^r(M,U) \to C^r(M,R^p)$ is C^s by Theorem 3.1. Q.E.D.

Recall that we identify the tangent space to $C^r(M,N)$ at g with $\Gamma_g{}^r$ and $T(C^{r+s}(N,P))_f$ with Γ_f^{r+s}. So D comp(f,g) should be a map from $\Gamma_f^{r+s} \times \Gamma_g^r$ to $\Gamma_{f\circ g}^r$. We can give an explicit formula for it.

COROLLARY 4.2. *$D\,\text{comp}(f,g): \Gamma_f^{r+s} \times \Gamma_g{}^r \to \Gamma_{f\circ g}^r$ is given by*

$$((D\,\text{comp}(f,g))(\gamma_1,\gamma_2))(x) = Df(g(x))(\gamma_2(x)) + \gamma_1(g(x)),$$

where $\gamma_1 \in \Gamma_f^{r+s}$ and $\gamma_2 \in \Gamma_g{}^r$.

Proof. This follows immediately from the proof of the previous theorem and Corollary 3.1.

Suppose $f \in C^{r+s}(N,P)$; we define $\omega_f: C^r(M,N) \to C^r(M,P)$ by $\omega_f(g) = f \circ g$.

COROLLARY 4.3. *ω_f is C^s and $D\omega_f(g): \Gamma_g{}^r \to \Gamma_{f\circ g}^r$ is given by $((D\omega_f(g)) \times (\gamma))(x) = Df(g(x))(\gamma(x))$, where $\gamma \in \Gamma_g{}^r$.*

Proof. This is an immediate consequence of Corollary 4.2.

PROPOSITION 4.2. *Let $g \in C^r(M,N)$ and define $\alpha_g: C^r(N,P) \to C^r(M,P)$ by $\alpha_g(f) = f \circ g$. Then α_g is C^∞ and $D\alpha_g(f): \Gamma_f{}^r \to \Gamma_{f\circ g}^r$ is given by*

$$(D\alpha_g(f))(\gamma) = \gamma \circ g, \qquad \text{where} \quad \gamma \in \Gamma_f{}^r.$$

Proof. Suppose N is a submanifold of R^n, P a submanifold of R^p, and $\pi: U \to N$ a tubular neighborhood of N in R^n. The map $\alpha_g: C^r(N,P) \to C^r(M,P)$ can be written as the composition

$$C^r(N,P) \xrightarrow{i} C^r(N,R^p) \xrightarrow{\alpha_\pi} C^r(U,R^p) \xrightarrow{\alpha_g} C^r(M,R^p),$$

where i is the inclusion and hence C^∞, α_π is continuous and linear by Lemma 4.1, and α_g is a linear map which is continuous as a consequence of the first part of Lemma 3.2.

Hence $\alpha_g: C^r(N,P) \to C^r(M,P)$ is C^∞. Then $D\alpha_g(f) = D_1\,\text{comp}(f,g)$, so by Corollary 4.2, $(D\alpha_g(f))(\gamma) = \gamma \circ g$.

LEMMA 4.2. *If $f \in C^1(M,M)$ is a diffeomorphism and $F: C^0(M,M) \to C^0(M,M)$ is given by $F(h) = f \circ h \circ f^{-1}$, then F is C^1 and $DF(id): \Gamma_{id}^0 \to \Gamma_{id}^0$ is given by $DF(id)(\gamma) = Df \circ \gamma \circ f^{-1}$.*

Proof. This is immediate from Corollary 4.3 and Proposition 4.2 when one notes $F = \omega_f \circ \alpha_{f^{-1}}$.

DEFINITION 4.3. The map $f^{\#}: \Gamma_{id}^{0} \to \Gamma_{id}^{0}$ given by $f^{\#}(\gamma) = Df \circ \gamma \circ f^{-1}$ is called the adjoint map of f.

We have shown in Lemma 4.2 that $f^{\#} = DF(id)$.

5. ANOSOV DIFFEOMORPHISMS

DEFINITION 5.1. If E is a Banach space and $L: E \to E$ is a continuous isomorphism, then L is called *hyperbolic* if its spectrum is disjoint from the set of complex numbers of absolute value one.

DEFINITION 5.2. A C^1 diffeomorphism $f: M \to M$ with M compact is called an Anosov diffeomorphism if $f^{\#}: \Gamma_{id}^{0} \to \Gamma_{id}^{0}$ is hyperbolic. Recall $f^{\#}(\gamma) = Df \circ \gamma \circ f^{-1}$.

For a different definition, examples, and discussion of the properties of Anosov diffeomorphisms see Franks [4] or Smale [13]. Mather [10] proves that the previous definition is equivalent to the definition in Franks [4] or Smale [13].

LEMMA 5.1. *If f is an Anosov diffeomorphism, then f is expansive. That is, there exists a number $\epsilon > 0$ such that x, $y \in M$ and $x \neq y$ implies $d(f^n(x), f^n(y)) > \epsilon$ for some integer n, where $d(\ ,\ )$ is a metric on M. The number ϵ is called an expansive constant for f.*

Proof. See (8.7) of [13].

THEOREM 5.1. (Structural stability of Anosov diffeomorphisms). *If f: $M \to M$ is a C^1 Anosov diffeomorphism, then there is a neighborhood N of f in $C^1(M, M)$ and a C^1 function $\varphi: N \to C^0(M, M)$ such that*

(1) *$\varphi(f) = id$ and $\varphi(g)$ is a homeomorphism of M for each g in N.*
(2) *$g \circ \varphi(g) = \varphi(g) \circ f$ for each $g \in N$.*

Proof. Choose a chart $\psi: U \to E$ on $C^0(M, M)$ such that $id \in U$ and $\psi(id) = o \in E$; E a Banach space. Choose an open neighborhood V of id in $C^0(M, M)$ and an open neighborhood N_1 of f in $C^1(M, M)$ such that $(g, h) \in N_1 \times V$ implies $g \circ h \circ f^{-1} \in U$. This can be done by the continuity of composition.

Now define $H: N_1 \times V \to E$ by $H(g, h) = \psi(g \circ h \circ f^{-1}) - \psi(h)$. The map $(g, h) \to g \circ h \circ f^{-1}$ is C^1 because it can be written as the composition of comp and $(\mathrm{id}, \alpha_{f^{-1}})$. comp: $N_1 \times V \to C^0(M, M)$ is C^1 by Corollary 4.3 and $\alpha_{f^{-1}}$ is C^∞ by Proposition 4.2. Thus since ψ is C^∞ we know H is C^1.

We note that $H(f, h) = \psi(F(h)) - \psi(h)$, where $F(h) = f \circ h \circ f^{-1}$, so

$$D_2H(f, id) = D\psi(id) \circ DF(id) - D\psi(id) = D\psi(id) \circ (f^{\#} - I)$$

by Lemma 4.2, where $I: \Gamma_{id}^0 \to \Gamma_{id}^0$ is the identity. Since f is Anosov, 1 is not in the spectrum of $f^{\#}$ so $(f^{\#} - I)$ is an isomorphism.

Thus by the Implicit function theorem there is a neighborhood N_2 of f in N_1 and a C^1 function $\varphi: N_2 \to V$ such that $H(g, \varphi(g)) = 0$ for all $g \in N_2$. Thus $\psi(g \circ \varphi(g) \circ f^{-1}) = \psi(\varphi(g))$ so $g \circ \varphi(g) \circ f^{-1} = \varphi(g)$ and $g \circ \varphi(g) = \varphi(g) \circ f$.

From this it follows easily that $g^n \circ \varphi(g) = \varphi(g) \circ f^n$ for any integer n, so if $x, y \in M$ and $\varphi(g)(x) = \varphi(g)(y)$, then $\varphi(g)(f^n(x)) = \varphi(g)(f^n(y))$.

Now choose an open neighborhood N of f in N_2 sufficiently small that if $g \in N, \varphi(g)$ is homotopic to id and $d(\varphi(g)(x), x) < \frac{1}{2}\epsilon$ for all $x \in M$, where ϵ is an expansive constant for f.

Suppose that $x, y \in M$, $g \in N$, and $\varphi(g)(x) = \varphi(g)(y)$. Then

$$\begin{aligned} d(f^n(x), f^n(y)) &\leqslant d(f^n(x), \varphi(g)(f^n(x))) + d(\varphi(g)(f^n(x)), \varphi(g)(f^n(y))) \\ &\quad + d(\varphi(g)(f^n(y)), f^n(y)) \\ &\leqslant \tfrac{1}{2}\epsilon + \tfrac{1}{2}\epsilon = \epsilon. \end{aligned}$$

Hence $x = y$, so $\varphi(g)$ is injective. Since $\varphi(g)$ is homotopic to id, it is surjective and since M is compact, $\varphi(g)$ must be a homeomorphism. Q.E.D.

6. The Stable Manifold Theorem

Suppose that $E = E_1 \times E_2$ is a Banach space and $T = T_1 \times T_2$ a continuous linear isomorphism of E, where $T_1: E_1 \to E_1$ and $T_2: E_2 \to E_2$ satisfy $\max\{\|T_1\|, \|T_2^{-1}\|\} < \tau < 1$. The object of this section is to show that if T is perturbed slightly, then the points of E that stay near 0 under positive iteration of the perturbed map form a submanifold of E.

We assume the norm on E is given by $\|(x_1, x_2)\| = \max\{\|x_1\|, \|x_2\|\}$. We shall need the following lemma from Hirsch and Pugh [5].

LEMMA 6.1. *Let B be an open ball centered at 0 in E and let $f: B \to E$ be a C^1 map which satisfies $\sup_{z \in B} \|Df(z) - T\| < \epsilon < (1 - \tau)/(1 + \tau)$. Suppose $z = (x, y)$ and $z' = (x, y')$ are elements of B such that $f^n(z) \in B$ and $f^n(z') \in B$ for all $n \geqslant 0$. Then $y = y'$.*

Proof. First note that an application of the Mean value theorem shows that for any $v, w \in B$

$$\|(f - T)(v - w)\| \leqslant \epsilon\|v - w\|.$$

Let $f_i: E \to E_i$, $i = 1, 2$, denote f followed by projection onto E_i. Similarly, if $v \in E$, let v_i, $i = 1, 2$, denote the component of v in E_i.

Suppose now that $v, w \in B$ and $\|v_2 - w_2\| \geqslant \|v_1 - w_1\|$. Then

$$\begin{aligned} \|f_2(v) - f_2(w)\| &= \|T_2(v - 2) + (f_2 - T_2)(v) + (f_2 - T_2)(w)\| \\ &\geqslant \|T_2(v_2 - w_2)\| - \epsilon\|v - w\| \\ &\geqslant \tau^{-1}\|v_2 - w_2| - \epsilon\|v_2 - w_2\| \\ &\geqslant (\tau^{-1} - \epsilon)\|v_2 - w_2\|, \end{aligned}$$

since $\|v_2 - w_2\| = \|v - w\|$. Similarly,

$$\begin{aligned} \|f_1(v) - f_1(w)\| &\leqslant \tau\|v_1 - w_1\| + \epsilon\|v - w\| \\ &\leqslant (\tau + \epsilon)\|v - w\|. \end{aligned}$$

Since $\epsilon < (1 - \tau)/(1 + \tau)$, we have $\epsilon < 1 - \tau$ and $\epsilon < \tau^{-1} - 1$ which shows $(\tau + \epsilon) < 1 < (\tau^{-1} - \epsilon)$. Putting these inequalities together gives

$$\|f_2(v) - f_2(w)\| \geqslant (\tau^{-1} - \epsilon)\|v_2 - w_2\| \geqslant \|f_1(v) - f_1(w)\|.$$

We apply this inequality to z and z'. Suppose B is the ball of radius r. Then

$$\begin{aligned} 2r &\geqslant \|(f^n)_2(z) - (f^n)_2(z')\| = \|f_2(f^{n-1}(z)) - f_2(f^{n-1}(z'))\| \\ &\geqslant (\tau^{-1} - \epsilon)\|(f^{n-1})_2(z) - (f^{n-1})_2(z')\| \\ &\vdots \\ &\geqslant (\tau^{-1} - \epsilon)^n\|z - z'\| = (\tau^{-1} - \epsilon)^n\|y - y'\|. \end{aligned}$$

But as $n \to \infty$, $(\tau^{-1} - \epsilon)^n$ also tends to infinity so we must have $y = y'$. Q.E.D.

We can now give a proof of the Stable manifold theorem due to Irwin [6].

THEOREM 6.1. (Stable manifold theorem for a fixed point). *Let $E = E_1 \times E_2$ be a Banach space and let $T_1: E_1 \to E_1$ and $T_2: E_2 \to E_2$ be isomorphisms with* $\max\{\|T_1\|, \|T_2^{-1}\|\} < \tau < 1$. *Let U be an open neighborhood of 0 in E and let $g: U \to E$ be a C^r map ($r \geqslant 1$) with $g(0) = 0$ and $Dg(0) = 0$. Let $f = T_1 \times T_2 + g$. Then there exist open balls C and D centered at 0 in E_1 and E_2, respectively, and a unique map $h: C \to D$ such that* $f(\text{graph}(h)) \subset \text{graph}(h)$. *The map h is C^r and for all $z \in C \times D$, $f^n(z) \to 0$ as $n \to \infty$ if and only if* $z \in \text{graph}(h)$.

Proof. Let $B = C_1 \times D$ be an open ball centered at 0 in $E_1 \times E_2$ such that $\sup\{\|Dg(z)\| \mid z \in B\} < \epsilon < (1 - \tau)/(1 + \tau)$. Let $\sup\{\|g(z)\| \mid z \in B\} = b$. Let X denote the set of integers $\{0, 1, 2, \ldots\}$ with the discrete topology compactified with a point ∞.

We are looking for points in B that tend to 0 under positive iteration by f or, equivalently, for sequences $\{z_n\}$ in B converging to 0 such that $f(z_i) = z_{i+1}$ for all i. Let $H = \{\gamma \in C^0(X, E) | \gamma(\infty) = 0\}$ so H is a closed subspace of $C^0(X, E)$ and is really just the space of sequences in E converging to 0. Let $G = \{\gamma \in C^0(X, B) | \gamma(\infty) = 0\}$ so G is an open subset of H.

Suppose that $\gamma \in G$ is, for some $z \in B$, the sequence $f^n(z)$. Then

$$\begin{aligned}\gamma(n) &= T \circ (T+g)^{n-1}(z) + g(\gamma(n-1))\\ &= T^2 \circ (T+g)^{n-2}(z) + T \circ g(\gamma(n-2)) + g(\gamma(n-1))\\ &= T^n(z) + \sum_{i=0}^{n-1} T^{n-1-i} \circ g(\gamma(i)).\end{aligned}$$

The second component of this is

$$T_2{}^n\left(z_2 + \sum_{i=0}^{n-1} T_2^{-1-i} \circ g_2(\gamma(i))\right).$$

Since T_2 is an expansion, we must have that $\sum_{i=0}^{\infty} T_2^{-1-i} \circ g_2(\gamma(i))$ exists and equals $-z_2$.

This motivates the definition of a function $F: C_1 \times G \to H$ given by

$$F(x, \gamma)(n) = \gamma(n) - \left(T_1{}^n(x) + \sum_{i=0}^{n-1} T_1^{n-1-i} \circ g_1(\gamma(i)), -\sum_{i=n}^{\infty} T_2^{n-1-i} \circ g_2(\gamma(i))\right)$$

and $F(x, \gamma)(\infty) = 0$. We want to show this is well defined, and then for each x find a γ such that $F(x, \gamma) = 0$.

To see that F is well defined, note that the infinite sum converges because $\|g_2(z)\| < b$ for $z \in B$ and $\|T_2^{-1}\| < \tau < 1$. Also it is clear that $\gamma(n)$, $T_1{}^n(x)$ and the infinite sum tend to 0 as n tends to infinity.

Note too that for all m with $0 \leqslant m \leqslant n$,

$$\begin{aligned}\left\|\sum_{i=0}^{n-1} T_1^{n-1-i} \circ g_1(\gamma(i))\right\| &\leqslant \sum_{i=0}^{m-1} \tau^{n-1-i}\|g_1(\gamma(i))\| + \sum_{i=m}^{n-1} \tau^{n-1-i}\|g_1(\gamma(i))\|\\ &\leqslant (1-\tau)^{-1}\tau^{n-m}b + (1-\tau)^{-1} \sup_{i \geqslant m} \|g_1(\gamma(i))\|.\end{aligned}$$

Given $\epsilon > 0$, we can fix m large enough for the second summand to be $< \epsilon/2$ and then for sufficiently large n the first term is also $< \epsilon/2$. This completes the proof that $F(x, \gamma)$ is a well-defined element of H.

For fixed $\gamma \in G$ the map $C_1 \to H$ given by $x \mapsto F(x, \gamma)$ is continuous affine. By Corollary 3.2 the map $\omega_{g_1}: G \to H$ is C^r. The map $L: H \to H$ given by $L(\gamma)(n) = \sum_{i=0}^{n-1} T_1^{n-1-i}(\gamma(i))$ is continuous linear. Thus $L \circ \omega_{g_1}$ is C^r. Similarly the map $G \to H$ which takes γ to the function whose value at n is $\sum_{i=n}^{\infty} T_2^{n-1-i} \circ g_2(\gamma(i))$ is C^r. It follows that for fixed x the map $G \to H$ given by $\gamma \to F(x, \gamma)$ is C^r and hence F is C^r.

Now $F(0,0) = 0$ and if $u \in H$, then from Corollary 3.2,

$$D_2F(x,\gamma)(u)(n)$$
$$= u(n) + \left(\sum_{i=0}^{n-1} T_1^{n-1-i} \circ Dg_1(\gamma(i))(u(i)), -\sum_{i=n}^{\infty} T_2^{n-1-i} \circ Dg_2(\gamma(i))(u(i))\right).$$

So $D_2F(0,0)$ is the identity map on H.

By the Implicit function theorem 1.6 there is a ball C in E_1 centered at 0 and a C^r map $\varphi: C \to G$ such that $\varphi(0) = 0$ and for all $x \in C$, $F(x, \varphi(x)) = 0$. This implies that

$$\varphi(x)(0) = \left(T_1{}^0(x), -\sum_{i=0}^{\infty} T_2^{-1-i} \circ g_2(\varphi(x)(0))\right) = (x, h(x))$$

if we define $h(x) = -\sum_{i=0}^{\infty} T_2^{-1-i} \circ g_2(\varphi(x)(0))$. The map $h: C \to D$ is C^r since φ is C^r and the map $\theta: G \to B$ taking γ to $\gamma(0)$ is continuous linear. Furthermore, for any positive integer n,

$$\begin{aligned}\varphi(x)(n+1) &= \left(T_1^{n+1}(x) + \sum_{i=0}^{n} T_1^{n-i} \circ g_1(\varphi(x)(i)), -\sum_{i=n+1}^{\infty} T_2^{n-i} \circ g_2(\varphi(x)(i))\right)\\ &= T(\varphi(x)(n)) + (g_1(\varphi(x)(n)), g_2(\varphi(x)(n)))\\ &= f(\varphi(x)(n)).\end{aligned}$$

Induction gives $\varphi(x)(n) = f^n(x, h(x))$. Thus it follows that $z \in \text{graph}(h)$ implies $f^n(z)$ tends to 0 as n tends to infinity. Lemma 6.1 shows that if $z \in C \times D$ and $f^n(z) \in B$ for all $n \geqslant 0$, then $z \in \text{graph}(h)$. Hence $f(\text{graph}(h)) \subset \text{graph}(h)$.

All that remains is to show $Dh(0) = 0$. This follows because $\text{graph}(Dh(0))$ is the tangent space of $\text{graph}(h)$ at 0 and hence is invariant under $Df(0) = T$. Thus $\text{graph}(Dh(0))$ must be E_1 so $Dh(0) = 0$. Q.E.D.

THEOREM 6.2. *Suppose $T: E \to E$ and U are as in Theorem 6.1. Then there is a ball C centered at 0 in E_1, a neighborhood V of 0 in $C_b{}^r(U, E)$, and a continuous function $W: V \to C_b{}^r(C, E_2)$ such that if $g \in V$, then $\text{graph}(W(g))$ is a stable manifold for $T + g$.*

Proof. Let $K = \{g \in C_b{}^r(U, E) \mid g(0) = 0\}$. Then for some neighborhood V_1 of 0 in K and some ball C_1 centered at 0 in E_1, the function $F: C_1 \times G \times V_1 \to H$ given by

$$F(x, \gamma, g)(n) = \gamma(n) - \left(T_1{}^n(x) + \sum_{i=0}^{n-1} T_1^{n-1-i} \circ g_1(\gamma(i)), -\sum_{i=n}^{\infty} T_2^{n-1-i} \circ g_2(\gamma(i))\right)$$

is well defined.

As in Theorem 6.1 it is not difficult to show, using Corollary 3.1, that F is C^r. Also it is clear that $D_2F(0,0,0)$ is the identity map on H. Applying the Implicit function theorem 1.6 gives a ball C_2 centered at 0 in E_1, a ball V_1 centered at 0 in K, and a C^r function $\varphi: C_2 \times V_1 \to G$ such that $\varphi(0,0) = 0$ and $F(x, \varphi(x,g), g) = 0$ for all $x \in C$ and $g \in V$. If $\pi: E \to E_2$ is a projection and $h(x,g) = \pi \circ \varphi(x,g)(0)$, then by the proof of Theorem 6.1 graph$(h(\cdot, g))$ is a stable manifold for g. Clearly the map $W_1: V_1 \to C_b{}^r(C_2, E_2)$ given by $W_1(g)(x) = h(x,g)$ is continuous.

It is not difficult to show that there is a neighborhood V_2 of 0 in $C_b{}^r(U,E)$ such if $g \in V_2$, $T + g$ has a unique fixed point near 0 and also that the function $\beta: V_2 \to E$ which assigns to g this fixed point is a continuous function. Let $\Delta: V_2 \to K$ be defined by $\Delta(g)(z) = g(z + \beta(z)) - \beta(z)$. Clearly $\Delta(g)(0) = 0$.

Now choose a neighborhood V of 0 in V_2 and a ball C centered at 0 in E_1 such that $g \in V$ implies $\Delta(g) \in V_1$ and $x \in C$, $g \in V$ implies $x - \beta_1(g) \in C_2$.

Now define $W: V \to C_b{}^r(C, E_2)$ by $W(g)(x) = W_1(\Delta(g))(x - \beta_1(g)) + \beta_2(g)$. One checks easily that W has the desired properties. Q.E.D.

References

1. R. Abraham, Lectures of Smale on differential topology, Lecture Notes, Columbia Univ., New York, 1962.
2. R. Abraham and J. Robbin, "Transversal Mappings and Flows," Benjamin, New York, 1967.
3. J. Eells, A setting for global analysis, *Bull. Amer. Math. Soc.* **72** (1966), 751–807.
4. J. Franks, Anosov diffeomorphisms, Global Analysis, *Proc. Symp. Pure Math.*, Vol. 14, pp. 61–94, Amer. Math. Soc., Providence, Rhode Island, 1970.
5. M. W. Hirsch and C. Pugh, Stable manifolds for hyperbolic sets, Global Analysis, *Proc. Symp. Pure Math.*, Vol. 14, pp. 133–163, Amer. Math. Soc., Providence, Rhode Island, 1970.
6. M. C. Irwin, On the stable manifold theorem, *Bull. London Math. Soc.* **2** (1970), 196–198.
7. M. C. Irwin, On the smoothness of the composition map, *Quart. J. Math. Oxford Ser.* **23** (1972), 113–133.
8. S. Lang, "Analysis II," Addison–Wesley, Reading, Massachusetts, 1969.
9. S. Lang, "Introduction to Differentiable Manifolds," Wiley (Interscience), New York, 1962.
10. J. Mather, Characterization of Anosov diffeomorphisms, *Nederl. Akad. Wentensch. Proc. Ser. A* **30** (1968), no. 5.
11. E. Nelson, "Topics in Dynamics I: Flows," Princeton Univ. Press, Princeton, New Jersey, 1969.
12. R. S. Palais, "Foundations of Global Non-Linear Analysis," Benjamin, New York, 1968.
13. S. Smale, Differentiable dynamical systems, *Bull. Amer. Math. Soc.* **73** (1967), 747–817.

STUDIES IN ANALYSIS
ADVANCES IN MATHEMATICS SUPPLEMENTARY STUDIES, VOL. 4

Positive Definite Functions on Abelian Semigroups without Zero

PAUL RESSEL

Institut für Mathematische Statistik
Westfälische Wilhelms-Universität
Münster, Germany

0. INTRODUCTION

The important rule played by positive definite functions on abelian groups has been well known for many years. Recently it has been shown in a work of Berg *et al.* [1], that for abelian semigroups there also exists a natural concept of positive definiteness, strongly connected with an abstract form of the Laplace transformation. Whereas every positive definite function on a group is automatically bounded, this does not hold in general on semigroups and hence was made part of the definition in [1]. More precisely: Let S be an Abelian semigroup, written additively, with neutral element 0. A real-valued function $f: S \to \mathbb{R}$ is called *positive definite* if f is bounded and the matrix $(f(s_i + s_j))_{i,j=1,\ldots,n}$ is positive semidefinite for every finite subset $\{s_1, \ldots, s_n\}$ of S. In [1] we obtained integral representations for these functions as well as for the closely related class of negative definite functions. One of our main results was that the class of functions on S that are monotone of infinite order in the sense of Choquet, turns out to be an extreme subcone of the positive definite functions on S, and also a similar relation between alternating functions of infinite order and negative definite functions.

Now there are standard examples of completely monotone functions that are naturally defined only on a semigroup without zero element, such as $x \to 1/x$ on the open half-line $]0, \infty[$ or the function $n \to 1/n$ on $\mathbb{N} = \{1, 2, 3, \ldots\}$ under addition; although this last function is bounded on $\mathbb{N}$, it does not admit an extension to a completely monotone function on $\mathbb{N}_0 = \{0, 1, 2, \ldots\}$. However, it is easy to see that both functions are positive definite.

Motivated by these and similar examples we have found a definition of positive definiteness for real-valued functions on Abelian semigroups that agrees with that given in [1] if the semigroup contains a zero element, but

ISBN 0-12-599150-9

that also covers the above examples (Definition 1). The concept of negative definite function also requires some change: these functions are no longer necessarily bounded below. A standard example is given by $x \to \log x$ on $]0, \infty[$ (cf. Definition 2).

The main results are contained in Theorems 1 and 3 where uniquely determined integral representations for positive and negative definite functions on a suitable locally compact subspace of the dual semigroup are given. The relation to functions monotone (respectively, alternating) of infinite order is clarified in Theorem 6. The examples and applications are gathered together in Section 3.

1. Positive and Negative Definite Functions

Let $(H, +)$ be an abelian semigroup. We now have in mind a semigroup without zero, but formally this shall not be required. An element $h \in H$ is called *maximal* if it is not of the form $h_1 + h_2$ with $h_1, h_2 \in H$ (cf. Choquet [4]). $H' = H + H$ denotes the subsemigroup of nonmaximal elements of H. Note that $H' = H$ whenever H is 2-divisible, but, for example, $H' = H$ also for the non-2-divisible semigroup $H = (]-1, 1[, \cdot)$. Put $S = H \cup \{0\}$ with some symbol $0 \notin H$. Let $0 + h = h$, $\forall h \in H$, $0 + 0 = 0$; then S is an Abelian semigroup with neutral element 0, and $\hat{S}$ denotes the full semigroup of *characters* on S, i.e., the set of all $\rho: S \to [-1, 1]$, where $\rho(0) = 1$ and $\rho(s + t) = \rho(s)\rho(t)$ for all $s, t \in S$. In the topology of pointwise convergence $\hat{S}$ is a compact topological semigroup. It is obvious that $1_{\{0\}} \in \hat{S}$; it is the so-called *absorbing element* of $\hat{S}$ ($1_{\{0\}} \cdot \rho = 1_{\{0\}}$, $\forall \rho \in \hat{S}$) which in any abelian semigroup is (if existent) uniquely determined. $\check{H} = \hat{S} \setminus \{1_{\{0\}}\}$ is a locally compact subspace of $\hat{S}$, containing the neutral element, i.e., the function $s \mapsto 1$, $\forall s \in S$.

Definition 1. $f: H' \to \mathbb{R}$ is called *positive definite* iff

(1) $\sup_{s \in S} |f(h' + s)| < \infty$, $\forall h' \in H'$.
(2) $\sum_{i,j=1}^{n} c_i c_j f(h_i + h_j) \geqslant 0, \forall (h_1, \ldots, h_n) \in H^n, (c_1, \ldots, c_n) \in \mathbb{R}^n, n \in \mathbb{N}$.

$$P(H) := \{f: H' \to \mathbb{R} \mid f \text{ is positive definite}\}.$$

If H contains a zero, then $H' = H$ and the notion agrees with that in [1]. We state some simple properties of $P(H)$.

Proposition 1. (i) *$P(H)$ is a closed convex cone in $\mathbb{R}^{H'}$.*
(ii) *$f_1, f_2 \in P(H)$ implies $f_1 \cdot f_2 \in P(H)$.*

(iii) *$f \in P(H)$ implies that $h' \to f(h') \pm f(s + h')$ belongs to $P(H)$ for any $s \in S$.*

(iv) *$f \in P(H)$, $h_1, h_2 \in H$ implies that*

$$s \to f(s + 2h_1) + f(s + 2h_2) - 2f(s + h_1 + h_2)$$

belongs to $P(S)$.

Proof. (i) Note that $\sup_{s \in S} |f(2h + s)| = f(2h)$ and that

$$|f(h_1 + h_2 + s)| \leqslant [f(2h_1)f(2h_2 + 2s)]^{1/2}$$

for all $h, h_1, h_2 \in H$ (cf. [1, Proposition 2.2]).

(ii) This follows from the fact that positive semidefinite matrices are closed under multiplication.

(iii) If $h_1, \ldots, h_n \in H$ and $c_1, \ldots, c_n \in \mathbb{R}$, then $F: S \to \mathbb{R}$ defined by

$$F(s) = \sum_{i,j=1}^{n} f(s + h_i + h_j)c_i c_j$$

belongs to $P(S)$ (cf. [1, Lemma 2.4]), hence $|F(s)| \leqslant F(0)$ for all $s \in S$, i.e.,

$$\sum_{i,j=1}^{n} c_i c_j [f(h_i + h_j) \pm f(s + h_i + h_j)] \geqslant 0.$$

(iv) Take F as defined above for $n = 2$, $c_1 = 1$, and $c_2 = -1$. ∎

Let $f \in P(H)$ and put $f_{h,k}(s) = f(s + h + k)$, $s \in S$, $h, k \in H$; furthermore $f_h = f_{h,h}$. Of course f_h belongs to $P(S)$, hence there exists a uniquely determined positive measure $\nu_h \in M_+(\hat{S})$ [$M(\hat{S})$ denotes all real Radon measures on S] such that

$$f_h(s) = \int_{\hat{S}} \rho(s)\, d\mu_h(\rho), \qquad \forall s \in S$$

(cf. [1, Corollary 2.5]). From Proposition 1 we learn the existence of $\nu_{h,k} \in M_+(\hat{S})$ such that

$$f(s + 2h) + f(s + 2k) - 2f(s + h + k) = \int_{\hat{S}} \rho(s)\, d\nu_{h,k}(\rho)$$

for all $s \in S$. Define

$$\mu_{h,k} = \tfrac{1}{2}(\mu_h + \mu_k - \nu_{h,k}).$$

Then $\mu_{h,k} \in M(\hat{S})$, $\mu_{h,h} = \mu_h \in M_+(\hat{S})$, and

$$f(s + h + k) = \int_{\hat{S}} \rho(s)\, d\mu_{h,k}(\rho), \qquad \forall s \in S.$$

Our aim is to get an integral representation for the functions in $P(H)$. In order to avoid pathologies we have to restrict ourselves to the subcone

$$P_0(H) = \{f \in P(H) \,|\, \mu_{h,k}(\{1_{\{0\}}\}) = 0, \forall h, k \in H\}.$$

The example $H = ([1, \infty[, +)$ and $f = 1_{\{2\}}$ shows that $P_0(H) \subset P(H)$ is possible. However, the following lemma shows that for 2-divisible semigroups examples of this kind are not possible.

LEMMA 1. *Let f belong to $P(H)$. Then we have*:

(i) *If $2H + S = H'$ and $\mu_h(\{1_{\{0\}}\}) = 0$ for all $h \in H$, then also $\mu_{h,k}(\{1_{\{0\}}\}) = 0$ for all $h, k \in H$.*

(ii) *If $H = H'$, then $P_0(H) = P(H)$.*

Proof. (i) Let $h, k \in H$; we can find $h_0 \in H$ and $a \in S$ such that $h + k = 2h_0 + a$. For any $s \in S$ we have

$$\int_{\hat{S}} \rho(s)\, d\mu_{h,k}(\rho) = f(s + h + k) = f(s + 2h_0 + a)$$

$$= f_{h_0}(s + a) = \int_{\hat{S}} \rho(s)\rho(a)\, d\mu_{h_0}(\rho)$$

hence $d\mu_{h,k}(\rho) = \rho(a)\, d\mu_{h_0}(\rho)$ by Corollary 1.6 in [1]. This implies

$$\mu_{h,k}(\{1_{\{0\}}\}) = 1_{\{0\}}(a)\mu_{h_0}(\{1_{\{0\}}\}) = 0.$$

(ii) For given $h,\ k \in H$ we find $h_1,\ h_2 \in H$ such that $h = h_1 + h_2$. From

$$\int_{\hat{S}} \rho(s)\, d\mu_{h,k}(\rho) = f(s + h + k) = f_{h_1,h_2}(s + k)$$

$$= \int_{\hat{S}} \rho(s)\rho(k)\, d\mu_{h_1,h_2}(\rho)$$

for all $s \in S$ we deduce

$$d\mu_{h,k}(\rho) = \rho(k)\, d\mu_{h_1,h_2}(\rho)$$

and therefore

$$\mu_{h,k}(\{1_{\{0\}}\}) = 0. \quad \blacksquare$$

Remark 1. If H already contains a neutral element 0_H, then it is easily seen that

$$\hat{S}\backslash\{1_{\{0\}}\} \to \hat{H}, \qquad \rho \to \rho|_H$$

is a topological semigroup isomorphism, showing in particular that $1_{\{0\}}$ is an isolated point in $\hat{S}$.

Remark 2. $H'' = 2H + S$ is always a subsemigroup of H'. $H = (\mathbb{N}, +)$ is an example where $H'' = H' \subset H$. That $H'' \subset H'$ is possible is shown by the semigroup $H = (\mathbb{N}\backslash\{1\}, +)$.

THEOREM 1. *Let $f \in P_0(H)$ be given. Then there exists a uniquely determined Radon measure $\mu \in M_+(\check{H})$ such that*

$$f(h') = \int_{\check{H}} \rho(h')\, d\mu(\rho) \qquad \text{for all} \quad h' \in H'.$$

Proof. For any $h_1, h_2, k_1, k_2 \in H$ we have

$$\int_{\hat{S}} \rho(s + h_1 + h_2)\, d\mu_{k_1,k_2}(\rho) = f(s + h_1 + h_2 + k_1 + k_2)$$
$$= \int_{\hat{S}} \rho(s + k_1 + k_2)\, d\mu_{h_1,h_2}(\rho)$$

for all $s \in S$, giving us

$$\rho(h_1 + h_2)\, d\mu_{k_1,k_2}(\rho) = \rho(k_1 + k_2)\, d\mu_{h_1,h_2}(\rho).$$

Let $O_h = \{\rho \in \hat{S} \mid \rho(h) \neq 0\}$; O_h is open and $\bigcup_{h \in H} O_h = \check{H}$. Setting

$$d\mu(\rho) = (1/\rho^2(h))\, d\mu_h(\rho) \qquad \text{on} \quad O_h,$$

we obtain a well-defined nonnegative Radon measure μ on $\check{H}$. We claim that

$$\rho(h_1 + h_2)\, d\mu(\rho) = d\mu_{h_1,h_2}(\rho) \qquad \text{on} \quad \check{H}.$$

In fact, on O_k we have

$$\rho(h_1 + h_2)\, d\mu(\rho) = (1/\rho(k + k))\rho(h_1 + h_2)\, d\mu_k(\rho) = d\mu_{h_1,h_2}(\rho).$$

(This also shows that μ_{h_1,h_2} is concentrated on $O_{h_1} \cap O_{h_2}$.)

Therefore we obtain for $h' = h_1 + h_2 \in H'$,

$$\int_{\check{H}} \rho(h')\, d\mu(\rho) = \mu_{h_1,h_2}(\check{H}) = \mu_{h_1,h_2}(\hat{S}) = f(h'),$$

where we use that, by hypothesis, $f \in P_0(H)$.

The unicity of μ follows immediately from the fact that

$$f(2h + s) = \int_{\check{H}} \rho(s)\rho^2(h)\, d\mu(\rho) = \int_{\hat{S}} \rho(s)\, d\mu_h(\rho)$$

for all $s \in S$, for all $h \in H$. ■

Remark 3. Put $\varphi_{h'}(\rho) = \rho(h')$ for $h' \in H'$ and $\rho \in \check{H}$ and consider

$$\{\mu \in M_+(\check{H}) \mid \varphi_{h'} \in L^1(\check{H}, \mu)\, \forall h' \in H'\}.$$

Theorem 1 establishes a one-to-one mapping from $P_0(H)$ into this set of measures. This mapping is also onto. In fact, for μ belonging to this set put

$f(h') = \int \rho(h')\, d\mu(\rho)$. Condition (1) of Definition 1 follows trivially, and from

$$|f(h' + s)| \leqslant \int |\rho(h' + s)|\, d\mu(\rho) \leqslant \int |\rho(h')|\, d\mu(\rho)$$

one gets condition (2). Therefore $f \in P(H)$ and indeed $f \in P_0(H)$ as one can easily see.

A natural question arises at this point: Under what condition can a given function $f \in P_0(H)$ be extended to a function $\tilde{f} \in P(S)$? It is, of course, necessary that f be bounded, but as the example $H = (\mathbb{N}, +)$ and $f(n) = 1/n$ shows, this condition is not sufficient. An answer is given by

PROPOSITION 2. *Let f belong to $P_0(H)$ and let μ be the disintegrating measure given by Theorem* 1. *Then*:

(i) *f can be extended to a function $\tilde{f} \in P(S)$ if and only if $\mu(\check{H}) < \infty$;*
(ii) *If this is fulfilled and if furthermore $H' = H$, then the additional requirement $\tilde{f}(0) = \mu(\check{H})$ determines $\tilde{f}$ uniquely.*

Proof. (i) If $\mu(\check{H}) < \infty$, then $\tilde{f}(s) := \int_{\check{H}} \rho(s)\, d\mu(\rho)$ for all $s \in S$ defines an extension. On the other hand, suppose that $\tilde{f}(s) = \int_{\hat{S}} \rho(s)\, d\nu(\rho)$ with $\nu \in M_+(\hat{S})$ is an extension of f. Then by the unicity statement in Theorem 1 we obtain $\mu = \nu$ on $\check{H}$; in particular, $\mu(\check{H}) < \infty$.

(ii) We identify μ with a measure on $\hat{S}$ having the property that $\mu(\{1_{\{0\}}\}) = 0$. If $\tilde{f}(s) = \int_{\hat{S}} \rho(s)\, d\nu(\rho)$ is an extension of f, then $\tilde{f}(0) = \mu(\check{H})$ (as well as the condition $\nu(\{1_{\{0\}}\}) = 0$) implies that $\nu = \mu$. ∎

Remark 4. One might guess that in case H is 2-divisible, the boundedness of $f \in P_0(H)$ should imply that the representing measure μ is (totally) finite. For some semigroups this is true, but in general it does not hold.

Suppose that $f \in P_0(H)$ and $sup_{h \in H} f(h) = \alpha < \infty$, where H is 2-divisible. We know that with $O_h = \{\rho \in \check{H} \,|\, \rho(h) > 0\}$ we have $\check{H} = \bigcup_{h \in H} O_h$. Using the 2-divisibility of H we obtain

$$\int_{\check{H}} (\rho(h))^{2^{-n}}\, d\mu(\rho) \leqslant \alpha \qquad \text{for} \quad \text{all } n \in \mathbb{N},$$

hence $\mu(O_h) \leqslant \alpha$ for all $h \in H$. Let us define $d(H) = \inf\{n \in \mathbb{N} \,|\, \exists h_1, \ldots, h_n \in H$ such that $\check{H} = \bigcup_{i=1}^{n} O_{h_i}\}$ ($d(H) = \infty$ if no such n exists). Then for finite $d(H)$ we obtain $\mu(\check{H}) \leqslant \|f\|_\infty \cdot d(H) < \infty$. For example, $d(]0, \infty[, +) = 1$ and $d([0, \infty[^n \backslash \{0\}, +) = n$. Therefore it may come as no surprise that a counterexample to the above-mentioned guess is given by $H = [0, \infty[^{\mathbb{N}} \backslash \{0\}$. We identify $[0, \infty]^{\mathbb{N}} \backslash \{(\infty, \infty, \ldots)\}$ with a subset of $\check{H}$, let $\rho_i = (\infty, \infty, \ldots, 0, \infty, \infty, \ldots)$, zero standing at the ith coordinate, and put $\mu = \sum_{i=1}^{\infty} \epsilon_{\rho_i}$, ϵ_ρ being the one-point measure in ρ. It is easy to see that $f(h) = \int_{\check{H}} \rho(h)\, d\mu(\rho)$,

$h \in H$, is nothing but the indicator function of $\{h \in H \mid \#\{i \mid h_i > 0\} = 1\}$. In particular, $d(H) = \infty$. But $d(H)$ need not to be finite in order to have that $\|f\|_\infty < \infty$ is equivalent with $\mu(\check{H}) < \infty$. Take, for example, $H = ([0, 1[, \frown)$ (cf. Section 7 in [1]). Here $d(H) = \infty$, but there exists an increasing sequence $\{O_{h_n}\}$ with union $\check{H}$ that implies that $\mu(\check{H}) \leqslant \|f\|_\infty$ and hence $\mu(\check{H}) = \|f\|_\infty$ for all $f \in P_0(H)$.

DEFINITION 2. $\psi: H' \to \mathbb{R}$ is called *negative definite* iff

(1) $\inf_{s \in S} \psi(h' + s) > -\infty$, $\forall h' \in H'$,

(2) $\sum_{i,j=1}^n c_i c_j \psi(h_i + h_j) \leqslant 0$ for $(h_1, \ldots, h_n) \in H^n$ and $\forall (c_1, \ldots, c_n) \in \mathbb{R}^n$ such that $\sum_{i=1}^n c_i = 0$, $\forall n \in \mathbb{N}$.

Then let

$$N(H) := \{\psi: H' \to \mathbb{R} \mid \psi \text{ is negative definite}\}.$$

It follows immediately from this definition that if $\psi \in N(H)$ and $d_h = \inf_{s \in S} \psi(2h + s)$, then $\psi_h \in N(S)$, where

$$\psi_h(s) = \psi(2h + s) - d_h, \qquad s \in S, \quad h \in H.$$

In particular, $\psi_h(s) \geqslant \psi_h(0)$, cf. [1, Proposition 3.3], i.e., $d_h = \psi(2h)$. A slight inconvenience might be caused by the fact, that in [1] functions of $N(S)$ are nonnegative by definition. It is perhaps more natural to replace this condition by the requirement that $\inf_{s \in S} \psi(s) > -\infty$. The only change in the representation theorem 3.7 of [1] would be the replacement of the nonnegative constant by a real constant.

Some simple properties of $N(H)$ are contained in the following

PROPOSITION 3. (i) *$N(H)$ is a closed convex cone in $\mathbb{R}^{H'}$, containing all constant functions.*

(ii) $-P(H) \subseteq N(H)$.

(iii) $\psi \in N(H)$ *iff* $\exp(-t\psi) \in P(H)$, $\forall t > 0$.

(iv) *Put $\Delta_a \psi(h') = \psi(a + h') - \psi(h')$ for $h' \in H'$. Then if $\psi \in N(H)$, we obtain $\Delta_a \psi \in P(H)$ for all $a \in S$.*

Proof. Case (ii) is a trivial property. Case (iii) follows from a general fact about positive semidefinite matrices (see [5, Proposition 4.1]). Case (iii) together with Proposition 1 imply (i). To show (iv) we observe that if $\psi \in -P(H)$, then $\Delta_a \psi \in P(H)$ by Proposition 1, and because of

$$\psi = \lim_{t \to 0} (1/t)(1 - \exp(-t\psi))$$

we obtain $\Delta_a \psi \in P(H)$ for every $\psi \in N(H)$, using the fact that $P(H)$ is closed. ■

Let us for a moment consider the semigroup S with neutral element. If $\psi \in N(S)$ and

$$\psi(s) = c + g(s) + \int_{\hat{S}\setminus\{1\}} (1 - \rho(s))\, d\nu(\rho)$$

is the uniquely determined representation given by Theorem 3.7 in [1], where $c \in [0, \infty[$, $g: S \to [0, \infty[$ is additive, and $\nu \in M_+(\hat{S}\setminus\{1\})$, and if $\Delta_a\psi(s) = \psi(s + a) - \psi(s) = \hat{\sigma}_a(s)$, then

$$(1 - \rho(a))\, d\nu(\rho) = d\sigma_a(\rho) \qquad \text{on} \quad \hat{S}\setminus\{1\}.$$

Hence $\sigma_a(\{1_{\{0\}}\}) = 0$ for all $a \in S$ if $\nu(\{1_{\{0\}}\}) = 0$ and of course the converse is also true.

Motivated by this result and taking into account Proposition 3 (iv) we define

$$N_0(H) := \{\psi \in N(H) \mid \Delta_a\psi \in P_0(H),\ \forall a \in S\}.$$

Lemma 1 implies that $N_0(H) = N(H)$ for any semigroup H without maximal elements. It is easily seen that $N_0(H)$ is an extreme subcone of $N(H)$.

THEOREM 2. *Let $\psi \in N_0(H)$ be given. Then there exists an additive function $g: S \to [0, \infty[$ and a Radon measure $\nu \in M_+(\check{H}\setminus\{1\})$ such that*

$$\psi(h' + s) - \psi(h') = \int_{\check{H}\setminus\{1\}} (1 - \rho(s))\rho(h')\, d\nu(\rho) + g(s)$$

for all $h' \in H'$ and $s \in S$. The pair (g, ν) is uniquely determined by these properties and given in the following way.

$$g(s) = \lim_{n\to\infty} \psi(2h + ns)/n \qquad \text{independent of } h \in H,$$

$$d\nu(\rho) = (1 - \rho(a))^{-1}\, d\sigma_a(\rho) \qquad \text{on the open set } \{\rho \in H \mid \rho(a) < 1\},$$

where σ_a is the positive Radon measure on $\check{H}$ satisfying

$$\Delta_a\psi(h') = \int_{\check{H}} \rho(h')\, d\sigma_a(\rho), \qquad h' \in H'.$$

Proof. $O_a = \{\rho \in \check{H} \mid \rho(a) < 1\}$ is open and $\bigcup_{a\in H} O_a = \check{H}\setminus\{1\} = \hat{S}\setminus\{1, 1_{\{0\}}\}$. $\Delta_a\psi \in P(H)$ implies $-\Delta_a\psi \in N(H)$ and $\Delta_b(-\Delta_a\psi) \in P(H)$. We have

$$\begin{aligned}
[\Delta_b(-\Delta_a\psi)](h') &= -\Delta_a\psi(h' + b) + \Delta_a\psi(h') \\
&= -\psi(a + b + h') + \psi(h' + b) + \psi(h' + a) - \psi(h') \\
&= \int_{\check{H}} \rho(h')\, d\sigma_a(\rho) - \int_{\check{H}} \rho(h' + b)\, d\sigma_a(\rho) \\
&= \int_{\check{H}} \rho(h')(1 - \rho(b))\, d\sigma_a(\rho) = \int_{\check{H}} \rho(h')(1 - \rho(a)\, d\sigma_b(\rho)
\end{aligned}$$

for all $h' \in H'$, hence from the unicity statement in Theorem 1,

$$(1 - \rho(b))\, d\sigma_a(\rho) = (1 - \rho(a))\, d\sigma_b(\rho) \qquad \text{on} \quad \check{H}.$$

This shows that $d\nu(\rho) = (1 - \rho(a))^{-1}\, d\sigma_a(\rho)$ on O_a is a well-defined positive Radon measure on $\check{H}\backslash\{1\}$ satisfying

$$(1 - \rho(a))\, d\nu(\rho) = d\sigma_a(\rho) \qquad \text{on} \quad \check{H}\backslash\{1\}.$$

Therefore we have

$$\Delta_a\psi(2h) = \int_{\check{H}\backslash\{1\}} \rho^2(h)(1 - \rho(a))\, d\nu(\rho) + \sigma_a(\{1\})$$

for all $a \in S$, $h \in H$.

Now we observe that $\Delta_s\psi(2h) = \psi(2h + s) - \psi(2h) = \psi_h(s)$ and $\psi_h \in N(S)$, for which we know the unique representation

$$\psi_h(s) = g_h(s) + \int_{\hat{S}\backslash\{1\}} (1 - \rho(s))\, d\nu_h(\rho)$$

with additive functions $g_h : S \to [0, \infty[$ and $\nu_h \in M_+(\hat{S}\backslash\{1\})$. We obtain

$$\begin{aligned}\psi(2h + s) - \psi(2h) &= \int_{\check{H}\backslash\{1\}} (1 - \rho(s))\rho^2(h)\, d\nu(\rho) + \sigma_s(\{1\}) \\ &= g_h(s) + \int_{\hat{S}\backslash\{1\}} (1 - \rho(s))\, d\nu_h(\rho)\end{aligned}$$

for all $h \in H$, $s \in S$. To draw conclusions we are first to show that $s \to \sigma_s(\{1\})$ is additive; in fact we shall show that $g_h(s) = \sigma_s(\{1\})$ independent of $h \in H$. From the proof of Theorem 3.7 in [1] we learn that $g_h(a) = \tau_{a,h}(\{1\})$ where $\Delta_a\psi_h(s) = \hat{\tau}_a(s)$ $\forall s \in S$. Put $h' = 2h + s$. Then $\Delta_a\psi(h') = \Delta_a\psi_h(s) = \int_{\check{H}} \rho(h')\, d\sigma_a(\rho) = \int_{\check{H}} \rho(s)\rho^2(h)\, d\sigma_a(\rho)$ and therefore

$$d\tau_{a,h}(\rho) = \rho^2(h)\, d\sigma_a(\rho)$$

implying $g_h(a) = \tau_{a,h}(\{1\}) = \sigma_a(\{1\})$, independent of h and additive in $a \in S$. Put $g(s) = \sigma_s(\{1\})$. Then indeed

$$\begin{aligned}\psi(h' + s) - \psi(h') = \Delta_s\psi(h') &= \int_{\check{H}} \rho(h')\, d\sigma_s(\rho) \\ &= \int_{H\backslash\{1\}} \rho(h')\, d\sigma_s(\rho) + \sigma_s(\{1\}) \\ &= \int_{H\backslash\{1\}} \rho(h')(1 - \rho(s))\, d\nu(\rho) + g(s)\end{aligned}$$

for all $h' \in H'$ and $s \in S$. From [1] we know that

$$g(s) = g_h(s) = \lim_{n\to\infty} \psi(2h + ns)/n$$

independent of $h \in H$. Using this fact the unicity of the representation can be shown as in the proof of Theorem 3.7 in [1]. ■

Theorem 2 is not yet completely satisfactory, because it gives a representation only for the differences of ψ, and in fact only for the special differences $\psi(h' + s) - \psi(h')$. Of course two functions $\psi_1, \psi_2 \in N_0(H)$ with the same pair (g, ν) differ only by a real constant, and calling such functions *equivalent*, Theorem 2 establishes a one-to-one map from the space of all equivalence classes into the set of pairs (g, ν), where $g: S \to [0, \infty[$ is additive and where $\nu \in M_+(\check{H}\setminus\{1\})$ has the property that all functions $\rho \to (1 - \rho(s))\rho(h')$ are integrable. Before giving an improved version of Theorem 2, stating that this map is onto, we present a simple lemma.

LEMMA 2. (i) *Suppose that the function $f: H' \to \mathbb{R}$ fulfills condition* (2) *of Definition* 1 *but condition* (1) *only for $h' \in 2H$. Then f belongs to $P(H)$.*

(ii) *Suppose that the function $\psi: H' \to \mathbb{R}$ fulfills condition* (2) *of Definition* 2 *but condition* (1) *only for $h' \in 2H$. Then ψ belongs to $N(H)$.*

Proof. (i) For $h_1, h_2 \in H$ and $s \in S$, the 2×2 matrix $(f(k_i + k_j + s))_{i,\, j=1,\, 2}$ is positive semidefinite, where $k_1 = h_1$ and $k_2 = h_2 + s$; hence

$$|f(h_1 + h_2 + s)| \leqslant [f(2h_1) \cdot f(2h_2 + 2s)]^{1/2}.$$

(ii) This case follows by applying (i) to $\exp(-\psi)$. ■

THEOREM 3. (i) *Let $\psi \in N_0(H)$ and (g, ν) the pair determined by Theorem* 2. *Then for all $h', k' \in H'$*

$$\psi(h') - \psi(k') = \int_{\check{H}\setminus\{1\}} (\rho(k') - \rho(h'))\, d\nu(\rho) + g(h') - g(k'). \qquad (*)$$

(ii) *Conversely if $g: S \to [0, \infty[$ is additive and $\nu \in M_+(\check{H}\setminus\{1\})$ such that $\rho \to \rho(k') - \rho(h')$ belongs to $L^1(\nu)$ for all $h', k' \in H'$, then*

$$\psi(h') = \int (\rho(k') - \rho(h'))\, d\nu(\rho) + g(h')$$

($k' \in H'$ arbitrary, but fixed) belongs to $N_0(H)$ and satisfies equation $(*)$.

Proof. (i) This case follows easily by computation from Theorem 2.

(ii) Let $(h_1, \ldots, h_n) \in H^n$ and $(c_1, \ldots, c_n) \in \mathbb{R}^n$ such that $\sum_{i=1}^n c_i = 0$. Then we obtain

$$\sum_{i,\, j=1}^{n} c_i c_j \psi(h_i + h_j) = \int \sum_{i,j} c_i c_j [\rho(k') - \rho(h_i + h_j)]\, d\nu(\rho) + \sum_{i,j} c_i c_j [g(h_i) + g(h_j)]$$

$$= \int -[\sum c_i \rho(h_i)]^2\, d\nu(\rho) \leqslant 0.$$

Furthermore for $h' = 2h \in 2H$,

$$\psi(h' + s) \geqslant \int (\rho(k') - \rho(2h))\, dv(\rho)$$

for all $s \in S$, hence $\psi \in N(H)$ by Lemma 2. It is now immediately seen that ψ belongs to $N_0(H)$ and that condition (*) is fulfilled. ■

Again one may ask for conditions that a given function $\psi \in N_0(H)$ have an extension to $\tilde{\psi} \in N(S)$ or, to be precise, to a function $\tilde{\psi}$ in $N(S) \oplus \mathbb{R}$, because functions in $N(S)$ are nonnegative by definition.

PROPOSITION 4. *Let ψ belong to $N_0(H)$ and let (g, v) be the pair associated to ψ by Theorem 2 or 3. Then:*

(i) *ψ can be extended to a function in $N(S) \oplus \mathbb{R}$ if and only if*

$$\int_{\check{H}\setminus\{1\}} (1 - \rho(h))\, dv(\rho) < \infty$$

for all $h \in H$.

(ii) *If H is 2-divisible and*

$$d(H) := \inf\left\{n \,\middle|\, \exists h_1, \ldots, h_n \in H \text{ such that } \check{H} = \bigcup_{i=1}^{n} O_{h_i}\right\}$$

is finite, then $\inf_{h \in H} \psi(h) > -\infty$ iff ψ has an extension as in (i).

Proof. We only prove (ii), (i) being straight forward to verify. Fix $h \in H$. Then for all $k \in H$ we have

$$\psi(h) - \psi(k) = \int_{\check{H}\setminus\{1\}} (\rho(k) - \rho(h))\, dv(\rho) \leqslant \alpha < \infty$$

for suitable $\alpha \in \mathbb{R}$. By assumption

$$\int [(\rho(k))^{2^{-p}} - \rho(h)]\, dv(\rho) \leqslant \alpha$$

hence also

$$\int [1_{O_k}(\rho) - \rho(h)]\, dv(\rho) \leqslant \alpha.$$

If now $\check{H} = \bigcup_{i=1}^{n} O_{k_i}$ for some $n \in \mathbb{N}$, then we easily see

$$\int (1 - \rho(h))\, dv(\rho) < \infty$$

and we can apply part (i) to obtain the result. ■

2. Monotone and Alternating Functions of Infinite Order

For a function $f: H \to \mathbb{R}$ and elements $h \in H$, $s_1, \ldots, s_n \in S$ we define functions $\nabla_n f(\cdot; s_1, \ldots, s_n): H \to \mathbb{R}$ inductively by

$$\nabla_1 f(h; s_1) = f(h) - f(h + s_1),$$
$$\nabla_n f(h; s_1, \ldots, s_n) = \nabla_{n-1} f(h; s_1, \ldots, s_{n-1}) - \nabla_{n-1} f(h + s_n; s_1, \ldots, s_{n-1}).$$

Definition 3. A function $f: H \to \mathbb{R}$ is called *monotone of infinite order* if and only if

(1) f if nonnegative, and
(2) $\nabla_n f(h; s_1, \ldots, s_n) \geqslant 0$ for $\forall h \in H, \forall (s_1, \ldots, s_n) \in S^n, \forall n \in \mathbb{N}$.

The function f is called *alternating of infinite order* if and only if

$$\nabla_n f(h; s_1, \ldots, s_n) \leqslant 0 \qquad \text{for} \quad \forall h \in H, \quad \forall (s_1, \ldots, s_n) \in S^n, \quad \forall n \in \mathbb{N}.$$

$M(H)$ [respectively $A(H)$] denotes the set of all functions $f: H \to \mathbb{R}$ that are monotone [alternating] of infinite order.

This definition goes back to Choquet [4], the only difference being that we allow functions in $A(H)$ to have also negative values. If H already contains a neutral element, then the definition of $M(H)$ agrees with that given in [1], whereas $A(H)$ from the preceding definition coincides with $A(H) \oplus \mathbb{R}$ in [1]. Notice that in contrast with the positive and negative definite functions which in a natural way are only defined on the subsemigroup H' of nonmaximal elements of H, the monotone and alternating functions of infinite order are defined on all of H.

For $f: H \to \mathbb{R}$ and $h \in H$ let $f_h: S \to \mathbb{R}$ be defined by $f_h(s) = f(h + s)$ and for fixed $s \in S$ define $\Delta_s f: H \to \mathbb{R}$ by $\Delta_s f(h) = f(h + s) - f(h)$. Let us state some properties of $M(H)$ and $A(H)$.

Proposition 5. (i) *$M(H)$ and $A(H)$ are closed convex cones in $\mathbb{R}^H$.*
(ii) $-M(H) \subseteq A(H)$.
(iii) *A function $f: H \to \mathbb{R}$ belongs to $M(H)$ [respectively, $A(H)$] if and only if f_h belongs to $M(S)$ [respectively, $A(S)$] for all $h \in H$.*
(iv) *A function $\psi: H \to \mathbb{R}$ belongs to $A(H)$ if and only if for every $a \in S$ the function $\Delta_a \psi$ belongs to $M(H)$.*
(v) *$f, g \in M(H)$ implies $f \cdot g \in M(H)$.*
(vi) *$\psi \in A(H)$ iff $e^{-t\psi} \in M(H)$ for all $t > 0$.*

Proof. (i) and (ii) need no proof. (iii) follows from the identity

$$\nabla_n f_h(s; a_1, \ldots, a_n) = \nabla_n f(h+s; a_1, \ldots, a_n)$$

valid for all $h \in H$, $s \in S$, $(a_1, \ldots, a_n) \in S^n$, $n \in \mathbb{N}$. We obtain (iv) from

$$\nabla_n(\Delta_a \psi)(h; a_1, \ldots, a_n) = -\nabla_{n+1}\psi(h; a_1, \ldots, a_n, a)$$

which holds for all $h \in H$, $a \in S$, $(a_1, \ldots, a_n) \in S^n$, $n \in \mathbb{N}$.

Using (iii) and the fact that $M(S)$ is closed under pointwise multiplication we obtain property (v) observing that $(f \cdot g)_h = f_h \cdot g_h$. In the same way (vi) follows from Theorem 4.5 in Berg *et al.* [1]. ■

If f is monotone of infinite order on H, then we learn from Proposition 5 that for every $h \in H$ the function f_h belongs to $M(S)$; hence there exists a uniquely determined measure $\mu_h \in M_+(\hat{S}_+)$ such that

$$f_h(s) = f(h+s) = \int_{\hat{S}_+} \rho(s)\, d\mu_h(\rho)$$

for all $s \in S$, cf. Choquet [4]. Again we have to require that $\mu_h(\{1_{\{0\}}\}) = 0$ for all $h \in H$ in order to get an integral representation on $\check{H}$. We put

$$M_0(H) = \{f \in M(H) \mid \mu_h(\{1_{\{0\}}\}) = 0 \text{ for } \forall h \in H\}$$

which is an extreme subcone of $M(H)$.

LEMMA 3. *If $H = H'$ (in particular, if H is 2-divisible or if H contains already a neutral element), then $M(H) = M_0(H)$.*

Proof. Let $f \in M(H)$ and $h \in H$ have the form $h = h_1 + h_2$ with h_1, $h_2 \in H$. From

$$f(h+s) = \int_{\hat{S}_+} \rho(s)\, d\mu_h(\rho) = \int_{\hat{S}_+} \rho(s)\rho(h_2)\, d\mu_{h_1}(\rho)$$

for all $s \in S$, we conclude

$$d\mu_h(\rho) = \rho(h_2)\, du_{h_1}(\rho) \qquad \text{on} \quad \hat{S}_+$$

from which

$$\mu_h(\{1_{\{0\}}\}) = 1_{\{0\}}(h_2)\mu_{h_1}(\{1_{\{0\}}\}) = 0. \quad ■$$

THEOREM 4. *For every f in $M_0(H)$ there exists a uniquely determined Radon measure $\mu \in M_+(\check{H}_+)$ such that*

$$f(h) = \int_{\check{H}_+} \rho(h)\, d\mu(\rho) \qquad \textit{for all } h \in H.$$

Proof. It is easy to see that

$$\rho(h)\, d\mu_k(\rho) = \rho(k)\, d\mu_h(\rho)$$

for all $h, k \in H$; hence the Radon measure μ defined by $d\mu(\rho) = 1/\rho(h)\, d\mu_h(\rho)$ on $O_h = \{\rho \in \hat{S}_+ \mid \rho(h) > 0\}$ is well defined on $\check{H}_+$ and satisfies

$$\rho(h)\, d\mu(\rho) = d\mu_h(\rho)$$

on $\check{H}_+$. Furthermore

$$f(h+s) = \int_{\check{H}_+} \rho(s)\, d\mu_h(\rho) = \int_{\check{H}_+} \rho(s+h)\, d\mu(\rho)$$

gives, for $s = 0$,

$$f(h) = \int_{\check{H}} \rho(h)\, d\mu(\rho)$$

for all $h \in H$. This proves the existence of μ and the uniqueness follows in the usual way. ■

In analogy with the negative definite functions we introduce

$$A_0(H) = \{\psi \in A(H) \mid \Delta_a \psi \in M_0(H) \qquad \text{for} \quad \forall a \in S\}.$$

If $H = H'$, then $A_0(H) = A(H)$ by Lemma 3 and Proposition 5.

THEOREM 5. (i) *Let $\psi \in A_0(H)$ be given. Then there exist an additive function $g: S \to [0, \infty[$ and a Radon measure $\nu \in M_+(\check{H}_+ \backslash \{1\})$ such that for all $h, k \in H$*

$$\psi(h) - \psi(k) = \int_{\check{H}_+ \backslash \{1\}} (\rho(k) - \rho(h))\, d\nu(\rho) + g(h) - g(k). \qquad (**)$$

(ii) *If $g: S \to [0, \infty[$ is additive and $\nu \in M_+(\check{H}_+ \backslash \{1\})$ is a Radon measure such that $\rho \to \rho(k) - \rho(h)$ belongs to $L^1(\nu)$ for all $h, k \in H$, then*

$$\psi(h) = \int (\rho(k) - \rho(h))\, d\nu(\rho) + g(h)$$

($k \in H$ arbitrary, but fixed) belongs to $A_0(H)$ and fulfills $(**)$.

We do not give the proof, because it is very similar to those of Theorems 2 and 3.

From Theorem 4.2 in [1] we know that for an Abelian semigroup S with neutral element we always have $M(S) \subseteq P(S)$ and $A(S) \subseteq N(S)$ with equality in case H is 2-divisible. Similar relations are of course also to be expected on semigroups without zero. We state them in the following theorem.

THEOREM 6. (i) *The restriction mapping* $M_0(H) \to P_0(H)$ *given by* $f \to f|_{H'}$ *is one-to-one. If* $H' = H$, *then* $M_0(H)$ *is an extreme subcone of* $P_0(H)$. *For 2-divisible H we have* $M_0(H) = P_0(H)$.

(ii) *The restriction mapping* $A_0(H) \to N_0(H)$ *is injective. If* $H' = H$, *then* $A_0(H)$ *is an extreme subcone of* $N_0(H)$. *We have* $A_0(H) = N_0(H)$ *if H is 2-divisible.*

Proof. (i) Let $f \in M_0(H)$ be given. Then f has an integral representation $f(h) = \int \rho(h)\, d\mu(\rho)$ as stated in Theorem 4, from which it is immediately seen that $f|_{H'}$ is positive definite. If $f, g \in M_0(H)$ have the same restrictions to H', then their representing measures coincide on each

$$O_{2h}^{+} = \{\rho \in S_+ \mid \rho(2h) > 0\}$$

from which they are equal because $\check{H}_+ = \bigcup_{h \in H} O_{2h}^{+}$.

If $H = H'$, then $M_0(H) \subseteq P_0(H)$ and $M_0(H)$ consists exactly of those positive definite functions in $P_0(H)$ whose representing measure is concentrated on $\check{H}_+$. This shows that $M_0(H)$ is an extreme subcone of $P_0(H)$.

For 2-divisible H we obtain the equality $M(H) = P(H)$ from the fact that $\check{H}_+ = \check{H}$.

The proof of (ii) is very similar to that of the first part. We therefore omit it. ∎

3. Examples and Applications

3.1

Let us consider the open half-line $H =]0, \infty[$ with addition. We have $S = [0, \infty[$, $\hat{S} = [0, \infty]$ under the natural identification, $\check{H} = [0, \infty[$, and $\check{H} \backslash \{1\} =]0, \infty[$. By Theorem 6 the cones $P(H)$, $P_0(H)$, $M(H)$, and $M_0(H)$ coincide and the same holds for $N(H)$, $N_0(H)$, $A(H)$, and $A_0(H)$. As a consequence of Theorem 1 and 3 we obtain

THEOREM 7. (i) *For every function* $f \in P(]0, \infty[)$ *there exists a uniquely determined nonnegative Radon measure* μ *on* $[0, \infty[$ *such that*

$$f(t) = \int_{[0,\infty[} e^{-\lambda t}\, d\nu(\lambda) \qquad \text{for} \quad \text{all } t \in]0, \infty[.$$

(ii) *For every function* $\psi \in N(]0, \infty[)$ *there exists a uniquely determined nonnegative Radon measure* ν *on* $]0, \infty[$ *and a constant* $\alpha \in [0, \infty[$ *such that*

$$\psi(s) - \psi(t) = \int_{]0,\infty[} (e^{-\lambda t} - e^{-\lambda s})\, d\nu(\lambda) + \alpha(s - t)$$

for all $s, t \in]0, \infty[$.

(iii) *If* $f \in P(]0,\infty[)$, *then* $\psi:]0,\infty[\to \mathbb{R}$, *defined by* $\psi(x) = \int_1^x f(t)\,dt$ *belongs to* $N(]0,\infty[)$ *and the derivative of any function in* $N(]0,\infty[)$ *belongs to* $P(]0,\infty[)$.

Proof. Only part (iii) needs to be proved and we could base our argument on known facts about completely monotone and Bernstein functions (cf. Berg and Forst [2, §9]). We prefer, however, to give a direct proof.

Let $f \in P(]0,\infty[)$ have the representation given in (i) and put $\psi(x) = \int_1^x f(t)\,dt$. Let $x_1, \ldots, x_n \in]0,\infty[$ and $c_1, \ldots, c_n \in \mathbb{R}$ such that $\sum_{i=1}^n c_i = 0$. Then

$$\begin{aligned}\sum_{i,j=1}^{n} c_i c_j \psi(x_i + x_j) &= \sum_{i,j} c_i c_j \int_{[0,\infty[} \int_1^{x_i + x_j} e^{-\lambda t}\,dt\,d\mu(\lambda) \\ &= \int_{]0,\infty[} \sum_{i,j} c_i c_j (1/\lambda)(e^{-\lambda} - e^{-\lambda(x_i + x_j)})\,d\mu(\lambda) \\ &= \int_{]0,\infty[} -1/\lambda \left(\sum_{i=1}^{n} c_i e^{-\lambda x_i} \right)^2 d\mu(\lambda) \leqslant 0.\end{aligned}$$

Hence ψ, being increasing, is negative definite.

Now suppose that ψ belongs to $N(]0,\infty[)$ and let ψ have the representation given in (ii). We have

$$\frac{\psi(s) - \psi(t)}{s - t} = \int_{[0,\infty[} \frac{e^{-\lambda t} - e^{-\lambda s}}{s - t}\,d\nu(\lambda) + \alpha, \qquad s \neq t.$$

If $t_n \uparrow s$, the quotient $(e^{-\lambda t_n} - e^{-\lambda s})/(s - t_n)$ decreases to $\lambda e^{-\lambda s}$ and if $t_n \downarrow s$, this quotient increases to $\lambda e^{-\lambda s}$. Using the fact that ψ is increasing we may conclude that ψ is differentiable and that $\psi'(s) = \int_{[0,\infty[} e^{-\lambda s}\,d\mu(\lambda)$, where $d\mu(\lambda) = \lambda\,d\nu(\lambda)$ on $]0,\infty[$ and $\mu(\{0\}) := \alpha$. Therefore ψ' belongs to $P(]0,\infty[)$. ∎

As an example we choose $f(x) = 1/x, x \in]0,\infty[$. From the well-known equation

$$f(x) = \int_{[0,\infty[} e^{-\lambda x}\,dm(\lambda),$$

where m denotes Lebesgue measure, we see that $f \in P(]0,\infty[)$. By Theorem 7(iii) the function $s \to \int_1^s f(x)\,dx = \log s$ is negative definite and therefore $\exp(-t \log s) = s^{-t}$ is positive definite for all $t > 0$ by Proposition 3. The integral representation for $\log s$ from Theorem 7(ii) is given by

$$\log s - \log t = \int_{]0,\infty[} (e^{-\lambda t} - e^{-\lambda s})\,d\lambda/\lambda, \qquad s, t \in]0,\infty[.$$

The function f is an infinitely divisible positive definite function and the corresponding one-parameter convolution semigroup $(\mu_t)_{t>0}$ of Radon measures on $[0, \infty[$ is easily seen to be

$$d\mu_t(\lambda) = (1/\Gamma(t))\lambda^{t-1}\, d\lambda$$

(cf. Example 14.16 in [2]). The negative definite functions we get by integrating $s \to s^{-t}$ are given by $s \to s^d$ for $0 < d \leqslant 1$ and $s \to -s^{-d}$ for all $d > 1$—this last family of functions belonging even to $-P(]0, \infty[)$.

3.2

There are several possibilities generalizing $]0, \infty[$ to higher dimensions. One example is given by $[0, \infty[^n \backslash \{0\}$, but in contrast with $[0, \infty[^1 \backslash \{0\} =]0, \infty[$ for $n \geqslant 2$ there exist discontinuous positive definite functions. For example, if $n = 2$, then

$$f_1(x, y) = \frac{1}{y} \cdot 1_{\{0\} \times]0,\infty[}(x, y) \qquad \text{and} \qquad f_2(x, y) = \frac{1}{x} \cdot 1_{]0,\infty[\times \{0\}}(x, y)$$

are positive definite and $f_1 f_2 \equiv 0$. Phenomena like that do, however, not occur if we choose $H =]0, \infty[^n$ instead of $[0, \infty[^n \backslash \{0\}$. Here $\hat{S}$ is easily seen to be the one-point compactification of $[0, \infty[^n$ and $\check{H} = [0, \infty[^n$. We obtain a complete generalization of Theorem 7(i) and (ii) just by replacing $[0, \infty[$ and $]0, \infty[$ by their respective powers. Instead of (iii) we can state that all first partial derivatives of a function in $N(]0, \infty[^n)$ belong to $P(]0, \infty[^n)$.

Before giving examples we state the following simple lemma.

LEMMA 4. *Let H be any Abelian semigroup and let ψ be a negative definite function on H such that $\psi(h) > 0$ for all $h \in H$. Then the following holds.*

$$\text{(i)} \quad \varphi \in P(]0, \infty[) \Rightarrow \varphi \circ \psi \in P(H).$$
$$\text{(ii)} \quad \varphi \in N(]0, \infty[) \Rightarrow \varphi \circ \psi \in N(H).$$

Proof. (i) This case follows immediately from Theorem 7(i), Lemma 2, and Proposition 3(iii).

(ii) We have to show that $e^{-t(\varphi \circ \psi)} \in P(H)$ for all $t > 0$. However $e^{-t(\varphi \circ \psi)} = (e^{-t\varphi}) \circ \psi$ and this is positive definite by (i). ■

Now we choose $\alpha_1, \ldots, \alpha_n \geqslant 0$ not all zero and $\{d_1, \ldots, d_n\} \subseteq]0, 1]$. The function $x \in]0, \infty[^n \to \sum_{i=1}^n \alpha_i x_i^{d_i}$ is negative definite, therefore, applying Lemma 4,

$$x \to \log\left(\sum_{=1}^{n} \alpha_i x_i^{d_i} \right)$$

is negative definite on $]0, \infty[^n$ and

$$x \to 1/(\sum \alpha_i x_i^{d_i})^t$$

is positive definite for any $t > 0$. In the particular case where $\alpha_1 = \alpha_2 = \cdots = \alpha_n = d_1 = d_2 = \cdots = d_n = 1$, the corresponding convolution semigroup of measures on $[0, \infty[^n$ is given by the image measures of $(\mu_t)_{t>0}$ from Section 3.1 under the diagonal mapping $\lambda \to (\lambda, \lambda, \ldots, \lambda)$ from $[0, \infty[$ to $[0, \infty[^n$. One might be tempted to believe that $\|x\|^{-1}$ is positive definite on $]0, \infty[^n$. This, however, is not true, and this can be seen by an application of the following characterization of functions in $P(]0, \infty[^n) = M(]0, \infty[^n)$: f belongs to $P(]0, \infty[^n)$ if and only if $f \in C^\infty(]0, \infty[^n)$ and

$$(-1)^{k_1 + \cdots + k_n} \partial^{k_1 + \cdots + k_n} f / \partial x_1^{k_1} \cdots \partial x_n^{k_n} \geqslant 0$$

for all $(k_1, \ldots, k_n) \in \mathbb{N}_0{}^n$ (cf. [3, Theorem 4.2.1]).

Another negative definite function on $]0, \infty[^n$ is given by $x \to \log(\prod_{i=1}^n x_i)$, it corresponds to the convolution semigroup $(\mu_t \otimes \cdots \otimes \mu_t)_{t>0}$.

Finally we mention that for any probability measure σ on $[0, 1]$ the function

$$x \to \left[\int_0^1 (1/(x_1 + \cdots + x_n))^\alpha \, d\sigma(\alpha)\right]^{-1}$$

is negative definite. This follows from a theorem of Ito (cf. Ito [7] and Hirsch [6, p. 195]).

3.3

Now we consider the semigroup $\mathbb{N} = \{1, 2, 3, \ldots\}$ of natural numbers with addition. We have $S = \mathbb{N}_0, \hat{\mathbb{N}}_0 = [-1, 1]$, $\check{\mathbb{N}} = [-1, 0[\cup]0, 1]$, and $\check{\mathbb{N}}_+ =]0, 1]$. The well-known fact

$$1/n = \int_0^1 t^{n-1} \, dt = \int_{]0,1]} t^n \, dt/t, \qquad \forall n \in \mathbb{N}$$

shows that $n \to 1/n$ is monotone of infinite order; although this function is bounded on $\mathbb{N}$, it has no extension to a function of $M(\mathbb{N}_0)$, as one can see from Proposition 2. From Section 3.1 we know that $n \to -\log(1/n) = \log n$ belongs to $A(\mathbb{N})$, hence $n \to n^{-d}$ is monotone of infinite order for all $d > 0$. The corresponding measures $(\mu_d)_{d>0}$ are a convolution semigroup on the multiplicative semigroup $]0, 1]$ and $d\mu_1(t) = dt/t$.

An example of a function $f \in P_0(\mathbb{N}) \backslash M_0(\mathbb{N})$ is given by

$$f(n) = (-1)^n \cdot (1/n).$$

Its representing measure on $\check{\mathbb{N}}$ is $(1/|t|) \cdot 1_{[-1,0[}(t) \, dt$.

One might ask if there exists a theorem analogous to part (iii) of Theorem 7 for the discrete case. Surprisingly enough this holds to be true, not on $\mathbb{N}$ but on $\mathbb{N}_0$. We state it here without proof (which is not difficult). For that reason $N^*(\mathbb{N}_0)[A^*(\mathbb{N}_0)]$ denotes all $\psi \in N(\mathbb{N}_0)[A(\mathbb{N}_0)]$ with $\psi(0) = 0$.

PROPOSITION 6. *For any function $f \in P(\mathbb{N}_0)$ define $\psi_f : \mathbb{N}_0 \to \mathbb{R}$ by $\psi_f(0) = 0$, $\psi_f(n) = \sum_{j=0}^{n-1} f(j)$ for all $n \geqslant 1$. Put $\Psi(f) = \psi_f$. Then the following holds:*

(i) *Ψ is a bijective and affine mapping from $P(\mathbb{N}_0)$ onto $N^*(\mathbb{N}_0)$.*
(ii) *$\Psi(M(\mathbb{N}_0)) = A^*(\mathbb{N}_0)$.*

Again there exist several possibilities for dimensions >1 to generalize the semigroup $\mathbb{N}$. Choosing $H = \mathbb{N}^k$ has the advantage that $\hat{S}$ is again a one-point compactification, namely of $([-1, 0[\cup]0, 1])^k$ which then is just $\check{H}$. As an example of a function in $M(\mathbb{N}^k)$ we take

$$n = (n_1, \ldots, n_k) \to 1/(n_1 + \cdots + n_k)$$

and

$$n \to (-1)^{n_1 + \cdots + n_k}/(n_1 + \cdots + n_k)$$

belongs to $P(\mathbb{N}^k) \backslash M(\mathbb{N}^k)$.

3.4

Let K be a compact Hausdorff space and let H be the set of all continuous real-valued functions on K that are strictly positive. Under pointwise addition H is an Abelian semigroup. Adding to H the zero function we obtain $S = H \cup \{0\}$. If κ is a nonnegative Radon measure on K, we obtain a character $\rho_\kappa \in \check{H}$ by defining

$$\rho_\kappa(h) = \exp\left(-\int h \, d\kappa\right), \qquad h \in S.$$

Of course $\rho_{\kappa_1} = \rho_{\kappa_2}$ implies $\kappa_1 = \kappa_2$. Suppose on the other hand that ρ belongs to $\check{H}$. Then $\lambda = -\log \rho$ is an additive function from S to the extended half-line $[0, \infty]$ and $\lambda(0) = 0$. It is easy to see that λ is monotone and therefore also positive homogeneous. By assumption there exists $h_0 \in H$ such that $\lambda(h_0) < \infty$; hence $\lambda(h) < \infty$ for all $h \in H$. We extend λ to $C_+(K)$ by $\lambda(f) = \lim_{n \to \infty} \lambda(h_n)$ for $h_n \in H$ with $h_n \downarrow f$; this definition does not depend on the choice of $\{h_n\}$. Now we can apply the Riesz representation theorem to find a nonnegative Radon measure κ on K such that $\lambda(f) = \int f \, d\kappa$ for all $f \in C_+(K)$. This leads to an identification of $\check{H}$ with the cone $M_+(K)$ of all nonnegative Radon measures on K which furthermore is, given in this way, a locally compact topology. Then $\hat{S}$ is the one-point compactification of $M_+(K)$. The particular case where K is a finite set leads us back to the semigroup $]0, \infty[^n$ we considered in Section 3.2.

Examples of negative definite functions on H are given by

$$h \to \int_K \alpha(x)[h(x)]^{d(x)}\, d\kappa(x),$$

where $\alpha: K \to [0, \infty[$ is measurable and bounded, $d: K \to [0, 1]$ is measurable, and $\kappa \in M_+(K)$. These functions, taken to the power $-t$, $t > 0$, are positive definite.

3.5

As the last example we mention the semigroup $H = [0, 1[$ with "$\wedge$" as semigroup operation. It is easy to identify $\check{H}$ as

$$\check{H} = \{1_{[a,1]} | 0 \leqslant a < 1 \cup \{1_{]a,1]} | 0 \leqslant a < 1\}$$

(cf. [1, §7]). With almost the same proof we obtain that $P(H)$ consists of all nonnegative increasing functions on $[0, 1[$ and that $N(H)$ is the set of all decreasing real-valued functions on $[0, 1[$.

REFERENCES

1. C. BERG, J. P. R. CHRISTENSEN, AND P. RESSEL, Positive definite functions on abelian semigroups, *Math. Ann.* **223** (1976), 253–274.
2. C. BERG AND G. FORST, "Potential Theory on Locally Compact Abelian Groups," Ergebnisse, Vol. 87, Springer-Verlag, Berlin and New York, 1975.
3. S. BOCHNER, "Harmonic Analysis and the Theory of Probability," Univ. of California Press, Berkeley and Los Angeles, 1955.
4. G. CHOQUET, Theory of capacities, *Ann. Inst. Fourier* (*Grenoble*) **5** (1954), 131–295.
5. A. GUICHARDET, "Symmetric Hilbert Spaces and Related Topics," Springer Lecture Notes, Vol. 261, Berlin and New York, 1972.
6. F. HIRSCH, Familles résolvantes, générateurs, cogénérateurs, potentiels, *Ann. Inst. Fourier* (*Grenoble*) **22** (1972), 89–210.
7. M. ITO, Sur les sommes de noyaux de Dirichlet, *C.R. Acad. Sci. Paris* **271** (1970), Sér. A, 937–940.

AMS (MOS) subject classifications: 20M15, 42A76, 42A88, 43A30.

STUDIES IN ANALYSIS
ADVANCES IN MATHEMATICS SUPPLEMENTARY STUDIES, VOL. 4

Invariant Weights on Decomposable von Neumann Algebras

NGHIEM DANG-NGOC[†‡]

Laboratoire de Probabilités
Université de Paris
Paris, France.

We prove in this chapter that a decomposable dynamical system (by "dynamical system" we mean an action of a group on a von Neumann algebra) is semifinite if and only if almost all component systems are semifinite. This result implies a classical theorem of Schwartz on semifinite von Neumann algebras and the existence of σ finite-invariant measures on certain Borel G-spaces (our proof is new even in these particular cases). In fact, we are able to prove a more general result on the integration of relatively invariant weights. We also give a characterization of semifinite systems by the existence of faithful invariant operatorial normal weights, solving a question of Guichardet. The first result allows us to deduce a theorem of Guichardet and Nest from a theorem of Størmer on the existence of invariant traces for certain dynamical systems.

1. DECOMPOSABLE SYSTEMS

We use the notations of [5]. Let $\mathscr{F} = (\mathscr{Q}, G, \tau, \mathscr{H}, U)$ be a concrete dynamical system, i.e., $\mathscr{Q}$ is a von Neumann algebra acting on a Hilbert space $\mathscr{H}$, G a locally compact group, τ a continuous representation of G into the group $\mathrm{Aut}(\mathscr{Q})$ of all automorphisms of $\mathscr{Q}$, and U a continuous unitary representation of G into $\mathscr{H}$ such that $\tau_g(x) = U_g x U_g{}^*$, $x \in \mathscr{Q}$, $g \in G$. We say that $\mathscr{F}$ is decomposable if there exist a standard measure space (Z, ν) and measurable fields $z \to \mathscr{H}(z)$, $z \to \mathscr{Q}(z)$, $z \to U(z)$ [$U(z)$ is a continuous unitary representation of G into $\mathscr{H}(z)$] such that

$$\mathscr{H} = \int_Z^{\oplus} \mathscr{H}(z)\, d\nu(z), \qquad \mathscr{Q} = \int^{\oplus} \mathscr{Q}(z)\, d\nu(z), \qquad U = \int^{\oplus} U(z)\, d\nu(z),$$

$$U_g(z)\mathscr{Q}(z)U_g(z)^* = \mathscr{Q}(z), \qquad \forall z \in Z, \quad \forall g \in G.$$

† Équipe de recherche n⁰ I, "Processus stochastiques et applications," dépendant de la Section n⁰ I, "Mathématiques-Informatique," associée au C.N.R.S.

‡ Present address Université de l'État à Mons (Belgium).

ISBN 0-12-599150-9

For every $z \in Z$, we have a concrete dynamical system

$$\mathscr{F}(z) = (\mathscr{Q}(z), G, \tau, \mathscr{H}(z), U(z))$$

and we write $\mathscr{F} = \int^{\oplus} \mathscr{F}(z)\, dv(z)$.

We say that $\mathscr{F}$ is semifinite if there exists a faithful normal semifinite weight on $\mathscr{Q}_+$ invariant by G.

We know that a dynamical system $\mathscr{F} = (\mathscr{Q}, G)$ with $\mathscr{Q}_*$ separable and G separable can be realized as a concrete dynamical system in a separable Hilbert space and if $\mathscr{C}^0$ is a von Neumann subalgebra of $\mathscr{C}^G$, the set of invariant elements of the center $\mathscr{C}$ of $\mathscr{Q}$, we can identify $\mathscr{C}^0 = L^\infty(Z, v)$ with (Z, v) standard and find a decomposition $\mathscr{F} = \int_Z^{\oplus} \mathscr{F}(z)\, dv(z)$ over Z (cf. [7, p. 14]).

THEOREM 1. *Let $\mathscr{F} = (\mathscr{Q}, G)$ by a dynamical system with $\mathscr{Q}_*$ separable and G a locally compact separable group. Consider a decomposition $\mathscr{F} = \int_Z^{\oplus} \mathscr{F}(z)\, dv(z)$ of $\mathscr{F}$ over a standard Borel space (Z, v). Then $\mathscr{F}$ is semifinite if and only if $\mathscr{F}(z)$ is semifinite for v almost all z in Z.*

Remarks. (a) If we take $\{u_n\}_{n \in \mathbb{N}}$, a countable group of unitaries of $\mathscr{Q}$ generating $\mathscr{Q}$ such that $u_n = \int^{\oplus} u_n(z)\, dv(z)$, with $\{u_n(z)\}_{n \in \mathbb{N}}$ unitary and generating $\mathscr{Q}(z)$, $\forall z \in Z$. A normal semifinite weight on $\mathscr{Q}_+$ (respectively, $\mathscr{Q}(z)_+$) is a trace if and only if it is invariant by the group of inner automorphisms determined by the u_n's (respectively, the $u_n(z)$'s) (cf. [3, Th. 4]). Theorem 1 implies the theorem of Schwartz (cf. [10], [5]) that a decomposable von Neumann algebra is semifinite if and only if almost all component algebras are semifinite.

(b) If $\mathscr{Q}$ is Abelian, we obtain Theorem IV-1 of [4] on decomposable Borel G-spaces implying the existence of σ finite-invariant measures on regular G-spaces when G is nilpotent (that is an extension of a result of Dixmier [6]).

(c) The methods used in (a) and (b) are based on notions of induced systems on Hermitian projections and the fact that a semifinite system possesses sufficiently many finite-induced systems and conversely. For general systems, these notions are not defined. To prove the theorem we construct directly a measurable field of invariant weights with the use of the cocycles of Connes [2, Theorem 1.2.1, p. 144]. First we prove some preliminary results.

LEMMA 1. *If a normal semifinite weight φ on $\mathscr{Q}_+$ is invariant by a dense subgroup G_0 of G, then it is invariant by G.*

Proof. Let $x \in \mathcal{Q}_+$, $g \in G$, let $g_\alpha \to g$ with $g_\alpha \in G_0$. We have $\varphi(\tau_{g_\alpha} x) = \varphi(x)$, $\forall \alpha$. As φ is lower semicontinuous for the weak topology, we have

$$\varphi(\tau_g x) = \varphi(\lim \tau_{g_\alpha} x) \leqslant \varphi(x), \qquad \varphi(x) = \varphi(\lim \tau_{g_\alpha^{-1}}(\tau_g x)) \leqslant \varphi(\tau_g x).$$

Therefore $\varphi(\tau_g x) = \varphi(x)$, $\forall g \in G$, $\forall x \in \mathcal{Q}_+$. Hence φ is G-invariant.

Let $\mathscr{C}^0 = L^\infty(Z, \nu)$ be the von Neumann algebra of diagonal operators in $\mathscr{H}$. We have $\mathscr{C}^0 \subset \mathscr{C}^G$, where $\mathscr{C}^G$ is the set of invariant elements of the center $\mathscr{C}$ of $\mathcal{Q}$.

We recall some results and notations of Combes [1]. Let φ be a faithful normal semifinite weight on $\mathcal{Q}_+$, let $\mathcal{N}_\varphi = \{x \in \mathcal{Q} \mid \varphi(x^*x) < \infty\}$, let $\mathscr{H}_\varphi$, π_φ, Λ_φ be the objects of the canonical representation defined by φ. The set $\mathscr{U} = \Lambda_\varphi(\mathcal{N}_\varphi \cap \mathcal{N}_\varphi^*)$ provided with the scalar product, the product and the involution deduced from that of $\mathcal{N}_\varphi \cap \mathcal{N}_\varphi^*$ is a left Hilbert algebra. Let $\mathscr{L}(\mathscr{U})$ be the associated left von Neumann algebra where $\mathscr{L}(\mathscr{U}) = \pi_\varphi(\mathcal{Q})$ [1].

The continuous action τ of G on $\mathcal{Q}$ induces an action of G on $\pi_\varphi(\mathcal{Q})$ that we denote again by τ. Following Haagerup [8, Lemma 3.9] there exists a continuous unitary representation U^φ of G into $\mathscr{H}_\varphi$ such that $\tau_g x = U_g{}^\varphi x U_g^{\varphi *}$, $x \in \pi_\varphi(\mathcal{Q})$, $g \in G$. As $\pi_\varphi(\mathscr{C}^0)$ is in the center of $\pi_\varphi(\mathcal{Q})$ and is composed of G-invariant elements of $\mathcal{Q}$, we can identify $\pi_\varphi(\mathscr{C}^0)$ with $L^\infty(Z, \nu)$ and obtain a decomposition of $\mathscr{H}_\varphi$, $\pi_\varphi(\mathcal{Q})$, and U^φ following $\pi_\varphi(\mathscr{C}^0)$:

$$\mathscr{H} = \int^\oplus \mathscr{H}_\varphi(z)\, d\nu(z), \qquad \pi_\varphi(\mathcal{Q}) = \int^\oplus \pi_\varphi(\mathcal{Q})(z)\, d\nu(z), \qquad U^\varphi = \int^\oplus U^\varphi(z)\, d\nu(z),$$

where $g \to U_g{}^\varphi(z)$ is a continuous unitary representation of G into $\mathscr{H}_\varphi(z)$.

Following [5, Proposition II.3–11, p. 185], we can suppose that $\mathscr{F}$ is realized in this way in $\mathscr{H}$, i.e., $\mathscr{H} = \mathscr{H}_\varphi$, $\mathcal{Q} = \pi_\varphi(\mathcal{Q})$, $U = U^\varphi$, etc. By a result of Sutherland [11, Th. 4.1] there exists a measurable field of left Hilbert algebras $z \to \mathscr{U}(z)$, $z \in Z$, such that $\mathscr{U} = \int^\oplus \mathscr{U}(z)\, d\nu(z)$ and if $\varphi(z)$ is the canonical faithful normal semifinite weight on $\mathscr{L}(\mathscr{U}(z))$, the field $z \to \varphi(z)$ is measurable and $\varphi = \int^\oplus \varphi(z)\, d\nu(z)$. Let $V_t^\varphi = \Delta_\varphi^{it}$, $V_t^{\varphi(z)} = \Delta_\varphi^{it}(z)$ and let $\sigma_t{}^\varphi$ and $\sigma_t^{\varphi(z)}$ be the modular automorphism group of φ and $\varphi(z)$, respectively, with $\sigma_t{}^\varphi(\cdot) = V_t^\varphi \cdot V_t^{\varphi *}$, $\sigma_t^{\varphi(z)}(\cdot) = V_t^{\varphi(z)} \cdot V_t^{\varphi(z)*}$, $\forall t \in \mathbb{R}$. By [10, Th. 2.1], we have

$$\sigma_t{}^\varphi = \int^\oplus \sigma_t^{\varphi(z)}\, d\nu(z), \qquad V_t^\varphi = \int^\oplus V_t^{\varphi(z)}\, d\nu(z), \qquad \forall t \in \mathbb{R}.$$

In this paragraph we fix a faithful weight φ and we write $\sigma_t = \sigma_t{}^\varphi$, $V_t = V_t^\varphi$, $V_t^z = V_t^{\varphi(z)}$, etc. (By "weight" we mean normal semifinite weight.) Consider now another faithful weight ψ on $\mathcal{Q}_+$. By Sutherland [11, Th. 3.4 and Th. 4.1] we can write $\psi = \int^\oplus \psi(z)\, d\nu(z)$, where $\psi(z)$ is a faithful weight on $\mathcal{Q}(z)$ (we suppress if needed a ν-null subset of Z).

Let $(D\psi : D\varphi)_t$ and $(D\psi(z) : D\varphi(z))_t$ be the Radon–Nikodym derivatives following Connes [2, Theorem 1.2.1].

LEMMA 2. *We have*

$$(D\psi : D\varphi)_t = \int^{\oplus} (D\psi(z) : D\varphi(z))_t \, d\nu(z), \qquad \forall t \in \mathbb{R}.$$

Proof. We come back to the construction of $(D\psi : D\varphi)$ by Connes [2, Th. 1.2.1]: let F_2 be the type I_2 factor acting on $\mathbb{C}^4$, let $(e_{ij})_{i,j=1,2}$ be a system of matrix units of F_2. By Dixmier [5, Prop. II.3–3 p. 177], we have

$$\mathscr{H} \otimes \mathbb{C}^4 = \int^{\oplus} \mathscr{H}(z) \otimes \mathbb{C}^4 \, d\nu(z) \qquad \text{and} \qquad \mathscr{Q} \otimes F_2 = \int^{\oplus} (\mathscr{Q}(z) \otimes F_2) \, d\nu(z).$$

As in [2, Lemma 1-2-2], let θ (respectively, $\theta(z)$) be the faithful weight defined on $\mathscr{Q} \otimes F_2$ (respectively, $\mathscr{Q}(z) \otimes F_2$) by

$$\theta(\textstyle\sum x_{ij} \otimes e_{ij}) = \varphi(x_{11}) + \psi(x_{22}) \qquad \text{for} \quad \textstyle\sum x_{ij} \otimes e_{ij} \geqslant 0$$

(respectively, $\theta(z)(\sum x_{ij}(z) \otimes e_{ij}) = \varphi(z)(x_{11}(z)) + \psi(z)(x_{22}(z))$ for $\sum x_{ij}(z) \otimes e_{ij} \geqslant 0$. We have $\theta = \int^{\oplus} \theta(z) \, d\nu(z)$. By Connes [2, Lemma 1.2.2] we have

$$\sigma_t^{\theta}(1 \otimes e_{21}) = (D\psi : D\varphi)_t \otimes e_{21}, \qquad \forall t \in \mathbb{R},$$
$$\sigma_t^{\theta(z)}(1_z \otimes e_{21}) = (D\psi(z) : D\varphi(z))_t \otimes e_{21}, \qquad \forall t \in \mathbb{R}, \quad \forall z \in Z,$$

and since

$$\sigma_t^{\theta} = \int^{\oplus} \sigma_t^{\theta(z)} \, d\nu(z) \qquad \text{(cf. [11, Th. 2.1])},$$

we have

$$(D\psi : D\varphi)_t = \int^{\oplus} (D\psi(z) : D\varphi(z))_t \, d\nu(z), \qquad \forall t \in \mathbb{R}.$$

Let $u_t^g = (D\tau_g\varphi : D\varphi)_t$, $u_t^g(z) = (D\tau_g\varphi(z) : D\varphi(z))_t$, $\forall z \in Z, \forall t \in \mathbb{R}, \forall g \in G$.

LEMMA 3. *Let ψ be another faithful weight on $\mathscr{Q}_+$, let $v_t = (D\psi : D\varphi)_t$, $t \in \mathbb{R}$. The following conditions are equivalent.*

(i) *The weight ψ is invariant by G.*
(ii) $1 = \tau_g v_t u_t^g v_t^*$, $\forall t \in \mathbb{R}, \forall g \in G$.
(iii) $1 = \tau_g v_t u_t^g v_t^*$, $\forall t \in \mathbb{Q}, \forall g \in G_0$, *where $\mathbb{Q}$ is the set of rational numbers and G_0 is a dense subgroup of G.*

Proof. The relation $\tau_g\psi = \psi$, $\forall g \in G$, can be written

$$1 = (D\tau_g\psi : D\psi), \qquad \forall g \in G, \quad \forall t \in \mathbb{R}$$

or

$$1 = (D\tau_g\psi : D\tau_g\varphi)_t(D\tau_g\varphi : D\varphi)_t(D\varphi : D\psi)_t, \qquad \forall t \in \mathbb{R}, \quad \forall g \in G.$$

or

$$1 = \tau_g v_t u_t^g v_t^*, \qquad \forall t \in \mathbb{R}, \quad \forall g \in G.$$

We have proved the equivalence of (i) and (ii). The last part of the lemma results from above, the lemma 1, and the fact that the variable t is continuous in the above equality.

Proof of Theorem 1. We fix a countable dense subgroup G_0 of G, $G_0 = \{g_n\}_{n \in \mathbb{N}}$.

(1) Suppose that $\mathscr{F}$ is semifinite. Let ψ be a faithful weight on $\mathscr{Q}_+$ invariant by G_0. Let $v_t = (D\psi : D\varphi)_t$, $v_t(z) = (D\psi(z) : D\varphi(z))_t$. We have

$$1 = \tau_g(v_t)u_t^g v_t^*, \qquad \forall g \in G_0, \quad \forall t \in \mathbb{Q}.$$

By Lemma 2, for almost all $z \in Z$, we have

$$1 = \tau_g(v_t(z))u_t^g(z)v_t(z)^*, \qquad \forall g \in G_0, \quad \forall t \in \mathbb{Q}.$$

Lemma 3 implies that $\psi(z)$ is G-invariant for almost all $z \in Z$. Therefore $\mathscr{F}(z)$ is semifinite for almost all $z \in Z$.

(2) Suppose now that $\mathscr{F}(z)$ is semifinite for almost all $z \in Z$. We are going to construct a continuous application $v: \mathbb{R} \to \mathscr{U}(\mathscr{Q})$ [$\mathscr{U}(\mathscr{Q})$ is the set of unitaries of $\mathscr{Q}$ provided with the strong topology] such that

$$v_{t+s} = v_t\sigma_t(v_s), \qquad \forall t, s \in \mathbb{R}, \tag{*}$$

$$1 = \tau_g v_t u_t^g v_t^*, \qquad \forall g \in G_0, \quad \forall t \in \mathbb{Q}. \tag{**}$$

It results from Lemma 3 that the canonical faithful weight ψ associated with $\{v_t\}_{t \in \mathbb{R}}$ by Connes (i.e., such that $(D\psi : D\varphi)_t = v_t, \forall t \in \mathbb{R}$) is invariant by G, which implies that $\mathscr{F}$ is semifinite.

By a standard method (cf. [5, Ch. II]) we can suppose that:

(a) The field $z \to \mathscr{H}(z)$ is a constant field, $\mathscr{H}(z) = \mathscr{H}_0, \forall z \in Z$.

(b) The space Z is metric and compact, $\mathscr{F}(z)$ is semifinite for all $z \in Z$, and there exist *continuous* mappings $z \to T_n(z)$, $z \to T_n'(z)$, $z \to U_{g_n}(z)$, $z \to V_t(z)$, $z \to u_t^{g_n}(z)$ (with $n \in \mathbb{N}$, $t \in \mathbb{Q}$) of Z into the unit ball of $\mathscr{L}(\mathscr{H}_0)$ provided with the strong topology such that for all $z \in Z$ we have

(A) $\{T_n(z)^*\}_{n \in \mathbb{N}} = \{T_n(z)\}_{n \in \mathbb{N}}$, $\{T_n'(z)^*\}_{n \in \mathbb{N}} = \{T_n'(z)\}_{n \in \mathbb{N}}$.
(B) $\{T_n(z)\}_{n \in \mathbb{N}}$ generates $\mathscr{Q}(z)$, $\{T_n'(z)\}_{n \in \mathbb{N}}$ generates $\mathscr{Q}(z)'$
(C) $U_{g_n}(z)$ and $V_t(z)$ are unitary, $\forall n \in \mathbb{N}, \forall t \in \mathbb{Q}$.
(D) $u_t^{g_n}(z)$ is a unitary of $\mathscr{Q}(z), \forall n \in \mathbb{N}, \forall t \in \mathbb{Q}$.

Let $C(R, \mathscr{U}_0)$ be the set of all *continuous* mappings of R into the set $\mathscr{U}_0$ of all unitaries of $\mathscr{H}_0$ provided with the strong topology. Then $C(R, \mathscr{U}_0)$ is a polonais space for the topology of uniform convergence on compact subsets of R. We provide $C(R, \mathscr{U}_0)$ with this topology.

Consider the set X of couples $(z, \alpha) \in Z \times C(R, \mathscr{U}_0)$ such that

(i)′ $\alpha(t)T_n'(z) = T_n'(z)\alpha(t), \forall n \in \mathbb{N}, \forall t \in \mathbb{R}.$
(ii)′ $\alpha(r + s) = \alpha(r)V_r(z)\alpha(s)V_r(z)^*, \forall r \in \mathbb{Q}, \forall s \in \mathbb{Q}.$
(iii)′ $1 = U_{g_n}(z)\alpha(r)U_{g_n}(z)^*u_r^{g_n}(z)\alpha(r)^*, \forall n \in \mathbb{N}, \forall r \in \mathbb{Q}.$

The set X is Borel in $Z \times C(R, \mathscr{U}_0)$; by the hypothesis and the Lemma 3, for every $z \in Z$ there exists $\alpha \in C(R, \mathscr{U}_0)$ such that $(z, \alpha) \in X$.

By [5, Appendix V, p. 332], there exists a ν-measurable mapping $z \to v(z)$ from Z into $C(R, \mathscr{U}_0)$ such that $(z, v(z)) \in X$, $\forall z \in Z$. For each t in R the field $z \to v(z)_t$ is measurable and essentially bounded. Let

$$v_t = \int^{\oplus} v(z)_t \, d\nu(z), \qquad \forall t \in \mathbb{R}.$$

The mapping $t \to v(z)_t$ is strongly continuous for every z in Z and the Dominated convergence theorem implies that the mapping $t \to v_t$ is also continuous.

The continuity of the mapping $t \to v_t$ and properties (i)′–(iii)′ show that v is a continuous mapping from R into the set of unitaries of $\mathscr{Q}$ satisfying (*) and (**). The theorem is proved.

Let χ be a positive real character of G; a weight ψ is said to be *χ-relatively invariant by G* if $\tau_g \psi = \chi(g)\psi, \quad \forall g \in G$.

PROPOSITION 1. *Suppose that G is a separable locally compact group, $\mathscr{F} = \int^{\oplus} \mathscr{F}(z) \, d\nu(z)$, let χ be a positive real character of G. The system $\mathscr{F}$ possesses a faithful weight χ-relatively invariant by G if and only if $\mathscr{F}(z)$ possesses a faithful weight χ-relatively invariant by G for almost all z in Z.*

Idea of the Proof. The proof remains practically the same except for two modifications:

(1) If $\varphi(\tau_g x) = \chi(g)\varphi(x), \forall g \in G_0, \forall x \in \mathscr{Q}_+,$ then $\tau_g \varphi = \chi(g)\varphi, \forall g \in G.$

We fix $x \in \mathscr{Q}_+$ and consider the mapping $g \to f(g) = \chi(g^{-1})\varphi(\tau_g x)$. It is lower semicontinuous and $f(g_0 g) = f(g)$, $\forall g_0 \in G_0$, $\forall g \in G$; a proof similar to that of Lemma 1 shows that $f(g) = f(e)$, with e the unit of G, hence $\varphi(\tau_g x) = \chi(g)\varphi(x)$, $\forall g \in G$, $\forall x \in \mathscr{Q}_+$.

(2) The equality in Lemma 3 is replaced by

$$\chi(g)^{it} = \tau_g v_t u_t^g v_t^*, \qquad \forall t \in \mathbb{R}, \quad \forall g \in G.$$

Remark. For $\chi \not\equiv 1$ the method of induced systems in [10] cannot be used to prove this proposition even when τ_g is inner for all g in G.

PROPOSITION 2. *The set of points z in Z such that $\mathscr{F}(z)$ is semifinite (respectively, finite, discrete, continuous* (cf. [3]), *admitting a faithful χ-relatively invariant weight) is ν-measurable.*

Proof. Let Z_{sf} be the set of $z \in Z$ such that $\mathscr{F}(z)$ is semifinite; by a standard method (cf. Dixmier [5, Ch. II]) we can suppose that we have the situation of (2) of the proof of Theorem 1 except for the hypothesis that $\mathscr{F}(z)$ is semifinite for all $z \in Z$. The set Z_{sf} is exactly the projection into Z of the Borel subset X of $Z \times C(R, \mathscr{U}_0)$. Therefore Z_{sf} is analytic, hence ν-measurable.

The other cases are treated similarly by using different proofs of [4].

Remark. Størmer has proved that if $(\mathscr{Q}, G, \tau)$ is a *finite* dynamical system with $\mathscr{Q}$ semifinite and $\mathscr{C}^G = \{\text{scalars}\}$, then there exists a faithful semifinite normal trace on $\mathscr{Q}_+$ invariant by G. If $\mathscr{Q}_*$ is separable, we can use Lemma 1 and suppose that G is countable. We can then decompose the system into centrally ergodic ones and apply the above result of Størmer to each component and obtain the Guichardet and Nest theorem: If $(\mathscr{Q}, G, \tau)$ is finite with $\mathscr{Q}$ semifinite and $\mathscr{Q}_*$ separable, then there exists a faithful semifinite normal trace on $\mathscr{Q}_+$ invariant by G.

2. INVARIANT OPERATORIAL WEIGHTS AND SEMIFINITE SYSTEMS

In this section we consider a general dynamical system $(\mathscr{Q}, G, \tau)$ with G a locally compact group (not necessarily separable). Let $\mathscr{C}^0$ be a von Neumann subalgebra in the center of $\mathscr{Q}$. We realize $\mathscr{C}^0 = L^\infty(Z, \nu)$ with Z compact and ν a locally finite measure (cf. [5]). Let $\hat{\mathscr{C}}_+{}^0$ be the set of positive real ν-measurable functions on Z.

We call *$\mathscr{C}^0$-weight* (or operatorial weight into $\hat{\mathscr{C}}_+{}^0$) every positive, positively linear mapping E from $\mathscr{Q}_+$ into $\hat{\mathscr{C}}_+{}^0$ such that

$$E(b^*ab) = b^*E(a)b, \qquad \forall a \in \mathscr{Q}_+, \quad \forall b \in \mathscr{C}^0 \tag{1}$$

Then E is said to be *finite* if $E(\mathscr{Q}_+) \subset \mathscr{C}_+{}^0$, *semifinite* if $E^{-1}(\mathscr{C}_+{}^0)$ is total in $\mathscr{Q}$, true if the relations $a \in \mathscr{Q}_+$, $E(a) = 0$ imply $a = 0$. A finite $\mathscr{C}^0$-weight E is said to be *normal* if $E(\lim_\alpha \uparrow a_\alpha) = \lim_\alpha \uparrow E(a_\alpha)$ for all increasing family $\{a_\alpha\}$ of $\mathscr{Q}_+$. A general $\mathscr{C}^0$-weight E is said to be *normal* if there exists a family E_α of normal finite $\mathscr{C}^0$-weights E such that $E(a) = \sup E_\alpha(a)$, $\forall a \in \mathscr{Q}_+$.

Suppose that $\mathscr{C}^0 \subset \mathscr{A}^G$. We say that E is *invariant* by G if $E \circ \tau_g = E$, $\forall g \in G$. We say that E is *χ-relatively invariant* by G if $E \circ \tau_g = \chi(g)E$, $\forall g \in G$.

PROPOSITION 3. *Let $\mathscr{F} = (\mathscr{Q}, G, \tau)$ be a dynamical system, χ a positive real character of G. The system $\mathscr{F}$ possesses a faithful weight χ-relatively invariant by G if and only if there exists a faithful normal semifinite $\mathscr{C}^G$-weight E χ-relatively invariant by G.*

Assume that such E exists and suppose furthermore that $\mathscr{C}^G = \mathscr{Q}^G$. Then the mapping $\varphi \to \psi = \varphi \circ E$ is a positive positively linear bijection of the set $\mathscr{P}(\mathscr{C}^G)$ of normal semifinite weights on $\mathscr{C}^G$ onto the set $\mathscr{P}(\mathscr{Q})^G$ of normal semifinite weights on $\mathscr{Q}_+$ χ-relatively invariant by G.

LEMMA 4. *Let $\mathscr{C}^0$ be a central von Neumann subalgebra of $\mathscr{Q}$, let ψ be a normal weight on $\mathscr{Q}_+$, φ a faithful weight on $\mathscr{C}_+{}^0$. There exists a unique normal semifinite $\mathscr{C}^0$-weight $E_\varphi{}^\psi$ on $\mathscr{Q}_+$ such that*

$$\varphi(E_\varphi{}^\psi a) = \psi(a), \qquad \forall a \in \mathscr{Q}_+. \tag{2}$$

The weight ψ is faithful if and only if $E_\varphi{}^\psi$ is. For fixed φ, the mapping $\psi \to E_\varphi{}^\psi$ is a bijection from the set of normal weights on $\mathscr{Q}_+$ onto the set of normal semifinite $\mathscr{C}^0$-weights of $\mathscr{Q}_+$.

Proof. (a) Suppose that ψ is finite. Let p be the support of ψ and let $\tilde{E}^\psi$ be the canonical conditional expectation of $\mathscr{Q}_p$ onto $\mathscr{C}_p{}^0$ such that $(\varphi|_{\mathscr{C}_p{}^0}) \circ \tilde{E}^\psi = (\psi|_{\mathscr{Q}_p})$ [12]. Let c be the $\mathscr{C}^0$-support of p. It is easy to see that the mapping $\mathscr{C}_c{}^0 \ni b \to bp \in \mathscr{C}_p{}^0$ is a $*$-isomorphism; let j be the inverse isomorphism. Let $E_\varphi{}^\psi$ be the mapping

$$\begin{array}{llllll} E: a \to pap & \to \tilde{E}^\psi(pap) & \to j(\tilde{E}^\psi(pap)) & \to hj(\tilde{E}^\psi(pap)) \\ \mathscr{Q}^+ \to \mathscr{Q}_p{}^+ & \to \mathscr{C}^0_{p+} & \to \mathscr{C}^0_{c+} & \to \hat{\mathscr{C}}_+{}^0 \end{array}$$

where h is the Radon–Nikodym derivative of $(\psi|_{\mathscr{C}^0})$ with respect to φ [8]. It is easy to see that $E_\varphi{}^\psi$ is a normal semifinite $\mathscr{C}^0$-weight satisfying (2). The uniqueness follows from (1) and (2).

(b) General case: Following Combes [1] there exists an increasing family (ψ_α) of positive normal linear forms on $\mathscr{Q}$ such that $\psi(a) = \lim_\alpha \uparrow \psi_\alpha(a)$ for $\forall a \in \mathscr{Q}_+$. Let $E_\varphi^{\psi_\alpha}$ be the $\mathscr{C}^0$-weight associated to ψ_α in (a). For every $a \in \mathscr{Q}_+$, the family $\{E_\varphi^{\psi_\alpha}(a)\}$ is increasing. Let $E_\varphi{}^\psi(a) = \lim_\alpha E_\varphi^{\psi_\alpha}(a) \in \hat{\mathscr{C}}_+{}^0$. For $a \in \mathscr{Q}_+$, $b \in \mathscr{C}^0$, we have

$$E_\varphi{}^\psi(b^*ab) = \lim_\alpha E_\varphi^{\psi_\alpha}(b^*ab) = \lim_\alpha (b^*E_\varphi^{\psi_\alpha}(a)b) = b^*(\lim_\alpha E_\varphi^{\psi_\alpha}(a)b),$$

hence

$$E_\varphi{}^\psi(b^*ab) = b^*E_\varphi{}^\psi(a)b.$$

We have

$$\begin{aligned}\psi(a) &= \lim_\alpha \uparrow \psi_\alpha(a) = \lim_\alpha \uparrow \varphi(E_\varphi^{\psi_\alpha}a) = \varphi\left(\lim_\alpha \uparrow E_\varphi^{\psi_\alpha}a\right)\\ &= \varphi \circ E_\varphi{}^\psi(a).\end{aligned}$$

The uniqueness of $E_\varphi{}^\psi$ follows from (1) and (2) and the semifiniteness of $E_\varphi{}^\psi$ results from that of ψ.

The last assertion of the lemma is immediate.

Proof of Proposition 3. (a) Let ψ be a faithful weight on $\mathscr{Q}_+$, χ-relatively invariant by G, let φ be a faithful weight on $\mathscr{C}^G$, let $E = E_\varphi{}^\psi$ be the $\mathscr{C}^G$-weight associated by Lemma 4.

For all $g \in G$, $a \in \mathscr{Q}_+$, $b \in \mathscr{C}^G$, we have

$$\varphi(b^*(E_0\tau_g(a))b) = \varphi(b^*(\chi(g)Ea)b).$$

As φ is faithful, we have $E_0\tau_g(a) = \chi(g)E(a)$.

(b) The converse is immediate.

(c) Suppose furthermore that $\mathscr{C}^G = \mathscr{Q}^G$. Let ψ' be another weight χ-relatively invariant by G, let p be its support. It is clear that p is G-invariant; therefore $p \in \mathscr{C}^G$. Let $\psi_p{}' = (\psi'|_{\mathscr{Q}_p{}^+})$, $\psi_p = (\psi|_{\mathscr{Q}_p{}^+})$ (they are faithful and semifinite since p is central). Let

$$u_t = (D\psi_p{}' : D\psi_p)_t, \qquad t \in \mathbb{R} \quad (\text{cf. } [2])$$

$$\tau_g u_t = (D\tau_g\psi_p{}' : D\tau_g\psi_p)_t = (D\chi(g)\psi_p{}' : D\chi(g)\psi_p)_t = u_t, \qquad \forall t \in \mathbb{R}, \quad \forall g \in G.$$

Therefore $u_t \in \mathscr{C}_p{}^G$, $\forall t \in \mathbb{R}$.

As $u_{t+s} = u_t\sigma_t^{\psi_p}$, $u_s = u_t u_s|_{t,s\in\mathbb{R}}$, there exists an inversible element h of of $\mathscr{C}_{p+}^G$ such that $\psi'(pap) = \psi(hpap)$, $\forall a \in \mathscr{Q}_{p+}$ [9]. Hence $\psi'(a) = \psi(ha)$, $\forall a \in \mathscr{Q}_+$.

Let φ' be the weight on $\mathscr{C}_+{}^G$ defined by $\varphi'(\cdot) = \varphi(h\cdot)$. We have $\psi' = \varphi' \circ E$.

The last assertion of the proposition follows immediately

In the case $\chi \equiv 1$, we obtain.

THEOREM 2. *Let* $\mathscr{F} = (\mathscr{Q}, G, \tau)$ *be a dynamical system. Then* $\mathscr{F}$ *is semifinite if and only if there exists a faithful normal semifinite* $\mathscr{C}^G$*-weight invariant* E *on* $\mathscr{Q}_+$.

Assume that E exists and suppose furthermore that $\mathscr{C}^G = \mathscr{Q}^G$. The mapping $\varphi \to \psi = \varphi \circ E$ is a positive affine bijection of the set of all weights φ on $\mathscr{C}^G$ and the set of all invariant weights ψ on $\mathscr{Q}_+$.

COROLLARY. *Let e_{sf} be the semifinite part of $\mathscr{F}$ (i.e., the greatest of all projections e such that $\mathscr{F}_e$ is semifinite), suppose that $\mathscr{Q}^G_{e_{sf}} \subset \text{center}(\mathscr{Q})$. There exists a normal semifinite G-invariant $\mathscr{C}^G$-weight E such that A weight ψ on $\mathscr{Q}_+$ is G-invariant if and only if it is of the form $\psi = \varphi \circ E$ with φ a weight on $\mathscr{C}^G$.*

Proof. We apply the theorem to the induced system $\mathscr{F}_e$.

Remarks. When $G = \text{Int}(\mathscr{Q})$, we obtain the classical results on operatorial traces (cf. [5, Chapter III-4]). Theorem 2 solves a question of [7, Remark II-3 p. 28].

REFERENCES

1. F. COMBES, Poids associés à une algebre Hilbertienne a gauche, *Compos. Math.* **23**(1), (1971), 49–77.
2. A. CONNES, Une classification des facteurs de type III, *Ann. Sci. École Norm. Sup., 4th Ser.* **6** (1973), 133–252.
3. N. DANG-NGOC, Sur la classification des systèmes dynamiques non commutatifs, *J. Functional. Anal.* **15** (1974), 188–201.
4. N. DANG-NGOC, Décomposition et classification des systèmes dynamiques, *Bull. Soc. Math. Fr.* **103** (1975), 149–175.
5. J. DIXMIER, "Les algèbres d'opérateurs dans l'espace Hilbertien," Gauthier-Villars, Paris, 1969.
6. J. DIXMIER, Sur la représentation regulière d'un groupe localement compact connexe, *Ann. Sci. École Norm. Sup., 4th Ser.* **2** (1969), 423–436.
7. A. GUICHARDET, Systemes dynamiques non commutatifs, *Asterisque* **13–14**, (1974).
8. U. HAAGERUP, The Standard Form of von Neumann Algebras, preprint, July 1973.
9. G. K. PEDERSEN AND M. TAKESAKI, The Radon–Nikodym theorem for von Neumann algebras, *Acta Math.* **130** (1973), 53–87.
10. J. SCHWARTZ, Type II Factors in a Central Decomposition, *Comm. Pure and Appl. Math.* **16** (1963), 247–252.
11. C. E. SUTHERLAND, Direct Integral Theory for Weights and the Plancherel Formula, preprint, 1973.
12. M. TAKESAKI, Conditional expectations in von Neumann algebras, *J. Functional Anal.* **9**, no. 3 (1973).

AMS (MOS) subject classifications: 46L10, 28A65.

STUDIES IN ANALYSIS
ADVANCES IN MATHEMATICS SUPPLEMENTARY STUDIES, VOL. 4

A Uniqueness Result for the Infinite Symmetric Group[†]

ROBERT R. KALLMAN[‡]

Department of Mathematics
University of Florida
Gainesville, Florida

Let S_∞ be the group of all permutations of the integers. Then S_∞ is a complete separable metric group in a natural topology. In this natural topology a basic neighborhood of the identity consists of all permutations that leave pointwise fixed a given finite subset of the integers. It is a simple matter to check that S_∞ is not locally compact in this topology. The purpose of this chapter is to give a very short proof of the following theorem, which asserts that this natural topology is the only topology in which S_∞ is a complete separable metric group. An immediate consequence of this theorem is that S_∞ cannot be made into a locally compact group with a countable basis for its topology. This answers a question asked by Ulam [3, p. 57]. This last result has already been obtained by Gaughan [1], who gave a more intricate proof based on other ideas.

THEOREM. *Let G be the group S_∞ with the natural topology described above and let G' be the group S_∞ with another topology in which it is a complete separable metric group. Let $\psi : G' \to G$ be the natural isomorphism. Then ψ is a topological isomorphism.*

Proof. Let B be a finite subset of the integers with three or more elements. Let $S(B)$ be the set of all permutations of the integers that map B into itself and are the identity outside of B. Let $S(B)'$ be the subgroup of S_∞ consisting of all elements that commute pointwise with the elements of $S(B)$. A simple computation shows that an element of S_∞ is an element of $S(B)'$ if and only if it leaves the elements of B pointwise fixed. Hence, as B ranges over the finite subsets of the integers with three or more elements, the $S(B)'$ form a neighborhood basis for G at the identity.

Since the set of elements that commute pointwise with a given set in any topological group is closed, we have that $\psi^{-1}(S(B)')$ is a closed subset of

[†] Received by the editors in January 1974.

[‡] Supported in part by NSF Grant GP-38023.

ISBN 0-12-599150-9

G'. Hence, ψ is a Borel mapping since G is separable. Hence, ψ is a Borel isomorphism by a well-known theorem of Souslin. But ψ and ψ^{-1} are then continuous on sets of the second category [2, p. 400]. We claim that ψ and ψ^{-1} are continuous everywhere. Choose a set P of first category in G' such that $\psi | G' - P$ is continuous. Let a_n $(n \geqslant 1)$ and a be elements of G' such that $a_n \to a$. Now if Q is the set that is the union of $a^{-1} \cdot P$ and $a_n^{-1} \cdot P$, then Q is again a set of the first category. Hence, $G' - Q$ is not empty. Let b be an element of $G' - Q$. Then ab is in $G' - P$ and $a_n b$ is in $G' - P$ for every integer n. But $a_n b \to ab$. Hence, $\psi(a_n b) \to \psi(ab)$ and so $\psi(a_n) \to \psi(a)$. Thus ψ is a continuous one-to-one mapping of G' onto G. One similarly proves that ψ^{-1} is also a continuous one-to-one mapping of G onto G'. Hence, G and G' are topologically isomorphic. Q.E.D.

References

1. E. D. Gaughan, Topological group structures of infinite symmetric groups, *Proc. Nat. Acad. Sci. U.S.A.* **58** (1967), 907–910.
2. K. Kuratowski, "Topology," Vol. I, Academic Press, New York, 1966.
3. S. M. Ulam, "A Collection of Mathematical Problems," Wiley (Interscience), New York, 1960.

AMS (MOS) subject classifications: primary, 22A05; secondary, 22D05